Incomplete Data in Sample Surveys

Volume 1

Report and Case Studies

Incomplete Data in Sample Surveys

Volume 1 Report and Case Studies
Volume 2 Theory and Bibliographies
Volume 3 Proceedings of the Symposium

Incomplete Data in Sample Surveys

Volume 1

Report and Case Studies

Panel on Incomplete Data
Committee on National Statistics
Commission on Behavioral and Social Sciences
and Education
National Research Council

PART I Report

PART II Case Studies

Edited by

William G. Madow
Committee on National Statistics
National Research Council
National Academy of Sciences
Washington, D.C.

Harold Nisselson
Westat, Inc.
Rockville, Maryland

Ingram Olkin
Department of Statistics and School of Education
Stanford University
Stanford, California

1983

ACADEMIC PRESS
A Subsidiary of Harcourt Brace Jovanovich, Publishers
New York London
Paris San Diego San Francisco São Paulo Sydney Tokyo Toronto

NOTICE: The project that is the subject of this report was approved by the Governing Board of the National Research Council, whose members are drawn from the councils of the National Academy of Sciences, the National Academy of Engineering, and the Institute of Medicine. The members of the committee responsible for the report were chosen for their special competences and with regard for appropriate balance.

This report has been reviewed by a group other than the authors according to procedures approved by a Report Review Committee consisting of members of the National Academy of Sciences, the National Academy of Engineering, and the Institute of Medicine.

The National Research Council was established by the National Academy of Sciences in 1916 to associate the broad community of science and technology with the academy's purposes of furthering knowledge and of advising the federal government. The council operates in accordance with general policies determined by the academy under the authority of its congressional charter of 1863, which establishes the academy as a private, nonprofit, self-governing membership corporation. The council has become the principal operating agency of both the National Academy of Sciences and the National Academy of Engineering in the conduct of their services to the government, the public, and the scientific and engineering communities. It is administered by both academies and the Institute of Medicine. The National Academy of Engineering and the Institute of Medicine were established in 1964 and 1970, respectively, under the charter of the National Academy of Sciences.

ACADEMIC PRESS, INC.
111 Fifth Avenue, New York, New York 10003

United Kingdom Edition published by
ACADEMIC PRESS, INC. (LONDON) LTD.
24/28 Oval Road, London NW1 7DX

Library of Congress Cataloging in Publication Data

Main entry under title:

Incomplete data in sample surveys.

Includes index.
Contents: v. 1. Report and case studies -- v. 2. Theory and bibliographies -- v. 3. Proceedings of the symposium.
1. Sampling (Statistics)--Addresses, essays, lectures. 2. Social surveys--Response rate--Addresses, essays, lectures. I. Madow, William G. (William Gregory), Date . II. Nisselson, Harold. III. Olkin, Ingram. IV. National Research Council (U.S.). Panel on Incomplete Data.
HA31.2.I52 1983 001.4'225 82-20591
ISBN 0-12-363901-8 (v. 1)

PRINTED IN THE UNITED STATES OF AMERICA

83 84 85 86 9 8 7 6 5 4 3 2 1

PANEL ON INCOMPLETE DATA IN SAMPLE SURVEYS

Ingram Olkin, Chair; Department of Statistics and School of Education, Stanford University

Barbara A. Bailar, Statistical Standards and Methodology, Bureau of the Census, U.S. Department of Commerce

Barbara A. Boyes, Office of Survey Design, Bureau of Labor Statistics, U.S. Department of Labor (deceased; member until March 1981)

Arthur P. Dempster, Department of Statistics, Harvard University

Robert M. Elashoff, Department of Biomathematics, University of California, Los Angeles

Robert L. Freie, Estimates Division, Statistical Reporting Service, U.S. Department of Agriculture

Louis Gordon, Office of Energy Information Validation, Energy Information Administration, U.S. Department of Energy (resigned, September 1979)

Robert M. Groves, Survey Research Center, University of Michigan

Morris H. Hansen, Westat, Inc.

Harold Nisselson, Westat, Inc.; Statistical Standards and Methodology, Bureau of the Census to 1979

Richard A. Platek, Census and Household Survey Methods Division, Statistics Canada

Donald B. Rubin, Department of Statistics and Department of Education, University of Chicago

Frederick J. Scheuren, Statistics of Income Division, Internal Revenue Service, U.S. Department of the Treasury

Joseph H. Sedransk, Department of Mathematics, State University of New York at Albany

Monroe G. Sirken, Office of Mathematical Statistics, National Center for Health Statistics, U.S. Department of Health and Human Services

William G. Madow, *Study Director*
M. Haseeb Rizvi, *Staff Officer*
Barbara A. (Booker) Malone, *Administrative Secretary*

(Biographical sketches of panel members and staff appear in the Appendix to Part I.)

COMMITTEE ON NATIONAL STATISTICS (1981–1982)

Stephen E. Fienberg, Chair; Departments of Statistics and Social Science, Carnegie-Mellon University

Jean D. Gibbons, Department of Management Science and Statistics, University of Alabama

Zvi Griliches, Department of Economics, Harvard University

Clifford Hildreth, Center for Economic Research, University of Minnesota

Nathan Keyfitz, Department of Sociology, Harvard University; Department of Sociology, Ohio State University

Leslie Kish, Institute for Social Research, University of Michigan

Gary G. Koch, Department of Biostatistics, University of North Carolina

Paul Meier, Department of Statistics, University of Chicago

Lincoln E. Moses, Department of Statistics and School of Medicine, Stanford University

Ingram Olkin, Department of Statistics and School of Education, Stanford University

Burton H. Singer, Department of Mathematical Statistics, Columbia University

Judith M. Tanur, Department of Sociology, State University of New York, Stony Brook

Edward R. Tufte, Departments of Political Science and Statistics, Yale University

William Gemmell Cochran
1909–1980

This report is dedicated to the memory of William Gemmell Cochran, a man of warmth, wit, and wisdom who enriched the lives of all who knew him and a major contributor to the development of sampling theory as well as other areas of statistics.

Bill Cochran was born in Rutherglen, Scotland, and received degrees from Glasgow University and Cambridge University. During and after his academic work, he served as a statistician at the Rothamsted Experimental Station in England. During World War II, he was on the faculty of Iowa State College, and he also worked with the statistical research group of Princeton University on problems of naval warfare. Beginning in 1946, he was, successively, associate director of the Institute of Statistics at the University of North Carolina, head of the Department of Biostatistics at The Johns Hopkins University, and professor of statistics at Harvard University. In 1976 he became professor emeritus at Harvard.

In addition to his full academic life, Bill was an active member of many professional organizations. He served as president of the Institute of Mathematical Statistics (1946), the American Statistical Association (1953), The Biometric Society (1954), and the International Statistical Institute (1967–1971). He was elected honorary fellow of the Royal Statistical Society in 1959, and he was elected to the National Academy of Sciences in 1974. Bill was awarded a Guggenheim Fellowship in 1964, and in 1967 he received the S. S. Wilks Medal of the American Statistical Association for his many contributions to the advancement of the design and analysis of experiments and their value for military research.

Bill was involved in the early work of the panel, and we acknowledge his contribution as a participant in our discussions and as a colleague in the field of survey research.

Contents

3 Measuring and Reporting Nonresponse

4 Review of Case Studies

5 Review of Theory

PART II Case Studies

3 USDA Livestock Inventory Surveys

Robert L. Freie

4 Report on the Survey Research Center's Surveys of Consumer Attitudes

Charlotte G. Steeh, Robert M. Groves, Robert Comment, and Evelyn Hansmire

7 Incomplete Data in the Survey of Consumer Finances

Rodger Turner and Murray Lawes

8 An Empirical Investigation of Some Item Nonresponse Adjustment Procedures

M. Haseeb Rizvi

Contributors

Numbers in parentheses indicate the pages on which the authors' contributions begin.

Ronald Andersen (391), Center for Health Administration Studies, Graduate School of Business, University of Chicago, Chicago, Illinois 60637

Martha J. Banks (391), Center for Health Administration Studies, Graduate School of Business, University of Chicago, Chicago, Illinois 60637

Barbara A. Boyes[1] (123), Office of Survey Design, Bureau of Labor Statistics, U.S. Department of Labor, Washington, D.C. 20212

David W. Chapman (435), Statistical Research Division, Bureau of the Census, U.S. Department of Commerce, Washington, D.C. 20233

Robert Comment (173), Survey Research Center, University of Michigan, Ann Arbor, Michigan 48106

Margaret E. Conlon (123), Bureau of Labor Statistics, U.S. Department of Labor, Washington, D.C. 20212

Martin R. Frankel (391), Bernard Baruch College, City University of New York, New York, New York 10021

Robert L. Freie (141), Statistical Reporting Service, U.S. Department of Agriculture, Washington, D.C. 20250

Robert M. Groves (173), Survey Research Center, University of Michigan, Ann Arbor, Michigan 48106

Morris H. Hansen (209), Westat, Inc., Rockville, Maryland 20850

Evelyn Hansmire (173), Survey Research Center, University of Michigan, Ann Arbor, Michigan 48106

Murray Lawes (269), Census and Household Survey Methods Division, Statistics Canada, Tunney's Pasture, Ottawa, Ontario, K1A OT6 Canada

[1]Deceased.

William G. Madow (237, 367), Committee on National Statistics, National Research Council, National Academy of Sciences, Washington, D.C. 20418

Harold Nisselson (107), Westat, Inc., Rockville, Maryland 20850

M. Haseeb Rizvi[2] (299), Committee on National Statistics, National Research Council, National Academy of Sciences, Washington, D.C. 20418

Charlotte G. Steeh (173), Survey Research Center, University of Michigan, Ann Arbor, Michigan 48106

Richard Tomasino[3] (209), Corporation Office of Market Research, Xerox Corporation, Rochester, New York 14644

Rodger Turner[4] (269), Statistics Canada, Tunney's Pasture, Ottawa, Ontario, K1A OT6 Canada

Richard Valliant (209), Westat, Inc., Rockville, Maryland 20852

[2]Present address: Sysorex International, Inc., Cupertino, California 95014.

[3]Present address: Continental Telephone Co., Atlanta, Georgia.

[4]Present address: Revenue Canada, Taxation, Government of Canada, Ottawa, Ontario K1A OL8.

Preface

Incompleteness of data occurs for a variety of reasons and in a variety of statistical contexts. When sample surveys are done, units that should be included among those from which a sample is selected may not be included; units selected for the sample may not respond; responding units may not respond to all items; and unit or item responses may be unusable. Each of these factors results in incompleteness. Although discussions of incompleteness have appeared in the sample survey literature for many years, concern with incompleteness and the volume of publications on theoretical aspects of nonresponse have been increasing.

The Panel on Incomplete Data, established by the Committee on National Statistics within the Commission on Behavioral and Social Sciences and Education of the National Research Council, was organized in 1977 because work in the field had reached a point at which it seemed desirable to make a comprehensive review of the literature on survey incompleteness in sample surveys and to explore ways of improving the methods of dealing with it. Initial planning for the study was undertaken by the Committee on National Statistics, with support provided by a consortium of federal statistical agencies. Funding for the work of the panel was provided by the National Science Foundation, the Social Security Administration, and the U.S. Department of Energy.

The panel's main task in the initial stages was to delineate the areas that should be included in the study. As a result of its deliberations, the panel decided that, in addition to a report, major parts of its work would consist of case studies and an exposition of theory dealing with incompleteness. The panel also planned a symposium to provide an opportunity to augment the range of topics discussed in the case studies and theory parts, which would permit new developments to be incorporated into its work. An annotated bibliography of important papers in the field was planned and a bibliography on nonresponse was later added. The panel's work is published in three volumes: the first includes the report of the panel and the case studies; the second includes the papers on theory and two bibliographies; and the third consists of the proceedings of the Symposium on Incomplete Data.

All members of the panel participated fully in the panel's work, including decisions about and preparation of parts of the report. In organizing its work, the panel designated three subpanels, on theory, on case studies, and on computer programs. The theory subpanel, chaired by Donald B. Rubin and with members Arthur Dempster, Robert M. Elashoff, Richard Platek, Ingram Olkin, Frederick Scheuren, and Joseph Sedransk, was responsible for preparing the theory part of Volume 2. The case studies subpanel, chaired by Harold Nisselson and with members Barbara Bailar, Barbara Boyes, Robert Elashoff, Robert Freie, Robert Groves, Morris Hansen, Ingram Olkin, Richard Platek, and Monroe Sirken, was responsible for preparing Part II of Volume 1. The computer programs subpanel, chaired by Joseph H. Sedransk and with members Ingram Olkin, Joseph I. Naus of Rutgers University, and Gordon T. Sande of Statistics Canada considered algorithms for dealing with missing data. The annotated bibliography was prepared by Lawrence V. Hedges of the University of Chicago and Ingram Olkin. The late William G. Cochran of Harvard University (to whom these volumes are dedicated) and Roderick J. A. Little of Datametrics Research, Inc. met with the panel on several occasions and contributed to the drafting of the panel's work. A steering committee consisting of Harold Nisselson, Ingram Olkin, Richard A. Platek, Donald B. Rubin, and Frederick J. Scheuren coordinated the project after the overall planning had been completed.

The panel relied heavily on its staff, William G. Madow, study director; M. Haseeb Rizvi, staff officer; and Barbara A. Malone, administrative secretary. In addition, the panel had the steady support and advice of Edwin D. Goldfield, executive director, and Margaret E. Martin, former executive director, of the Committee on National Statistics.

Drafts of the theory papers were reviewed by Professor Carl H. Särndal of the University of Montreal and Professor J. N. K. Rao of Carleton University; drafts of the case studies were reviewed by Professor Graham Kalton of the University of Michigan. I am indebted to them for their detailed and helpful suggestions.

A draft of the panel's report was also reviewed by three sets of reviewers within the National Research Council; I wish to express my thanks and appreciation for their comments.

In any venture of this magnitude the burden must naturally fall on a few shoulders. In the present instance, William G. Madow served not only as the study director, coordinating all the panel's activities and projects, but also provided intellectual guidance to authors on both theoretical and practical statistical issues. He was an author of many parts and a friendly critic of others. This project could not have been completed without his herculean efforts and constant help. It is hard to imagine how the project might have fared without him, and the panel wishes to acknowledge his efforts with special thanks.

Ingram Olkin, Chair
Panel on Incomplete Data

Contents of Volumes 2 and 3

Volume 2

Theory and Bibliographies

Volume 3

Proceedings of the Symposium

PART I

Report

CHAPTER **1**

Introduction and Recommendations

INTRODUCTION

Survey results are affected by errors arising from several sources, the most important of which are sampling error, error due to incomplete data, response error, and processing error. This report is concerned almost exclusively with incomplete data; errors due to other sources are discussed only when these are relevant to the handling of incomplete data.

Incomplete data resulting from three sources are of particular importance in sample surveys: undercoverage, unit nonresponse, and item nonresponse.

Undercoverage: units (e.g., households, persons, establishments, farms) that should be in the frames (or lists) from which a sample is selected are not in those frames, or units in the sample are mistakenly classified as ineligible or are omitted from the sample or from the units interviewed.

Unit nonresponse: units in the selected sample and eligible for the survey do not provide the requested information, or the provided information is unusable.

Item nonresponse: eligible units in the selected sample provide some, but not all, of the requested information, or the information provided for some items is unusable. (Survey takers sometimes define unit nonresponse as occurring if item nonresponse occurs for a specified set of items.)

Undercoverage is the most difficult type of incompleteness with which to deal, primarily because units not covered are hard to find and because undercoverage rates are rarely known. Sometimes, through analysis of data from outside sources, undercoverage rates may be estimated for a whole population;

INCOMPLETE DATA
IN SAMPLE SURVEYS
Volume 1, Part I

ISBN 0-12-363901-8

even then, however, undercoverage rates for important domains (subsets) of the population will usually be unknown. Also, when undercoverage occurs, units are rarely identifiable as omitted units.

With both unit and item nonresponse, at least the identification of the unit or item is almost always known. Usually, but not always, the locations of nonresponding units and of units with missing items are also known. An exception occurs in mail surveys when a nonresponding unit may no longer exist or may be ineligible for the survey. And sometimes a mail survey is conducted so that even units that reply are not identified.

Information other than identification may also be known for units that do not respond. Such information on units may be in the frame, be obtained during interviewing efforts, or have been obtained earlier or later, for example, in a panel or longitudinal survey. Items reported by a unit may be related to the items not reported by the unit; for example, rent or value of housing is related to income. Thus, a survey taker may have information for nonresponding units and missing items that is not available for units not covered.

The panel decided at an early stage that it would not attempt to make explicit recommendations on undercoverage. However, many of the recommendations given below are applicable to undercoverage, especially those dealing with survey operations. The particular problem of adjusting for undercoverage, namely, obtaining undercoverage rates for populations and domains in sufficient detail for adjustments to be made, is not discussed.

This study emphasizes estimates such as occur in statistical tables. It was felt that a review of the methods of dealing with the effects of incomplete data on complex statistical analyses should be left to a further study. However, the approaches in the theory papers in Volume 2, especially Part VI, are applicable to problems more general than the estimates of data appearing in statistical tables. In most parts of this study the statistical criteria used are those that are customary in sample surveys, namely, bias, variance, mean square error, and coefficient of variation. In addition, Bayesian and other model-based approaches leading to statistical distributions are discussed in the theory papers.

The remainder of this chapter consists of the conclusions of the panel regarding incomplete data, the recommendations of the panel for conducting and analyzing sample surveys and for further research, and a brief comment on the uses and limitations of the study.

Chapter 2 contains a discussion of the problems caused by incompleteness and how incompleteness due to various sources occurs in sample surveys. Some illustrative data on nonresponse are given, showing the levels and the changes in levels of response rates over surveys, over time, and for different domains and items. (Further data and discussions of these and related topics are presented in Chapter 4.) Simple expressions for the bias of a mean and ratio of two totals are presented in the last section of Chapter 2. These expressions both provide statistical motivation for the definition of the nonresponse rate and for formulations of models of nonresponse.

Chapter 3 presents a consideration of measures of nonresponse and some ways in which such measures can be used and misused. Chapter 4 contains summaries and reviews of the 10 case studies that appear in Part II of this volume. Chapter 5 presents a summary of the several theoretical points of view and approaches to statistical methods of dealing with nonresponse that are presented in the theory papers in Volume 2. Other statistical methods of dealing with nonresponse are discussed in Sessions II, III, and V of the symposium, Volume 3.

Except for the last sections of Chapters 2 and 4 and for parts of Chapter 5, this report assumes familiarity with sample surveys such as would result from a first nonmathematical course in sample surveys or experience in conducting surveys. The final sections of Chapters 2 and 4 require a little knowledge of survey or statistical theory. Chapter 5 makes a serious effort to provide the feeling of the theory without the mathematics, but does require more theoretical background than the earlier chapters.

CONCLUSIONS

No specific recommendation is made below on "acceptable" levels of nonresponse. Nonresponse rates should be "low," but whether 5%, 10%, 20%, or some other percentage should be an upper bound for acceptable nonresponse rates depends on the survey objectives and is difficult to specify even for a particular survey. Unit and item nonresponse rates must be low for a total survey and for significant domains and subpopulations of interest.

Few surveys can achieve 5% or smaller unit nonresponse rates without either accepting very high costs or defining the survey population especially to achieve that objective, i.e., excluding subsets of the population expected to have high nonresponse rates. Yet even with a low overall nonresponse rate, estimates for some items and domains may have much higher nonresponse rates, and estimates for items characterizing only a small proportion of the population (and other "low" estimates) may possibly be seriously affected when a nonresponse rate is as small as 5%. Thus, the acceptable rates of nonresponse depends on the inferences to be made and the procedures for making those inferences.

The major conclusions of the panel are presented in the rest of this section.

(1) Nonresponse must be dealt with in planning, in data collection, in preparing the data base, in analyzing data, and in reporting results of surveys.

(2) Nonresponse should be measured and reported not only overall but also by reason for nonresponse, by important domain if estimates are made for domains, and by item for at least important items. Efforts to estimate biases should also be made. These data will facilitate effective allocation of resources

in making and analyzing surveys and in studying trends in nonresponse rates and nonresponse biases among surveys and over time.

(3) Even with intensive efforts to collect data, the inevitable nonresponse requires the consideration of methods that improve analysis by statistical adjustment of the collected data. But no statistical methods will fully compensate for missing units and data. Biases will almost certainly remain. Good methods are chiefly aimed at reducing biases and mean square errors of estimators while reducing or at least not unduly increasing variances of estimators. The effects of thc methods used on covariances and distributions must also be considered. Evaluation of nonresponse procedures is rarely possible from survey data.

(4) The general methods currently most likely to reduce biases are those employing poststratification. The methods may utilize imputation or weighting techniques. In general, no statistical methodology for imputation or adjustment will reduce the need to attempt to collect data with high levels of response.

(5) It is desirable but difficult and often impossible to estimate biases both before and after adjustment. The effects on variances and covariances of nonresponse adjustments can usually be estimated, at least approximately. It is important to incorporate these effects in survey analyses and reports or to demonstrate that they may be neglected. Expected biases, conditional on assumed response mechanisms and the obtained data, can also be calculated and may be useful, but they are not a substitute for empirical estimates of actual biases.

(6) Alternative response mechanisms should be considered in order to attempt to improve estimation and to understand the effects of sources of biases. The use of individual response mechanisms for *each* item will usually be impractical; however, such use for selected items or classes of items may be practical.

(7) Special efforts must be made to obtain data sets that permit methods to be evaluated. The cost of obtaining such data sets and the need for their being generally available are so great that it may be necessary to plan and fund them by consortia.

RECOMMENDATIONS

Many of the recommendations that follow are concerned with statistical methods of dealing with nonresponse after data collection is completed. However, the statistical methods of dealing with nonresponse will not completely eliminate the effects of nonresponse. Biases and distortions of distributions will remain, and their effects cannot be estimated from the survey data themselves. Thus, the most important recommendation:

Recommendation 1. Collect the survey data as fully and accurately as possible, using callbacks and follow-ups as needed to do so.

The rest of the recommendations are classified into recommendations on survey operations and recommendations on research. The recommendations on survey operations are intended to supplement good survey practice. Thus, they omit recommendations desirable for some survey organizations and duplicate or be inferior to existing practices of some survey organizations.

If an extreme view is taken of certain recommendations, for example, those on documentation and reporting, it may appear that the costs of implementation are too high. The recommendations are intended to provide guides for achieving the goals of a survey, subject to available resources. In general terms, documentation should be sufficiently detailed to enable the replication of the essential steps in conducting the survey and should include an evaluation of the steps if possible. Also, the discussion of methodology in a survey report should include a statement of how all essential steps were performed and should contain information that enables survey users to judge the quality of at least the important statistical inferences that are or can be made.

No special recommendation has been made with respect to planning only because *every* recommendation should be considered at the planning stage. Almost every survey should be planned assuming nonresponse will occur, and at least informed guesses about nonresponse rates and biases based on previous experience and speculation should be made, not only for the survey as a whole but also for strata and poststrata and for important domains and items.

The panel, in making its recommendations, assumes that cost-effectiveness will be the basis for decision making and for implementing the recommendations both for each survey being taken and for the longer-range goals of the survey sponsor and survey taker.

Survey Operations

The recommendations on survey operations are intended to accomplish six objectives: to reduce nonresponse; to measure nonresponse; to reduce the effects of nonresponse on estimates; to improve analyses and reports; to propose the accumulation of information useful in survey design and evaluation; and to suggest further research.

Data Collection

Recommendation 2. Select and train interviewers to achieve low nonresponse rates as well as accurate interviews. Verify at least a sample of units classified as ineligible as well as a sample of completed interviews.

Recommendation 3. Begin early to interview sample units expected to be difficult to interview; attempt to convert noninterviews to interviews; and make a final interviewing effort with the help of supervisors and superior interviewers. (This is recommended in addition to making callbacks and follow-ups.)

Measurement of Nonresponse

Recommendation 4. Compute nonresponse rates, completion rates, and item coverage rates during as well as after the data collection effort for the sample, for important domains, and for important items.

Recommendation 5. Prepare one or more accountability tables and define the nonresponse and completion rates in terms of the entries in the accountability tables. Provide one or more tables containing the data used in calculating item coverage rates to avoid any misunderstanding of the definition.

Recommendation 6. Consider preparation of accountability tables for domains and specified numbers of callbacks as well as for the total sample.

Recommendation 7. Consider establishing a management information system that will provide the data on nonresponse for the survey and by selected important classifications, such as interviewers, interviewing areas, strata, domains, calls, and other important items in addition to the data required for survey management.

Sample Design

Recommendation 8. Design the sample so that if anticipated nonresponse rates within poststrata are approximately correct, the numbers of respondents in the poststrata will be adequate for imputation and for poststratification, weighting class, or other estimators to be used.

Recommendation 9. Consider whether to select subsamples of non-respondents for an intensive interviewing effort in order to reduce bias or to evaluate alternative nonresponse models or both.

Questionnaires, Control Forms, and Other Data Sources

Recommendation 10. Use questionnaire and control forms that include information useful in data collection, data base preparation, and analysis:

(1) The information on eligibility, callbacks, reasons for nonresponse, conversion efforts, and final data collection efforts should be in a form suitable for inclusion in the data record of each unit.

(2) Variables (items) should be included that are useful in poststratification and in adjustment for nonresponse.

(3) During conversion efforts or the final data collection, consideration should be given to using a shorter or simpler questionnaire that will facilitate analysis and increase the response rate.

(4) Data records should include variables related to items that are expected to have high nonresponse rates. (These variables may result from simpler or less threatening versions of questions expected to have high nonresponse rates. Examples of such variables are rent and a size classification of income for dealing with nonresponse to a detailed income question.)

Recommendation 11. Attempt to find data from censuses and administrative sources and from past, future, or other data records for units in the survey that may be helpful in approximating biases due to nonresponse before and after imputation or adjustment for nonresponse.

Data Base

Recommendation 12. Make sure the data base contains data records of nonresponding as well as responding units.

Recommendation 13. Include the following information in the data records:

(1) Outcome of each attempt to interview, whether by personal interview, mail, or telephone, including codes of reasons for ineligibility and nonresponse.

(2) Data that may be available for ineligible and nonresponding units as well as responding units.

(3) For each imputed or substituted data record, a code indicating that the record is imputed or substituted, using a different code for imputation and substitution.

(4) For important items, an identification of each imputed or substituted item and whether the item was missing or unusable. (Unusable items in a data record are often identified during editing. If so, "unusable" should be defined to exclude minor corrections made in the course of editing.)

(5) The original weights and the weights adjusted for nonresponse (if weights are entered on each data record).

(6) For each donor unit or item, the number of times the unit or item is used as a donor. (If this is impractical, a separate record should be maintained containing at least the identifications and numbers of times a data record is used as a donor for imputing entire data records and for important items.)

Analysis

Recommendation 14. Evaluate to the extent possible or at least speculate about the biases before and after imputation and adjustment for nonresponse.

Also include the component of variance due to nonresponse in the estimated variance of survey estimates.

Recommendation 15. Approximate or at least discuss the magnitudes of mean square errors as well as biases and variances.

Recommendation 16. Attempt to obtain data with which to evaluate imputation and adjustment procedures using data from administrative sources, censuses, other surveys, past or future waves from this survey, or subsamples of nonrespondents selected for intensive interviewing efforts.

Recommendation 17. Choose poststrata, weighting classes, and classes used for "hot-deck" imputations such that the differences between respondents and nonrespondents are relatively small within classes. Responding samples from these classes should also be large enough (20 or more units is often a satisfactory size) so that their contributions to variances are relatively small.

Recommendation 18. In imputation by hot-deck methods, try to avoid using the same donor record to impute values for more than one recipient record, provided that biases and variances are not unduly increased.

Recommendation 19. Consider the use of multiple imputation to improve estimates and estimates of variances and to compare the effects of alternative response mechanisms.

Recommendation 20. Prepare at least some analyses based on alternative reasonable assumptions concerning response mechanisms. One commonly used method of dealing with nonresponse is applied as though nonrespondents are random samples from the selected sample after poststratification. Other methods are applied as though a superpopulation distribution holds either for all units or for nonresponse units conditional on the responding units and their data. Objective information that indicates one set of assumptions is better than other reasonable alternatives is very rare. Hence the spread of estimates produced by such alternative methods is useful in making inferences.

Recommendation 21. Prepare tables that show the effects of nonresponse and the procedures used to deal with nonresponse. The tables might include

(1) Estimates for the responding population, i.e., using original weights

(2) Estimates for the entire population using the original weights adjusted for nonresponse overall or within strata (and, perhaps, within clusters)

(3) Estimates for the entire population using reweighting within poststrata

(4) Estimates after using other methods, e.g., raking procedures or procedures depending on assumed superpopulation distributions.

Include nonresponse rates in tables intended to show effects of nonresponse and procedures used in dealing with nonresponse.

Recommendation 22. Consider preparation of tables such as those discussed under Recommendation 21 for domains and subpopulations as well as for alternative nonresponse mechanisms (see Recommendation 20). The objectives would be to exhibit the effects of methods of dealing with nonresponse on individual estimates and on distributions.

Documentation and Reports

Recommendation 23. Include in both documentation and the final report of a survey approximately the information outlined in Table 1 including errors of response where possible. Documentation should preferably be done while a survey is in progress. Although not explicitly listed in Table 1, data evaluating the survey operations and outcomes should be included in both the documentation and the report whenever possible. For example, some tables should give the effects of alternative nonresponse procedures on important estimates (see Recommendations 14–22). Ideally, the methodological information in the report should be a summary of information taken from the documentation.

Recommendation 24. Include in press releases some information on nonresponse as well as other information on the survey. A good objective is to include brief descriptions of the target and survey populations and units, the sample size and design, how the survey was conducted, nonresponse rates, and, for important estimates, variances or coefficients of variations that include the effects of nonresponse. Also, discuss the possibility of nonresponse and other (e.g., response) biases and, if possible, give some information about them. (For some press releases, the above list may be too long, but in releases of important data, whether by government or nongovernment sources, the information provided should be helpful to the users.)

Research

The recommendations on research have three objectives: to provide a capital investment in computer programs and data sets that will make nonresponse methodology cheaper to implement and evaluate; to encourage research on and evaluation of theoretical response mechanisms; and to urge that long-term programs be undertaken by individual or groups of survey organizations and sponsors to provide for and accomplish cumulative survey research, including research on nonresponse.

Recommendation 1. General-purpose computer programs or modules should be developed for dealing with nonresponse. These programs and modules should include editing, imputing (single and multiple), and the calculation of estimators, variances, and mean square errors that, at least, reflect contributions due to nonresponse.

Recommendation 2. Current methods of improving estimates that take account of nonresponse, such as poststratification, weighting methods, and hot-deck imputation, especially hot-deck methods of multiple imputation, require further study and evaluation.

Recommendation 3. Theoretical and applied research on response mechanisms should be undertaken so that the properties and applicability of the models become known for estimates of both level and change.

TABLE 1

Recommended Content of Documentation and Survey Report on Methodology

Documentation and the report of a sample survey should contain statements of many of the following items, depending on cost and agreement between survey taker and survey sponsor. (The emphasis here is on content related to incomplete data; additional items should be added for response errors.)

(1) The objectives of the survey
  (a) The definition of the population to which these objectives relate (the target population)
  (b) The definitions of the population characteristics in terms of which the objectives are stated
(2) The questionnaire
  (a) The relation of the questions to the objectives of the study
  (b) Supplementary information needed for use in imputation or statistical adjustment for unit and item nonresponse
  (c) Choice of questions to reduce nonresponse
(3) Sample design selection
  (a) The survey population, i.e., the population from which the sample is to be selected
  (b) The frame population, i.e., the lists or sets from which the sample is actually selected, including the frames used at different stages of selection
  (c) The relation of the frame population to the target and survey populations, discussing whether the frame population has been defined in order to reduce nonresponse rates
  (d) The definition of which units in the frame are eligible for or within the scope of the survey
  (e) The selection procedure used at each stage of the survey, stated so that the selection procedure is repeatable
  (f) The definition of the probabilities of selection
  (g) The criteria to be satisfied if a purposive sample is to be selected
(4) Data collection methodology, especially steps taken to reduce incompleteness
  (a) Organization
  (b) Selection of interviewers
  (c) Training
  (d) Collection procedures, such as
    1. Introductory letter
    2. Incentives to respond, if offered
    3. Special techniques
  (e) Procedures for callbacks and followups
  (f) Attempts to convert nonresponse to response
  (g) Final collection procedures
  (h) Accountability tables
(5) Measures of incompleteness for the population, important domains, subpopulations, and items: for example,
  (a) Undercoverage rates, if possible
  (b) Unit nonresponse rates
  (c) Item nonresponse rates
  (d) Item coverage rates for important items
(6) Steps in preparing data records, such as
  (a) Field edit
  (b) Office edit
  (c) Imputation, including multiple imputation

TABLE 1 *(cont.)*

1. Unit imputation
2. Item imputation

(7) Statistical models, especially sampling and response mechanisms
(8) Estimators
 (a) Complete response estimators
 (b) Incomplete response estimators
 (c) Expected values of estimators
 (d) Biases
 (e) Distribution of estimators, if obtained
(9) Variances
 (a) Formulas and estimators of variance if complete response had been achieved
 (b) Formulas for and estimators of variances of the estimators actually used
 (c) Biases of estimates of variances
(10) Tables showing effects of
 (a) Editing, excluding identification and correction of minor errors identified during editing
 (b) Imputation and substitution
 (c) Response mechanisms
(11) Comparisons with data believed to be essentially complete
(12) Examples of inference illustrating effects of incomplete data on inference

Recommendation 4. A systematic summarization of information from various surveys should be undertaken on the proportions of respondents for specified parts of populations and for particular questions in stated contexts.

Recommendation 5. Research is needed to distinguish the characteristics of nonrespondents as opposed to respondents and to assess the impact of questionnaire design and data collection procedures on the level of nonresponse.

Recommendation 6. Data sets that permit good estimates of bias and variance to be made when various statistical methods of dealing with nonresponse are adopted should be made publicly available. Such data sets could be used for testing various methods of bias reduction and for assessing effects of the methods on variances. They could also be used for the evaluation of more general methods depending on models.

Recommendation 7. Theoretical and empirical research should be undertaken on methods of dealing with nonresponse in longitudinal and panel surveys.

Recommendation 8. Theoretical and empirical research on the effects of nonresponse on more complex methods of analysis of sample survey data, e.g., multivariate analysis, should be undertaken.

Recommendation 9. A consistent terminology should be adopted for descriptive parameters of nonresponse problems and for methods used to handle nonresponse in order to aid communication on nonresponse problems.

Recommendation 10. Research on response mechanisms that depend on reasons for nonresponse should be undertaken.

Recommendation 11. Data on costs should be obtained and analyzed in relation to nonresponse procedures so that objective cost-effective decisions may become increasingly possible.

COMMENT

Some overall recommendations may be given to those who use and take surveys. For any survey, survey users should learn the levels and distribution of nonresponse rates, learn how both nonresponse and techniques intended to reduce the effect of nonresponse may affect the data and analyses, including estimates of level and change, and require that documentation and the data base be available in a form that permits the user to make judgments on the survey and its results as well as on any desired further analyses.

Survey takers should attempt to predict the unit and item nonresponse rates and biases that may occur and prepare for nonresponse in data collection, questionnaire design, data base construction, and data analyses and documentation, including the preparation of computer programs intended to produce the necessary information; summarize information on nonresponse; identify and, if desired, impute for missing data; attempt to estimate biases and calculate estimates and estimates of variances that reflect the treatment and, often, the alternative treatments of nonresponse; document the steps taken to reduce the effects of nonresponse on the survey output; and ensure that the survey user has the information with which to make judgments on the analyses and uses of survey results.

To reduce the burdens that occur when documentation is prepared after a survey is done, survey organizations may find it desirable to prepare documentation on a current basis during a survey and may find it helpful to use a standardized format, possibly computerized, based on Table 1.

It is important to keep in mind that all statistical methods of dealing with nonresponse explicitly or implicitly impute for missing data. As a result of the explicit or implicit imputation, missing data are replaced by data that can be said to have response errors determined by the imputation and adjustment processes, the data provided by respondents, and any other data used.

The emphasis on incomplete data in this report should not be taken to imply that response errors are smaller or less important than errors due to sampling or incomplete data. Response errors often have greater effects on survey results and are far more difficult to measure and reduce than sampling or nonresponse errors.

CHAPTER **2**

Problems of Incomplete Data

This chapter is concerned with the causes and effects of incomplete data. First there is a brief discussion of the problems resulting from incomplete data, namely, biases and distortions of distributions occur and cannot be assumed to be reduced by increasing the size of the sample. Second, the problems that lead to incompleteness are discussed as well as some types of nonresponse and the fact that incompleteness may occur because respondents do not wish to provide the information that is requested. Third, illustrative data are presented on levels and trends of nonresponse and refusal rates, on differences in nonresponse rates among subgroups of the population, and on biases resulting from nonresponse.

Finally, an approach to obtaining definitions of nonresponse rates is given. Expressions for the bias due to nonresponse are obtained in two simple cases, the arithmetic mean and the ratio of two totals. In the case of the mean, the bias is the product of a nonresponse rate and the difference between the mean for nonrespondents and the mean for respondents. A similar result is obtained for the ratio of two totals. These expressions suggest formulas for both unit and item nonresponse rates. The approach can be applied to population characteristics and estimators based on samples.

EFFECTS OF INCOMPLETE DATA ON SURVEY RESULTS

The main problem caused by incomplete data in sample surveys is that estimators of population characteristics and relations must be assumed to be

INCOMPLETE DATA
IN SAMPLE SURVEYS
Volume 1, Part I

ISBN 0-12-363901-8

biased unless very convincing evidence to the contrary is provided. Also, the mean square errors of such estimators are likely to be larger than the mean square errors obtained with full response, but they may on occasion be smaller. Finally, the univariate and multivariate distributions of unit characteristics will be distorted. As a result, means, variances, covariances, and other statistical functions will be biased and have distributions affected by incompleteness.

To reduce biases, variances, and mean square errors that result from incompleteness, greater efforts are often made to collect data, missing data are imputed, and estimates are adjusted. Variances of estimators will usually be larger when nonresponse occurs because the responding sample is smaller than the total sample and because the data from respondents are used to estimate or impute the uncollected data. However, the distortion in the distributions caused by incompleteness may lead to smaller variances. An extreme example would occur if there is a symmetric distribution with equal nonresponse at the two tails, possibly leading to unbiased estimators that have variances with downward biases that may more than compensate for the reduced size of sample.

TYPES OF INCOMPLETENESS

Undercoverage

Undercoverage occurs if units that should be on the frames or lists from which a sample is selected are not on the lists, if units in the frame or sample are incorrectly classified as ineligible for the survey, or if units are omitted from the sample or skipped by the interviewer. The definition of eligibility should be a part of the definition of the population to be surveyed. Many frames contain units that are not eligible for the survey. It is rare that information in the frame can be used to determine eligibility. The decision on eligibility of sample units is often made by an interviewer on the basis of observation or a brief screening interview and sometimes is made from more detailed information on the questionnaire.

When eligibility of a unit is determined during data collection, an error may occur. If a unit is erroneously classified as eligible, the additional information obtained in the interview may lead to a later correction of the error. If a unit is erroneously classified as ineligible, no opportunity for correcting the error necessarily occurs unless the classification as ineligible is verified. For example, in a survey of occupied households, households classified as vacant or nonresidential by the interviewer may indeed be occupied or residential. In a survey of establishments in specified industries, establishments classified as ineligible because of not being in one of the specified industries may be found to be eligible if more detailed information is obtained. An erroneous classification as ineligible causes undercoverage. Since units incorrectly classified as ineligible are likely to differ from units classified as eligible, biases may result.

Different types of units may have different propensities to be covered in a survey. Skipped units are likely to be skipped because they are difficult to find, to identify, or to interview, and such units are likely to differ from the other units in the sample. In area surveys, undercoverage rarely occurs among large units such as counties. But smaller units, such as blocks, establishments, or farms, may sometimes be omitted from a frame because the smaller units, as of the date for which the frame is constructed, had no final units, namely, households in the case of blocks, employees at work in the case of establishments, or acreage in specified crops in the case of farms. Undercoverage and biases may result if the omitted smaller units contain final units, such as a newly constructed apartment house, as of the time of the survey.

Lists of large establishments are often used as frames. Such lists are almost always somewhat out of date: changes in establishment size may have occurred; establishments may have merged, split, or been bought or sold in whole or in part; establishments that no longer exist may be on the list; new establishments may not be on the list. Sometimes a supplementary sample, e.g., an area sample, is selected from the entire population of establishments in an effort to reduce or eliminate undercoverage due to use of existing lists of establishments. When an area sample is selected to supplement a list, considerations of cost may lead the survey taker to exclude some areas because no establishments are expected to be in those areas. Undercoverage and biases will occur if establishments not on the list are present in the excluded areas as well as if interviewers do not cover all units in areas in the supplementary sample.

Often in preparing for or conducting a sample survey, lists of units are prepared as part of the survey tasks: e.g., lists of residential units on sample blocks, lists of establishments in one or more specified industries in specified areas, lists of members of households in the sample, and lists of departments or divisions of establishments in the sample. Such lists may be incomplete because the lists are out of date by the time the survey is taken or because of errors of the interviewer or the person furnishing the information. When such lists are used for interviewing or to select a sample, undercoverage will occur unless the lists are checked and completed during survey operations.

Sometimes the lister of dwelling units may incorrectly omit units from a list because, as mentioned earlier, the lister classifies the building containing the units as out of the scope of the survey, e.g., the building is classified as a nonresidential building although it contains residential units. If the lists from which the final sample is selected are incomplete, perhaps because the lister missed one or more residential units in a basement, a penthouse, or elsewhere in the building, undercoverage and biases will result.

Interviewers may receive assignments of areas that have boundaries defined on maps and may be instructed either to list or to interview all units in the area or a sample of such units. The interviewer then either explicitly or implicitly constructs a list of units as part of the assignment. Problems often occur in finding units, e.g., in hilly terrain, or in identifying the boundaries of the areas. Either undercoverage or overcoverage (caused by units considered by the interviewer to be in the interviewer's area that are not in that area) may then occur.

All telephone surveys are subject to biases because of incompleteness. Telephone books exclude households that do not have telephones and households that have unlisted telephones. Hence, samples selected from telephone books are subject to undercoverage. If random-digit dialing is used, households with unlisted telephones will be covered, but not households without telephones. Households with more than one telephone may be overrepresented in samples and estimates unless care is taken.

Various types of undercoverage and overcoverage may occur in mail surveys. Lists may be out of date and thus incomplete. Mail may not be delivered, or it may be delivered to incorrect addresses. A unit that does not reply may have moved, no longer exist, be ineligible for the survey, or be a nonrespondent. Thus, classification of a unit as not covered or as a nonrespondent may not be possible without additional information. The same units may appear on mailing lists with different names and addresses. Also, in mail surveys those to whom the particular survey is important are more likely to respond than are others, and biases are likely to result.

The examples discussed suggest that it is desirable to analyze every survey to determine the possibilities and sources of undercoverage. Any information available from other studies or administrative sources should be used to estimate or speculate on the size and direction of biases due to undercoverage. Sometimes better information on coverage will be available for items rather than units because totals of items may be available from administrative sources. Also, the exclusion from estimates of units that interviewers incorrectly call ineligible and units that are omitted from the sample or during interviewing makes it important to check interviewers' work for completeness as well as for accuracy.

Unit Nonresponse

Unit nonresponse occurs if a unit is selected for the sample and is eligible for the survey, but no response is obtained for the unit or the obtained response is unusable.

There are four primary reasons for unit nonresponse in household surveys:

(1) No one is at the unit when the efforts are made to interview.

(2) The interviewer cannot communicate with the persons in the unit, e.g., because of illness or a language problem.

(3) Total refusal occurs or the interview is broken off by the respondent and the partial response prior to breakoff is classified a refusal.

(4) The responses given by the unit are later classified as unusable.

Nonresponse may also occur for other reasons. For example, the interviewer fakes or "curbstones" the interview (i.e., does not conduct the interview but fills in a schedule or questionnaire as though the interview had been conducted) and the faking becomes known and the questionnaire is discarded but the real interview is not obtained.

Tables of frequencies and percentages of nonresponse by type of nonresponse are often included in survey reports. (Such tables are discussed in more detail in Chapter 3.) Categorization by type of nonresponse is subject to measurement error. For example, some units classified as "nonresponse, no one at home, repeated calls" may really be refusals by the resident not answering a doorbell.

Some refusals occur because the respondent does not wish to give certain information, such as income. Other refusals occur for reasons probably unrelated to the subject matter of the survey: for example, a person may refuse an interview because the time is inconvenient, or the doorman of an apartment house may refuse entry to an interviewer.

The type of nonresponse may be related to the nature of the biases resulting from nonresponse. For example, the responding sample in surveys in which any adult is an acceptable respondent is likely to contain higher proportions of households with a housewife or unemployed adult than in the complete sample since such units (households) are more likely to have a respondent available when an attempt at interviewing is made. In some surveys, units that refuse may be more nearly like responding units at least for some demographic items than are units "not at home, repeated calls." Sometimes, rather than refusing or continuing to refuse to respond, respondents may supply inaccurate information.

At present commonly used statistical methods of dealing with nonresponse do not depend on type of nonresponse; such methods need to be more fully developed in the future, particularly for panel or other surveys in which some or all units continue in a survey over time.

In obtaining information for people, a proxy respondent is a person who according to survey rules may provide information concerning another person. Thus, a respondent is a self-respondent for information about the respondent but a proxy respondent for other persons. For example, an adult member of a household may be a respondent for information on the respondent or on the total household but a proxy respondent for any member of the household other than the respondent.

Surveys often specify those who will be acceptable as respondents for the survey as well as the units for which information is desired. Sometimes the respondent is uniquely identified, e.g., the oldest female in a household or the specified officer of an establishment. More often a category of respondents is specified, such as any person aged 18 or over. Nonresponse increases if the specified respondent is uniquely designated. The designated respondent may refuse or not be contacted even though others in the unit might be available or willing to respond.

Nonresponse may occur at more than one level in a survey. For example, a sample of households may be selected and information may be requested both for the household and for designated persons in the household. A respondent may then report information for the household but not for one or more of the designated persons in the household. Nonresponse rates (see Chapter 3) may be calculated for both the primary (household) units and the secondary (designated person) units.

Item Nonresponse

Item nonresponse occurs if questions that should be answered are not answered or if the answers are classified as unusable. Sometimes item nonresponse occurs because the interview is broken off after being partly completed, but the partial response is not classified as a unit nonresponse. Item nonresponse may occur because the respondent (e.g., a proxy respondent) does not have the information needed for one or more questions, because the respondent refuses to answer specific questions, or because the interviewer or respondent skips the question. If an interviewer fakes the data, the faked items are, if detected, unusable.

Item nonresponse may occur for blocks of questions: an interviewer may miss a branch point in an interview, or a respondent may refuse to answer all questions on a specific subject, say, income or the final questions in an interview.

LEVELS AND CHANGES IN LEVELS OF NONRESPONSE RATES

Nonresponse rates vary widely from survey to survey and from survey organization to survey organization. These rates change over time and change even for repetitions of the same survey. Two compilations of nonresponse rates (Bailar and Lamphier, 1978, pp. 36, 37 and National Research Council, 1979, p. 140) illustrate the variability in nonresponse rates that is reported for various surveys. Response rates in the reported surveys vary from 13 to 95%, largely as a result of differences in the efforts made to collect the data, and for some items, such as income, of the questions asked. Apparent differences in nonresponse rates may also be due to the use of different definitions of nonresponse rates in different surveys. (Measurement of nonresponse is discussed in more detail in Chapter 3.)

Table 2 presents a summary of response rates reported in the case studies (in Part II). These rates do not differ as widely as do those in the compilations mentioned earlier, but they also show marked differences.

TABLE 2

Response Rates Reported by Selected Surveys[a]

Survey	*Unit response percentage*
Employment Cost Index, Bureau of Labor Statistics, Phase I	70.7
Livestock Inventory Surveys, Department of Agriculture, List Frame, Cattle Survey	86.5
Economic Surveys, Institute for Survey Research, University of Michigan	70.0
Office Equipment Survey, Westat	88.0
Annual Survey of Manufactures, Bureau of the Census[b]	84.2
Survey of Consumer Finances, Statistics of Canada[c]	83.5
Readership of Ten Major Magazines, Audits and Surveys	81.5

[a] The rates are based on unweighted counts of units in the survey, sometimes not including late returns. The data are from the case studies in Part II; nonresponse rates may be differently defined in the surveys.

[b] In 1976 the ASM was based on a sample of companies. Single establishment companies had a response rate of 73.4%; multiestablishment companies had a response rate of 89.7%; the response rate for all establishments was 84.2%. Since 1979 the ASM is based on a sample of establishments; the response rates are greater than 84%.

[c] The percentage of eligible responding units that had earlier responded in the Canadian Labour Force Survey.

Nonresponse and refusal rates have appeared to increase in some surveys, with refusal rates increasing relative to total nonresponse rates. [See, for example, Thomsen and Siring (Session I, Volume 3) and chapter by Steeh, Groves, Comment, and Hansmire (Part II this volume).] Even in surveys that have been able to keep total nonresponse low and relatively constant (see Table 3), refusal rates have tended to increase.

Table 3 shows nonresponse rates for two major surveys for the period 1968–1980. For both surveys total nonresponse has been maintained at a low level. However, during most of that period the proportion of total nonresponse accounted for by refusals has been increasing. The proportion seems to have leveled out during the past 4 years. If, for any reasons, the refusal ratios were to continue to increase as they have during most of the period covered, total nonresponse rates would eventually increase. On the other hand, stability in rates may occur.

Nonresponse rates and their components depend on how a survey is done and on the resources devoted to reducing them. Thus, those rates are affected by changes in the methods of conducting surveys (mail, telephone, personal interviews) and changes in the resources used for reducing nonresponse, e.g., increased numbers of callbacks, transfer of assignments to more experienced

TABLE 3

Nonresponse Rate (%) for Two Federally Conducted, Continuing Household Surveys, 1968–1980

	Current Population Survey			*Household Health Interview Survey*		
Year	*Total nonresponse*	*Refusals*	*Refusal (%)*[a]	*Total nonresponse*	*Refusals*	*Refusal (%)*[a]
1968	4.6	1.8	39	4.7	1.2	26
1969	4.6	1.8	39	4.7	1.3	28
1970	4.0	1.6	40	4.2	1.1	26
1971	3.7	1.6	43	3.6	1.1	31
1972	4.0	1.8	45	3.9	1.4	36
1973	4.3	1.9	44	3.6	1.5	42
1974	4.1	2.0	49	3.2	1.5	47
1975	4.1	2.2	54	3.1	1.6	52
1976	4.5	2.5	56	3.8	2.1	55
1977	4.1	2.5	61	3.3	1.9	58
1978	4.6	2.6	57	3.8	2.1	55
1979	4.6	2.7	59	3.9	2.2	56
1980	4.1	2.4	59	3.9	2.2	56

Source: Published and unpublished data from the Bureau of the Census for the Current Population Survey and from National Center for Health Statistics for the Household Health Interview Survey.

[a] Ratio of refusal rate to nonresponse rate (in percent).

and "better" interviewers, use of incentives to respond, use of methods of guaranteeing privacy, and so on.

VARIABILITY OF RESPONSE RATES BY SUBGROUPS

In almost all surveys important estimates are provided for subgroups (categories, domains, or areas) of the population as well as for the entire population. These estimates also are subject to the biases and increases in variance resulting from nonresponse. Often such estimates may be sufficiently important for the purposes of the survey that special efforts will be made to increase response within the particular domains. Thus, it is useful to know nonresponse rates by subgroups.

When information from the full sample is available that enables nonresponse rates to be calculated by subgroups, that information can be used during the data collection process to allocate resources in order to improve data collection and to define estimators that may reduce the effects of nonresponse. Such information also increases the understanding of the survey estimates.

Two illustrations of such data are presented below. In the first, shown in Table 4, the Employment Cost Index Survey (see chapter by Boyes and Conlon, Part II, this volume) data on the number of employees in each establishment in the sample were available from a base period survey (Phase I). Then, in the quarterly survey that constituted Phase II, nonresponse rates were obtained by employment size class of establishment.

The data in Table 4 suggest that if it had been intended to publish data for the size class consisting of units having 1–7 employees (it was not intended), careful consideration would have been necessary since the possible bias for that size class could be very large, depending on the differences, usually unknown, between the respondents and nonrespondents in that class. In estimates of population totals, not by size class, the proportion of total employees, wages, etc., in the 1–7 size class may be small enough for the totals not to be badly biased by this size class, but the proportion in the size class will vary by industry. If data are to be published by industry, there may be industries for which biases due to nonresponse in the smallest size class result in appreciable biases in industry totals.

The second example deals with item nonresponse. In March of each year the Bureau of the Census's Current Population Survey (CPS) questionnaire includes a supplement in which detailed data on income are requested for each member aged 14 and over of the sample households. In March 1978, 11 questions on income were asked for each member. The nonresponse rate on income (one or more income items not reported) for persons aged 14 and over who were

TABLE 4

Nonresponse Rates by Employment Size Class of Establishments

Number of employees	*Establishment nonresponse rate* (%)
1–7	67.5
8–19	6.0
20–49	6.2
50–99	9.7
100–249	5.9
250–499	6.9
500–999	.9
1000 and over	4.8

Source: Chapter by Boyes and Conlon, Part II, this volume.

TABLE 5

Person Nonresponse Rates for Income

Males: age in years	*Nonresponse rate for one or more income items (%)*	*Nonresponse rate[a] for all income items (%)*
14–19	17	8
20–24	16	7
25–34	14	6
35–44	20	8
45–54	25	11
55–64	27	11
65 and over	26	10

Source: Letter from John Coder, Bureau of the Census, April 1980.
[a] The nonresponse rate for all items is part of the response rate for one or more items.

respondents for the CPS was 20%. As shown in Table 5, one of the variables with respect to which nonresponse varied most for males was age. A similar table for females shows minor differences from Table 5. For some purposes, the nonresponse rates for one or more income items suggest that care is required in using the data for older ages.

DATA ON BIASES

Relatively few sources exist for data on biases due to nonresponse. Some data are found in Chapter 4 of this part and in Tables 8 and 9 of the chapter by Thomsen and Siring (Session I, Volume 3). The tables by number of calls also suggest the nonresponse biases that would occur if interviewing stopped after a given number of calls.

Biases are minor for many items, but when the item, for example, size of family, is related to one of the demographic items associated with the cause for nonresponse, say, "No one at home—repeated calls," the bias may be large. Similarly, the bias may be large for an item such as income or illness when refusals occur because the respondent does not wish to report the income or illness that would be reported. Many items are not particularly associated with reasons for nonresponse. For such items nonresponse may occur approximately at random, at least within poststrata, but some bias is likely to remain.

Whether nonresponse biases and mean square errors are causes for concern depends on the uses to be made of a survey, knowledge of how large the errors may be, and the ways in which data are summarized.

BIASES AND NONRESPONSE RATES

As suggested, the differences between respondents and nonrespondents result, in general, in biases. In this section expressions for biases are obtained that provide statistical motivation for the commonly used definition of a nonresponse rate by showing the dependence of those biases on the nonresponse rate. The expressions also suggest the need for joint modeling of nonresponse and the relations of variables for respondents with variables for nonrespondents, or the modeling of the relation conditional on the nonresponse rate and the respondents and data about them. The formulation thus implies the need to consider how efforts to increase response may affect the differences between respondents and nonrespondents since both affect the biases. (Other models are discussed in Chapter 5 and elsewhere in this report.)

Suppose that the survey population consists of N units, identified by $1, 2, \ldots, N$, and Y_i is associated to unit i ($i = 1, 2, \ldots, N$), where Y is ordinarily a vector of many components but is here treated as a single variable. For simplicity, $Y_1, \ldots, Y_N$ will be assumed to be fixed values. (The discussion, with minor changes, also holds when the Y_i are assumed to be random variables.)

Suppose that the objective of a survey is to estimate a population total $T = Y_1 + \cdots + Y_N$ or the arithmetic mean $\bar{T} = T/N$. The entire population is, hypothetically, surveyed under the conditions of the sample survey that is being made. It is assumed that responding units have no item nonresponse. (This approach is conditional on the values of Y and on the nonrespondents that would result from a particular hypothetical survey of the population.)

Denote by T_{R} the total of Y for the N_{R} responding units and by T_{NR} the total of Y for the N_{NR} nonresponding units: $N_{\mathrm{R}} + N_{\mathrm{NR}} = N$. Denote by $\bar{T}_{\mathrm{R}}$ and $\bar{T}_{\mathrm{NR}}$ the corresponding arithmetic means. Then if $\bar{T}_{\mathrm{R}}$ is used as an approximation to $\bar{T}$, the bias is $\bar{T} - \bar{T}_{\mathrm{R}}$ (see Cochran, 1977, p. 361), and

$$\bar{T} - \bar{T}_{\mathrm{R}} = \frac{N - N_{\mathrm{R}}}{N}(\bar{T}_{\mathrm{NR}} - \bar{T}_{\mathrm{R}}).$$

Define the population unit nonresponse rate $\tilde{R}$ by

$$\tilde{R} = \frac{N - N_{\mathrm{R}}}{N}.$$

Then the bias $\bar{T} - \bar{T}_{\mathrm{R}}$ is the product of the nonresponse rate and the difference between the means for nonrespondents and respondents.

Suppose the values of Y_i for the $N - N_{\mathrm{R}}$ nonresponding units are estimated by $\hat{Y}_i$. The estimates may be imputations or any other estimators. An estimator $\hat{T}$ of $\bar{T}$ may then be defined by

$$\hat{T} = \frac{N_{\mathrm{R}}}{N}\bar{T}_{\mathrm{R}} + \frac{N - N_{\mathrm{R}}}{N}\hat{T}_{\mathrm{NR}},$$

where $\hat{T}_{\mathrm{NR}}$ is the mean of the $\hat{Y}_i$ for the $N - N_{\mathrm{R}}$ nonresponding units. Then

$$\bar{T} - \hat{T} = \tilde{R}(\bar{T}_{\mathrm{NR}} - \hat{T}_{\mathrm{NR}}).$$

In the special case where each $\hat{Y}_i$ is defined by

$$\hat{Y}_i = \bar{Y}_{\mathrm{R}},$$

it follows that $\hat{T} = \bar{T}_{\mathrm{R}}$.

Thus, the reduction in bias, if any, is due to using $\hat{Y}_i$ as an alternative to imputing $\bar{Y}_{\mathrm{R}}$ for each nonresponding unit i and is caused by the reduction, if any, of $|\bar{T}_{\mathrm{NR}} - \bar{T}_{\mathrm{R}}|$ to $|\bar{T}_{\mathrm{NR}} - \hat{T}_{\mathrm{NR}}|$. Imputation does not alter $\tilde{R}$.

The unit nonresponse rate $\tilde{R}$ also occurs in population biases of totals since $N(\bar{T} - \hat{T}) = N[\tilde{R}(\bar{T}_{\mathrm{NR}} - \hat{T}_{\mathrm{NR}})]$.

Two comments should be made. First, N is the number of *eligible* units in the surveyed population. If measurement errors in determining eligibility are considered, then N is a random variable. Second, $\tilde{R}$ is the population nonresponse rate. Corresponding to $\tilde{R}$ are $\tilde{r}$, the nonresponse rate for the sample, and $\hat{R}$, the estimated population nonresponse rate based on the obtained sample. In general, $\tilde{r}$ is defined by $\tilde{r} = (n - m)/n$ when n is the number of eligible units in the sample and m is the number of eligible responding units. Also, $\hat{R}$ will usually be a weighted nonresponse rate reflecting, in probability sample designs, the probabilities of selection of sample units, responding or not.

The sample nonresponse rate $\tilde{r}$ measures one of the several aspects of the effectiveness of a date collection effort. The estimated population nonresponse rate $\hat{R}$ estimates a factor $\tilde{R}$ of the bias due to nonresponse. Increased data collection efforts should reduce both $\tilde{R}$ and $|\bar{T}_{\mathrm{NR}} - \hat{T}_{\mathrm{NR}}|$, while statistical methods of dealing with nonresponse will reduce only $|\bar{T}_{\mathrm{NR}} - \hat{T}_{\mathrm{NR}}|$.

The unit nonresponse rate $\tilde{R}$ has the same value for all characteristics. The definition of an item nonresponse rate is the same as $\tilde{R}$ except that now N is the number of eligible population units that should respond to the particular item, and N_{R} is the number of such units that do respond to that item. Also, $\tilde{R}$ is the proportion of units in the population that should, but do not, respond to the particular item. It is often necessary to define item nonresponse relative to the population of respondents because of the difficulty of knowing whether a nonrespondent should have responded to the particular item. Then for item k the nonresponse rate is

$$\tilde{R}_k = \frac{N_{\mathrm{R}k} - N_{\mathrm{RR}k}}{N_{\mathrm{R}k}},$$

where $N_{\mathrm{R}k}$ is the number of eligible responding units that should respond to item k and $N_{\mathrm{RR}k}$ is the number of eligible responding units that should and do respond to item k.

As discussed, the values of $\tilde{R}$ and $\bar{T}_{\mathrm{R}} - \bar{T}_{\mathrm{NR}}$ will be affected by how a survey is conducted. For example, interviewers and their assignments influence both nonresponse and the biases resulting from nonresponse. So also do questionnaires, including the information requirements of various statistical procedures for treating nonresponse. Another factor influencing both occurrence of and bias due to nonresponse is the designation of allowable respondents.

Sometimes survey procedures are changed in order to decrease $\tilde{r}$ without sufficient concern for the effects on $\overline{T}_{\mathrm{NR}} - \hat{T}_{\mathrm{NR}}$. Increased efforts to obtain responses after noting that $\tilde{r}$ appears to be large may effectively increase response from only one part of the survey population and may not decrease $|\overline{T}_{\mathrm{NR}} - \hat{T}_{\mathrm{NR}}|$ for some important characteristics.

Ratios of characteristics are often estimated in sample surveys and rates, often called coverage rates but perhaps better called item or characteristic coverage rates, have been found to be useful.

Let X_i be the value of variable X associated to unit i of the population, $i = 1, 2, \ldots, N$, and suppose the population characteristic to be estimated is the ratio

$$Z = T/T_X,$$

where $T = Y_1 + Y_2 + \cdots + Y_N$ and $T_X = X_1 + X_2 + \cdots + X_N$. Define T_{R} and $T_{X\mathrm{R}}$ to be the sums of Y and X for the responding population. Algebraically,

$$\frac{T}{T_X} - \frac{T_{\mathrm{R}}}{T_{X\mathrm{R}}} = \frac{T_{X\mathrm{NR}}}{T_X}\left(\frac{T_{\mathrm{NR}}}{T_{X\mathrm{NR}}} - \frac{T_{\mathrm{R}}}{T_{X\mathrm{R}}}\right). \tag{1}$$

Then the item or characteristic coverage rate C is defined by

$$C = T_{X\mathrm{R}}/T_X,$$

and the item or characteristic noncoverage rate $\tilde{C}$ is $1 - C$. It follows that the population bias for the ratio Z is the product of the characteristic undercoverage rate $\tilde{C}$ and the difference of the ratios Z for the responding and nonresponding parts of the population.

Item or characteristic coverage rates are of particular interest in sampling establishments or farms, where X may be a variable such as number of employees or amount of sales or acreage in crops perhaps for an earlier time period. In establishment surveys, unit nonresponse is often much greater for small than for large establishments (see Table 4). Thus a survey with a large nonresponse rate may have high coverage rates for certain items or characteristics—which may be the important items in the survey. Thus, both nonresponse and characteristic coverage rates provide useful information.

If T_X is known and if the $\hat{Y}_i$ are estimated for nonresponding units, then

$$\frac{T}{T_X} - \frac{T_{\mathrm{R}} + \hat{T}_{\mathrm{NR}}}{T_X} = \frac{T_{X\mathrm{NR}}}{T_X}\left(\frac{T_{\mathrm{NR}}}{T_{X\mathrm{NR}}} - \frac{\hat{T}_{\mathrm{NR}}}{T_{X\mathrm{NR}}}\right), \tag{2}$$

which reduces to (1) if $\hat{Y}_i$ is calculated by

$$\hat{Y}_i = X_i \frac{T_{\mathrm{R}}}{T_{X\mathrm{R}}}.$$

Thus the expression (2) is the product of the characteristic noncoverage rate for X and the difference of actual and estimated ratios for the nonresponding units.

Characteristic coverage rates may be computed for the sample. The population characteristic coverage rate may be estimated from the sample.

When the X_i are not known for all units in the populations, T_X may be obtained from another source, perhaps an administrative source, or it may be estimated. Then the coverage rates are computed subject to (usually unknown) error due to the obtained T_X and are conditional on the value of T_X.

The previous results generalize when the type of nonresponse is considered. Suppose, for example, that there are $K - 1$ types of nonresponse. Then

$$\bar{T} = \sum_{k=1}^{K} \frac{N_k}{N} \bar{T}_k,$$

where $k = 1$ denotes response, $k = 2, \ldots K$ denote the types of nonresponse. Also, N_k and T_k are the numbers of units and the mean values of units in category k, if those values were known, $k = 1, \ldots, K$. Let

$$\hat{T} = \sum_{k=1}^{K} \frac{N_k}{N} \hat{T}_k,$$

where $\hat{T}_1 = T_1$ and $\hat{T}_2, \ldots, \hat{T}_k$ are the results of imputing or estimating the Y_i in response categories $2, \ldots, K$.

Then the bias is

$$\bar{T} - \hat{T} = \sum_{k=2}^{K} \frac{N_k}{N} (\bar{T}_k - \hat{T}_k), \tag{3}$$

where N_k/N is the nonresponse ratio for type k. Also

$$\bar{T} - \hat{T} = \tilde{R} \sum_{k=2}^{K} \frac{N_k}{N - N_k} (\bar{T}_k - \hat{T}_k), \tag{4}$$

where

$$\sum_{k=2}^{K} N_k = N - N_R.$$

The expression (4) both suggests the importance of estimating its components and the possibility of conjecturing biases based on those estimates.

CHAPTER **3**

Measuring and Reporting Nonresponse

In this chapter nonresponse rates are defined for the sample rather than for the population (as in the last section of Chapter 2). Completion rates and item or characteristic coverage rates are also defined. Uses of nonresponse rates are considered and problems of measuring and reporting nonresponse are discussed. Definitions of sample nonresponse, completion rates, and item coverage rates are given in the first section of this chapter and discussed both for household and establishment surveys. Limitations of the usefulness of nonresponse rates are discussed in the second section of this chapter.

One suggestion for increasing the understanding and usefulness of nonresponse rates is an accountability table, discussed in the third section of this chapter. Accountability tables disaggregate the difference between the size of a selected sample and the number of units in the responding subset of the sample, by the reason that sample units do not respond and perhaps by other variables as well. Accountability tables clarify the meaning of nonresponse rates, since the components used in the rates appear in the tables, and also make it possible for users to select alternative definitions that they prefer.

The report of a survey should provide information on nonresponse as well as on other aspects of the quality of the survey. Standards for providing information on nonresponse are discussed in the final section of this chapter. These standards should cover not only final reports but also preliminary and intermediate reports.

INCOMPLETE DATA
IN SAMPLE SURVEYS
Volume 1, Part I

ISBN 0-12-363901-8

DEFINITIONS AND USES OF RESPONSE AND NONRESPONSE RATES

Perhaps the most frequently used *nonresponse rate* is the ratio $\tilde{r}$ of the number of nonresponding eligible units $n - m$ to the number of eligible units n in the sample:

$$\tilde{r} = 1 - (m/n). \tag{1}$$

The *response rate* r is defined by $r = 1 - \tilde{r}$. The nonresponse rate $\tilde{r}$ is similar to the population nonresponse rate $\tilde{R}$ appearing in the expression for the population bias in Chapter 2.

Although the nonresponse rate $\tilde{r}$ as defined by (1) is used very widely, it is not appropriate for all survey designs. For example, suppose a sample containing n_1 eligible units is first selected of which m_1 units respond. Then a sample containing n_2 eligible units is selected from the $n_1 - m_1$ eligible nonrespondents and m_2 units respond. Then the nonresponse ratio

$$\tilde{r} = \tilde{r}_1\tilde{r}_2, \tag{2}$$

where $\tilde{r}_i = (n_i - m_i)/n_i$, $i = 1, 2$, is suggested by bias analyses similar to those in Chapter 2 performed both for the population and the sample.

Alternatives to (1) can be similarly suggested for stratified and other designs when the objective is to find a nonresponse ratio related to, although not uniquely determining, the bias due to nonresponse.

Completion rates are often prepared in addition to, and sometimes in place of, response rates. Many definitions of completion rates exist (see Wiseman and McDonald, 1980, and CASRO Task Force on Completion Rates, 1982). Completion rates are intended to measure completion of various tasks associated with a survey. One completion rate, for example, is

$$c_1 = \frac{\text{number of households contacted}}{\text{number of households in sample}},$$

where a contacted household might be defined to be one where the interviewer speaks with a person who would be an acceptable respondent. But "contacted" could have other definitions related to how well the data collection tasks have been fulfilled. (This emphasizes how important it is for a survey report to contain precise definitions of words such as contacted.)

Another example of a completion rate is

$$c_2 = \frac{\text{number of interviews with eligible households}}{\text{number of households in sample}}.$$

The only difference between c_2 and $\tilde{r}$ when the survey unit is defined to be a household is that the denominator is the sum of the numbers of eligible and ineligible households in the sample, thus avoiding the difficult task of determining eligibility of units that are not interviewed.

A third rate, often prepared, is the *item* or *characteristic coverage rate*

$$c = X_R/X_T,$$

where X_R is the total of some variable, e.g., employment or acreage for eligible responding units in the sample, and X_T is the corresponding total for all eligible units in the population (sometimes for the sample).

The item undercoverage rate is $\tilde{c} = 1 - c$. The coverage rate is most often used for establishments or farms. It is usually defined in terms of some variable that is fundamental to many of the analyses to be made and *is known* for all units in the population. Sometimes, however, the denominator of an item or characteristic coverage rate is known, at least approximately, from outside sources, and the numerator is based on the sample.

It is important to note that *item* or *characteristic undercoverage* just defined does not differ from *undercoverage* defined earlier to be the type of incompleteness occurring when *units* that should be in a frame are not in the frame or when units are incorrectly classified as ineligible or skipped during the interviewing process. In the earlier definition the reference is to units, i.e., the definition is the same if an item is defined as 1 if the unit is in the population, is eligible, and should be counted, and the item is defined as 0 otherwise. Then X_R is the number of units counted and X_T is the number that should be counted if the entire population (or sample) were counted without error.

It is often useful to compute nonresponse rates $\tilde{r}_i$, where i denotes an interviewer, a geographic area, a domain, or a category, e.g., three-person households, of the population; $\tilde{r}_i$ is defined as the ratio of $n_i - m_i$ (the number of responding eligible units in a category i) to n_i (the number of eligible sample units in category i), $r_i + \tilde{r}_i = 1$.

If nonresponse rates by interviewer or by interview area are among the data produced by a management information system before final data collection, the rates may identify interviewers and areas that need additional support in order to achieve satisfactory response rates. Also, preliminary tabulations may indicate that the responding sample is too small or too unequally distributed for satisfactory estimates of important population characteristics to be made. Final data collection and conversion efforts may be made in order to achieve response rates that are satisfactory for interviewers, areas, and important population characteristics.

An important type of nonresponse rate is one that corresponds to the *reason* for nonresponse. For example, the *refusal rate* may be defined to be the ratio of the number of eligible units that refuse to be interviewed to the number of eligible sample units. Other nonresponse rates by reason for nonresponse are similarly defined. A difficulty in computing nonresponse rates is that the nonresponse may occur without information on eligibility being provided.

Refusal rates may also be defined as completion rates, i.e., with the number of sample units as a denominator. Hence it is important in any survey to state the definition used and to provide the data with which alternative definitions may be computed.

Response rates by number of attempts to complete an interview, time since a mailing, date of telephone, or wave of a survey are very useful, particularly when the characteristics of eligible units interviewed are correspondingly summarized (see, for example, Thomsen and Siring, Session I, Volume 3, and Drew and Fuller, 1980).

Unweighted unit nonresponse rates provide an indication of how well the data collection effort was carried through for the entire sample and for specified subsets of the sample defined in terms of interviewers and their assignments, areas, and categories. They also provide a means of managing the data collection effort to obtain not only a good overall response rate but also a data collection effort distributed to achieve satisfactory response rates in strata and important domains.

Weighted unit nonresponse rates provide estimates of the proportion of the population or of specified totals or other population characteristics for the subpopulation that would have responded to the survey. Both weighted and unweighted unit nonresponse rates provide data useful in adjustment for nonresponse.

In practice, *item response rates* can provide information on item nonresponse only for *responding* units. Thus, it becomes practical to calculate item nonresponse rates for the entire eligible sample only if the number and characteristics of nonresponding units that should respond to a particular item are known. Hence, the item response rate is usually calculated as the ratio of the number of eligible units responding to an item to the number of *responding* eligible units that should respond to that item.

In most surveys, item nonresponse rates for different items do not differ much from item to item except for items such as illness and income, which involve privacy, confidentiality, or memory. Also complex items, i.e., items consisting of several component items, will have different item response rates from single items. Income is an example of an item in which an item nonresponse may be defined to occur if, say, there is a nonresponse to one or more of several sources of income.

Sometimes an item, e.g., number of employees, is such that the total of the item, perhaps estimated, may be available for the population. In such cases, item coverage rates as defined earlier may be calculated for the particular item. Sometimes, in establishment or farm surveys, the unit response rate may be fairly low, but the important population characteristics to be estimated are totals, and the low rate of unit response may be primarily due to nonresponse of small establishments or farms that cumulatively contribute little to the estimate of the population total. The item coverage rates may then be high even with a relatively low unit response rate if the sample includes all large units.

Nonresponse rates may often be computed for several types of units. For example, in household surveys, information may be sought for the household and also for one or more individuals within the household. Then nonresponse rates may be computed both for households and for individuals.

A questionnaire may have a section to be completed only for specified units, e.g., only if there are two adults in the household. Thus, there may be completed interviews for part of the questionnaire, but the unit may be ineligible or nonresponding for the part on which information is to be obtained only for units with two adults. Similarly, part of the questionnaire may be completed for a designated individual or subunit, perhaps chosen at random, perhaps not. The designated individual may not respond because the individual is not at home, refuses, or does not communicate. Undercoverage for subunits or people occurs if the respondent for the unit is asked to provide a list of subunits but provides an incomplete list. Thus, undercoverage and unit or item nonresponse may occur not only for the primary unit but also for secondary units within the primary unit, such as subfamilies or persons within a household. Hence, separate nonresponse rates and undercoverage rates, if available, for primary units and secondary units within primary units may be desirable in certain surveys.

Nonresponse rates may be defined for an entire questionnaire, for blocks of questions within the questionnaire, and for individual items. Sometimes an entire questionnaire may be classified as nonresponse because specified items are missing or unusable and are important to the survey objectives.

During data processing, missing or unusable entries or both may be replaced by usable entries determined from the questionnaire and other information, perhaps information obtained from other respondents in the survey or respondents in another survey. Such information is said to be *imputed.* The procedure for replacing missing or unusable items is called an imputation procedure. *Imputation does not alter the response rate.*

When unit nonresponse occurs, another population unit is sometimes substituted for the nonresponding unit. Responding substitute units, even when designated during selection of the original sample for use if needed, are units from the *responding subpopulation.* Also, the possibility of making a substitution may affect the interviewer's willingness to accept an originally selected unit as being a nonresponse unit. Finally, the probability of a substitute unit is not the same as that of the originally selected unit. Hence, in computing nonresponse rates, substitutions should be treated as though they are imputed units. *Nonresponse rates should not be decreased if substitutions or imputations are made.* However, a separate statement of numbers and rates of substituted units should be included in reports.

Interviews or items found to have been faked are unusable. Nonresponse rates that have been calculated before faked responses are identified should be recalculated unless faked responses are replaced by the actual responses. Good survey practice requires that field verification procedures keep faking at a negligible level.

Not only are many definitions of nonresponse possible, but also many implementations of those definitions of nonresponse are possible. For example, in panel surveys made in several waves of interviews, at least two relevant nonresponse rates for each wave may be defined: the *cumulative nonresponse rate*, in which the denominator is the total number of eligible sample units in the panel, and the *current nonresponse rate*, in which the denominator is the number of interviews or eligible interviews attempted in the current wave. Neither rate by itself provides an adequate measure of nonresponse. In longitudinal surveys in which the population is changing over time, several nonresponse rates may have meaning.

Nonresponse rates in different surveys are often compared. For such comparisons to be meaningful, the definitions should agree. Hence, it is desirable that a small number of nonresponse rates be defined in order that all surveys might include in their reports at least one of these rates (see CASRO Task Force on Completion Rates, 1982). Each survey might include other rates, in addition, if desired.

As has been suggested earlier in specific contexts, the different definitions of nonresponse and the problems of measuring nonresponse suggest that both numerators and denominators be *explicitly defined* and that the population from which the sample is selected be stated in detail. In order to define different rates in which different users may be interested, the information from which the ratios are or can be computed should be given in any survey report. One means of achieving this objective is to include accountability tables, which are discussed later.

Nonresponse rates for surveys of farms and establishments are defined in terms of units (as are those for surveys of households). Item or characteristic coverage rates are more often used for establishments or farms than for households. For example, a unit response rate of .80 for establishments may be associated with a coverage response rate of, say, .95, where the numerator is total sales (total employment) of responding establishments in the sample and the denominator is total sales (total employment) of all eligible establishments in the population; the denominator total may be obtained from an outside source or estimated in the survey, using nonresponse procedures.

Surveys of establishments or farms are often made using a list as the entire frame or as one of several frames. Nonresponse rates may be computed for the portion of the survey based on a list, as well as for the remainder of the survey and the total survey.

It is often useful to compute establishment nonresponse rates by size of establishment. In practice, nonresponse rates usually differ by size of establishment; often, an appreciable number of establishments in one or more size strata will not respond, thus creating the possibility of large biases and large mean square errors of estimates that depend on such strata.

PROBLEMS IN CALCULATING AND USING NONRESPONSE RATES

Problems in calculating and using nonresponse rates occur because

(1) Nonresponse rates depend on the definition of the survey population.
(2) It is often difficult to define and implement the definition of eligibility, especially for units that are not interviewed.
(3) Nonresponse rates may be difficult to define for complex surveys.
(4) Nonresponse rates do not measure the quality of survey estimators; they measure one aspect of the quality of data collection.

Consequently, measures of the quality of survey outcomes should not be limited to nonresponse rates but should include or at least discuss estimated biases, if available, variances, and mean square errors.

Nonresponse Rates and the Definition of the Survey Population

Sometimes, one can predict that nonresponse rates will be low for one definition of the survey population (the population of which a sample will be selected) and higher for another definition of the survey population. Thus, the relation of the survey population to the target population (the population for which inferences are to be made) must be considered, as well as the nonresponse rate for the survey population actually used. A survey with low nonresponse rate may appear to be of higher quality than a survey with higher nonresponse rate, but the appearance may be misleading if the survey population for the survey with the lower nonresponse rate differs more from the target population than does the survey population for the survey with the higher nonresponse rate.

In surveys with one or a small number of sponsors and not for public use (e.g., for use only by the sponsors and possibly a small number of other users), agreement may be reached to use a survey population for which a low nonresponse rate is anticipated, even though that survey population differs considerably from the target population. For example, survey takers expect that those who respond on one survey are more likely to respond on a second survey. Thus, if the sample for a second survey is a subsample of the respondents for the first survey, and the survey topics are similar, a rather high response rate may be expected for the second survey since it is a survey of respondents to the first survey. An alternative approach in the second survey would be to sample also the nonrespondents in the first survey, thus possibly reducing the bias and mean square error, even though the nonresponse rate may increase. Use of the first procedure is much easier to accept for a survey with few sponsors who share a known risk than for a survey with many possible users for which the risks of the users are unknown.

In general, different subsets of a population may have different proportions of respondents in a particular sample survey. To define a survey population by reducing or eliminating subsets having low response rates will lead to estimators having lower nonresponse rates in the survey population, but the estimators may have larger biases and possibly larger variances as estimators of characteristics of the target population. Inferences from the survey population to the target population may thus become very difficult.

A discussion of nonresponse rates should include clear definitions of survey and target populations, should discuss what is known about the nonresponse rates for the survey population and important domains, especially for items for which the response that would be made affects the probability of responding, and should discuss how to make inferences from the survey population to the target population.

Meaning and Implementation of a Nonresponse Rate

This section presents a discussion of the problems that occur in defining and calculating response rates in terms of the ratio of the number of nonresponding eligible sample units to the total number of eligible sample units. Very often, some of the units in a sample will be ineligible for inclusion in the survey. Also, the definition of eligibility will vary among surveys of the same types of units. For example, a vacant housing unit is ineligible for a survey of occupied households, but eligible for a survey of all housing units.

The determination of eligibility is often made by observation or a screening interview. Sometimes the determination of eligibility is made by inspection of records for the sample units, and sometimes interviewers are paid only for completed questionnaires, while time spent on nonresponse units is not paid. In the latter case, the number of units classified as ineligible is likely to be erroneously high. One biasing factor is that units likely to be difficult to interview may be classified as ineligible. Other sources of error with respect to eligibility occur when a sample unit is not contacted, the screening interview is refused, or access to the unit is inadequate for an inspection to determine eligibility.

Difficulties in assessing eligibility affect both the numerator and denominator of the nonresponse ratio. The number of nonresponses (numerator) may be too high because some nonresponses were really ineligible units but this could not be determined because of refusal or noncontact. Alternatively, or at the same time, the number of nonresponses may be too low because the interviewer classified some units as ineligible rather than as nonresponses. The number of eligible units (denominator) is similarly affected. One reason for preparing completion rates is to have rates with denominators not dependent on the estimation of eligibility for units not in the sample.

It is important not to change nonresponse rates because of imputation, substitution, or assumptions concerning the relations of data for respondents and nonrespondents. (As discussed earlier, the selection of a subsample of nonrespondents for intensive interview will affect the nonresponse rates computed for the first interviewing effort.) In general, callbacks and follow-ups result in reducing nonresponse rates.

Nonresponse and the Quality of Survey Estimators

It is possible to analyze the effects of nonresponse on totals, means, or ratios in such a way that the biases are products of nonresponse rates and the differences of characteristics of responding and nonresponding parts of the population (see Chapter 2). The nonresponse rate does not, therefore, uniquely determine the nonresponse bias. The differences in the characteristics of the responding and nonresponding units of the population also are important, and usually more important. Nonresponse rates may also be useful in estimating the variances of the estimators that take account of nonresponse. If reasonably small, nonresponse rates may be a minor part of such calculations. The formulas for biases and variances usually depend, not on the nonresponse rates for the entire sample, but on nonresponse rates for strata, including poststrata if these are used.

Thus, nonresponse rates do not uniquely determine the biases, the sampling errors, or the distributions of estimates. Large nonresponse rates only indicate the possibility of large biases and mean square errors of survey estimators or of badly distorted distributions of estimators. To measure fully the effects of nonresponse on estimators requires estimates of bias, variance, and mean square error and the distributions of estimators that reflect how the estimates were calculated in the presence of nonresponse.

ACCOUNTABILITY TABLES

Many surveys publish accountability tables or similar tables. The purpose of accountability tables is to account for the reduction from the number of units in the sample to the number of units for which responses are made by classifying units into categories for exclusion from the population, for ineligibility and for nonresponse by type of nonresponse and for response. Accountability tables should be constructed to enable alternate nonresponse or completion rates to be calculated. Illustrations of categories in accountability tables are given in Table 6. Accountability tables will vary from survey to survey, but the illustrations suggest the objectives of accountability tables.

An accountability table begins with the selected sample. For example, if the sample is a sample of households, then the accountability table begins with the number of households in the sample. The accountability table continues with a description of the units found at the household addresses by whether the units are eligible for the sample, e.g., the number of nonresidential units, the number of residential units, and the number of those residential units that are unoccupied. This classification could be used whether the survey is a survey of occupied units or of all units.

The next categories are relevant to determining the interview status of units in the sample. Units are classified as "not at home, repeated calls," "refusals," "language problems," "illness," and other categories into which the units are classified for the particular survey. As a result of the information obtained, there may be a further classification of whether the units are eligible for the survey according to the criteria set for the survey and also a determination of whether there are subunits within the unit, e.g., people within the household, for which information is desired. Thus, the number of primary units is determined. For each primary unit there may be a number of secondary units in the sample, and the questions of eligibility, contacting, refusal, etc., must be answered for the secondary units, and so on.

Accountability tables can be classified by stratum, by interviewing area, by interviewer, or by call number in order to facilitate the management of the survey, including making decisions as to whether additional support is required for certain interviewers or interviewing areas or whether a reassignment of units to other interviewers is desirable.

If accountability tables are prepared by stratum and summarized by call number within stratum, an indication of the total effort and total productivity with respect to response by stratum will be available. This information can be used in any decisions on a final effort intended to ensure that the sample should not become poorly distributed as a result of nonresponse. For example, it may be determined that one stratum has a very low response rate compared to another stratum and therefore that it may be necessary to make additional efforts in that stratum rather than to distribute the effort equally over all the strata.

An accountability table should be accompanied by descriptions of the survey population and the target population. The survey population should be defined in sufficient detail to show any limitations that may affect response rates. The target population may or may not be the survey population. If it is not the survey population, the differences between survey and target populations should be discussed, since direct inferences from the sample are to the survey population but the primary interest is for the target population. Possible

TABLE 6

Illustrative Accountability Tables

Occupied Household Survey: Personal Interview[a]

Households, total in sample
- Nonresidential
- Residential, vacant, nonseasonal
- Residential, vacant, seasonal
- Occupied households
- Refusal
- Not at home, repeated calls
- Illness
- Language problem
- Noninterviews, total
- Interviews, total
- Interview households not meeting requirements for secondary unit interviews
- Interview households meeting requirements for secondary unit interviews

Secondary units, total
- Refusal
- Not at home, repeated calls
- Illness
- Language problem
- Secondary units, noninterviews, total
- Secondary units, interviews, total

In household surveys, some basic possible categorizations are location (stratum or domain), interviewer, item, time period, call number, and wave.

Establishment Survey

Establishments in sample from address list
Establishments in sample not from address list
Establishments, total in sample
For establishments from list sample
- Establishments, ineligible (out of scope)
- Establishments, out of business
- Establishments, eligible (in scope)
- Refusal
- Unable to complete interview
- Noninterviews, total

For establishments not from list sample
- (Same categories as above)

For establishments, total, list and nonlist samples
- (Same categories as above)

In establishment surveys, some basic possible categories are size, standard industrial classification (SIC) category, domain, item, time period, call number, wave.

[a] The basic entry for each row is the number of households; percentages to various denominators are useful.

sources and sizes of undercoverage should also be discussed in an introduction to accountability tables.

REPORTING NONRESPONSES

The results of surveys are often made available in press releases, preliminary reports, and final reports, or their equivalent. Also, data bases are increasingly provided for the use of secondary analysts. This section discusses the information that should be given to users of survey results so that they can reach conclusions concerning the strengths and weaknesses of the information provided.

Users of a data base need full documentation of the survey more or less along the lines of the outline in Table 1 (in Chapter 1). For different surveys, the documentation outlined in that figure may be too long or too short, but the topics covered in the outline should all be discussed.

Ideally, the user of the tabular output of a survey should know as much about the survey as is included in the documentation to be provided to anyone using the data base, but the information may be more narrowly focused since the tables are known. Thus, the user of the survey should have available, in the final or preliminary survey report or elsewhere, at least knowledge of the population for which inferences are desired, the population from which the sample was selected, the types of sampling units used, the number of units in the sample, nonresponse rates, completion rates, and the coverage and nonresponse rates for at least the important items in the sample. A statement of the differences between the target and the survey populations, the methods of imputation, and the estimators used both for survey data and variances should be provided. Biases and their possible effects should be discussed. Finally, any evaluations of the possible effects of sampling, editing, undercoverage, nonresponse, and response errors on survey influences should be included.

For a press release to provide so much information would be impracticable, but even a press release should include information on the frames used, the survey and target populations, the size and nature of the sample, one or more unit nonresponse ratios and coverage ratios, and mean square errors for one or more of the important statistics included in the press release.

Any intermediate reports should contain information more similar to the information in the final report than to that in a press release. If such information is not available when an intermediate report or the final report is issued, at least a reference to the later availability of the information should be included.

The information suggested for inclusion in the final report may appear to be very detailed, but it can usually be summarized in relatively few pages depending on the complexity of the survey and on the kinds of information that the survey provides.

CHAPTER **4**

Review of Case Studies

INTRODUCTION

Ten studies of large-scale and significant survey efforts are included in Part II of this volume. (Seven were prepared by statisticians associated with the surveys; three, prepared by panel staff based on reports of the surveys, were reviewed by statisticians associated with the surveys.) The case studies describe the survey designs, the efforts undertaken to complete successfully the data collection, including control of nonresponse and other aspects of data quality, the methods adopted to deal with incomplete data in the survey, and such information as might be available for evaluation of the methods used.

This chapter presents a summary of the case studies in order to provide a broad overview of each of these aspects. It is intended to address the interests of those who sponsor surveys and use survey data, as well as those who conduct surveys or have theoretical statistical interests in survey methodology.

Table 7 lists the names of the case studies and the abbreviations for them used in the text. Table 8 summarizes the sample design, data collection unit, and primary method of data collection of the surveys covered in the case studies.

The next section outlines methods used to increase response in sample surveys and provides a summary description of each survey in the case studies, including how the surveys were conducted. Some information on the residual data incompleteness after data collection is given in Tables 2 and 3 (in Chapter 2 in this part). The methods adopted to make adjustments for unit and item nonresponse are described in the third section of this chapter. Several case studies present an analysis of alternative methods that were considered. In one or two

INCOMPLETE DATA
IN SAMPLE SURVEYS
Volume 1, Part I

ISBN 0-12-363901-8

TABLE 7

List of Case Studies and Abbreviations for Citations in Text

Part II Chapter No.[a]	*Case study*	*Abbreviation*
2.	The Employment Cost Index: A Case Study of Incomplete Data	ECI
3.	USDA Livestock Inventory Surveys	LIS
4.	The Survey Research Center's Surveys of Consumer Attitudes	SCA
5.	Treatment of Missing Data in an Office Equipment Survey	OES
6.	Annual Survey of Manufactures	ASM
7.	Incomplete Data in the Survey of Consumer Finances	SCF
8.	An Empirical Investigation of Some Item Nonresponse Adjustment Procedures—National Longitudinal Survey	NLS
9.	Readership of Ten Major Magazines	RTMM
10.	Total Survey Error—Center for Health Administration Studies	CHAS
11.	An Investigation of Nonresponse Imputation Procedures for the Health and Nutrition Examination Survey	HANES

[a] Chapter 1 of Part II is Overview.

instances it was feasible to evaluate the alternatives by comparison with respondent-supplied information or data from external sources. These are of particular interest.

The methods of imputation used in the case studies are summarized in the fourth section. The effects of the methods used in the case studies for dealing with nonresponse are summarized in the last two sections, for biases and variances, respectively. The last section ends with some guidelines for defining poststrata and estimating variances.

METHODS USED IN REDUCING NONRESPONSE

This section lists the various methods used in the surveys analyzed in the case studies to reduce nonresponse. Although data on the cost-effectiveness of the procedures are not available, the list should be useful and may encourage making studies of cost-effectiveness.

Studies concerned with reducing nonresponse are likely to include the following general steps in their survey methodology:

TABLE 8

Summary of Sample and Survey Designs of the Case Studies

Case study	*Sample design*[a]	*Data collection unit*	*Primary data collection method*
ECI	Two-phase stratified sample with controlled selection in 2nd phase	Sample occupation within sample "establishment"	Mail[b]
LIS	Combination of list and multistage area sample	Producer of cattle, hogs	List sample—mail Area sample—personal interview
SCA	Random digit dialing	Randomly selected adult household member[c]	Telephone interview
OES	Multistage area sample	Establishment	Personal interview
ASM	Single-stage Poisson sampling with variable probabilities	Each establishment of a sample company	Mail
SCF	Multistage area sample	Each individual in sample household	Personal interview
NLS	Stratified sample of secondary schools, sample of seniors within school[d]	Sample senior student in sample school	Mail[e]
RTMM	Multistage area sample	Sample individual in sample household	Personal interview
CHAS	Multistage area sample	Individuals in sample households	Personal interview
HANES	Multistage area sample	Sample person selected from screening interviews of household sample	Physical examination

[a] All the case studies are based on probability sample designs.
[b] Personal visit in base period when survey was initiated.
[c] For the index of consumer sentiment, unit is head of household or spouse.
[d] Sample design for base period when survey was initiated.
[e] Third follow-up.

(1) Prepare questionnaires that place as little burden on and reflect as much concern for the privacy of the respondent as possible.

(2) Train interviewers not only in methods of administering questionnaires but also in methods of obtaining cooperation.

(3) Send or deliver introductory letters and statements of survey objectives.

(4) Provide respondents with evidence that their responses will be held confidential.

(5) Identify unit and item nonresponse as early as possible in the data collection process in order to permit callbacks and follow-ups for unit and item nonresponse to be efficiently planned and made during the data collection period. In mail surveys, arrange to get changed addresses from the Postal Service, if they are available.

(6) Make telephone or personal callbacks or mail additional requests to units not initially responding, or use some other follow-up procedure.

(7) In mail surveys, include reminder postcards and make efforts to learn whether a survey unit has moved or is no longer in existence.

(8) Verify, at least for a sample, that inverviews were actually made and information correctly entered; that units classified as ineligible are in fact, ineligible; and that attempts were made to contact and interview units classified as nonresponses. In telephone surveys, monitor a subset of interviews in progress.

(9) Include items in the questionnaire that will be used in imputation or statistical adjustments for nonresponse.

(10) Attempt to convert refusals into responses in the final part of the data collection effort.

(11) In later interviewing efforts, assign the callbacks to the better interviewers or supervisors. (As the final date of the interviewing period approaches, a final intensive callback effort may be made using the best available interviewers.)

(12) Use a smaller questionnaire consisting of key items in the later or final data collection efforts to obtain the minimum data needed for the survey itself and/or for imputation or adjustment for missing data.

In addition to these general procedures, the following less generally adopted procedures occurred in the case studies or are in fairly general use.

(13) Increase both item response and the ability to adjust for item nonresponse by asking questions expected to have high item nonresponse in simpler or less threatening versions in addition to the desired version, e.g., a short classification of income by size as well as the detailed income question involving parts asking for information on amount of income or source.

(14) Offer incentives to respond to potential responders. (This is less generally done in government surveys than in nongovernment surveys.)

Many other survey methods are used to increase response, such as double sampling, network sampling, and randomized response methods (see Part III and the chapters by Sirkin and by Emrich in Part II Volume 2). [The use of substitute units is considered to be more nearly an alternative to imputation than a procedure for reducing nonresponse (see the chapter by Chapman in Volume 2, Part II).]

ADJUSTMENT FOR UNIT NONRESPONSE

No standard terminology has, as yet, been accepted by survey practitioners for types of nonsampling error and for methods of treating incomplete data (see, for example, Office of Federal Statistical Policy and Standards, 1978). The following definitions of adjustment, however, are essentially common in the panel's work.

Adjustment for unit nonresponse (e.g., complete nonresponse for a sample person, household, or business) refers to techniques of weighting up data for respondents to sample totals including nonrespondents (see Chapman, 1976).

Weighting-class adjustment refers to the technique in which estimates of the totals of some characteristic known for both respondents and nonrespondents are made using the full sample and the respondents and, in both cases, the survey sampling weights. Then the sampling weight of the respondents is multiplied by the ratio of the full-sample estimate to the respondent-based estimate (see chapter by Oh and Scheuren in Volume 2, Part IV). Specifically, let s_i denote the known characteristic and P_i the probability of selection of the ith unit. Then the inflation ratio would be

$$r = \frac{\sum_i w_i s_i}{\sum_j w_i s_i},$$

where $w_i = 1/P_i$, the numerator is the full sample estimate (the weighted sum over all units in the sample), and the denominator is the respondent-based estimate (the weighted sum over the respondents). Such an inflation ratio is useful when the known characteristic and the characteristic being estimated are highly positively correlated.

Suppose a simple random sample has been selected. Then $w_i = N/n$ for every unit, where N is the number of units in the population and n is the number in the sample. Let $s_i = 1$ for every unit. The inflation rate for nonresponse is then

$$\frac{n(N/n)}{m(N/n)} = \frac{n}{m},$$

where m is the number of respondents. Hence, the adjusted weight would be

$$\frac{n}{m} \times \frac{N}{n} = \frac{N}{m}.$$

This is the usual weighting adjustment for unit nonresponse when nothing is known of the characteristics of the nonrespondents.

If, within any group of sample units, the average values for respondents and nonrespondents of some characteristic to be estimated differ no more than would occur from random sampling, the weighting-class adjustment would yield approximately unbiased estimates of means or totals for the characteristic. Therefore, for purposes of adjustment, the sample is often divided into subclasses—weighting classes—on the basis of known characteristics of the sampling unit so that the units within a class are expected to be relatively homogeneous with regard to the characteristics of interest.

Poststratification adjustment refers to techniques of stratifying respondent units into poststrata for which a control total, independent of the sample, is available, making an estimate of the control total from the respondent units based on their sampling weights, and then inflating the sampling weight of each respondent unit by the ratio of the control figure for the poststratum to the estimate based on the respondents (see Cochran, 1977, Section 5A.9; Hansen, Hurwitz, and Madow, 1953, Chapter 9, Section 23).

Poststratification adjustment uses control totals that weight data from the respondents to the population totals independent of the sample. Weighting-class adjustment weights data from the respondents to totals for the weighting class, derived from data obtained for the selected sample. The difference in terminology may reflect the fact that frequently the classes defined for weighting-class adjustments use the strata employed in the sample design. Since post-stratification adjustment tends to reflect stratification not used in the sample design, it is also frequently used as a technique to try to reduce sampling errors of estimators.

Typically, in a population survey the poststrata are defined on the basis of such characteristics as age, sex, and race for which control totals in terms of number of persons are available from the Bureau of the Census (used in the RTMM) or an independent survey (used in the SCA). In establishment surveys, the preferred item to be used for poststratification adjustment is one that reflects economic activity—such as employment or total sales—but it may be the number of establishments, with the poststrata incorporating stratification by SIC codes or kinds of business and a size stratification (used in the OES). Because many surveys tend to be multipurpose, the poststrata should be defined to try to improve the key estimates to be made from the survey. In this sense they must be robust with regard to the different variables to be estimated. Additional gains for other variables may be sought through the estimation technique.

Poststratification is widely used for unit adjustment. It is, perhaps, the basic method in use. Poststratification adjustments may lead to reductions in the sampling errors of estimates, but the principal effect of such adjustments may be more important for reducing biases arising from differential coverage of units from poststratum to poststratum (see OES). In multistage samples, an initial weighting-class adjustment based on primary sampling unit characteristics may be used, followed by a poststratified adjustment based on the people, farms, or other final units in the sample (Hanson, 1978, Chapters V, VI). The potential bias improvement, of course, is affected by any bias in the control totals themselves. In practice, there may be a conviction—or, perhaps, even evidence—that an increased number of poststrata may be effective in reducing bias in that there is differential nonresponse or expected significant differences in means between poststrata, but control totals for such more detailed poststratification may be less accurate, which could increase the bias. If sample sizes within poststrata are small, they may unduly increase variance; if zero, they cause bias unless the poststrata have no elements in the population.

The variables used to define the poststrata must be known for both respondents and nonrespondents and should be correlated with the characteristics of major interest to be estimated. From this point of view, the use of more variables with a few strata on each is likely to be better than extensive stratification on a single variable (Cochran, 1977, Section 5.A.8; Hansen, Hurwitz, and Madow, 1953, Chapter 5, Section 11).

In the HANES case study, the application of a clustering technique to the problems of defining weighting classes was explored. It was found that a simple poststratification adjustment by age–sex–race groups was as effective as the more complex procedures examined. Because the sample sizes in the poststrata (or weighting classes) are random variables, the weights in the poststrata are ratio estimates.

If the sample sizes in the poststrata are very small, the effect may be to introduce large ratio-estimate biases and components of variance. This suggests that limits on the number of poststrata (weighting classes) should be established. The guidelines, similar to those for ratio estimates, suggest that poststrata with fewer than 20–30 respondents be pooled (Hansen *et al.*, 1953, Chapter 5, Section 16, Remark 3; Hanson, 1978).

Similarly, an upper bound is generally placed on the magnitude of the adjustment of the sampling weight in any stratum in order to control the magnitude of the differential in sampling weights caused by the nonresponse adjustment (SCA, Cochran, 1977, Section 5.A.9). This differential causes an increase in variance, which can be substantial in comparison with an optimal allocation over the strata. Keeping the adjustment of the sampling weights within the upper bound can be achieved by pooling of poststrata (SCF, Hansen, 1978, Chapters V, VI). However, the problem has generally been dealt with by arbitrarily limiting the maximum adjustment to some factor (2.3995 in CHAS) and distributing the remaining adjustment among similar strata.

The most effective trade-off between reducing bias and reducing variance is a matter for empirical study. As is illustrated in the case studies, a variety of methods have been used for selecting variables and defining strata for weighting-class or poststratification adjustments. In a large-scale study, several alternatives can be tried and the results examined for reasonableness from the point of view of subject matter expertise and awareness of problems in the data collection activities. A brief description of the methods illustrated in the case studies follows.

Employment Cost Index (ECI)

The ECI adjustment (for wages) for establishment response is essentially a weighting-class adjustment for active panel members. It is made for temporary nonresponse, seasonal nonresponse, and schedules still pending at the close of collection. This imputation approach reflects the fact that the index being estimated uses quarter-to-quarter link relatives[1] based on panels of identical establishments in the two quarters involved. For specified types of cases, an establishment's data for the previous quarter (reported or imputed) are moved by the average wage change of a matched sample of establishments/occupations at some "collapsed" cell (i.e., weighting class) level. Wages generally form a monotonically increasing function, and establishments that experience temporary or seasonal closings (such as amusement parks) do not reopen with wages at the level at the time of closing but with wages having changed in accordance with the wage movement during the period it was closed. Thus, temporary and seasonal responses are adjusted by the average change in wage rather than held constant. The "collapsed" cells for imputation are predetermined on the basis of SIC employment size class and other variables and follow a prescribed sequence in using those variables. Usually, cells are collapsed across 3-digit SIC codes and size class to maintain groups as homogeneous as feasible. The criterion for determining the "collapse" level is that cells are collapsed or combined (starting with the most detailed class classification) until the weighted employment of the respondents reaches at least 60% of the weighted employment of all "active" schedules in the cell. Thus, in effect, no sampling weight is adjusted by more than 5/3. An establishment/occupation can be assigned imputed data for no more than four consecutive quarters before being reviewed. If an establishment does not respond for four quarters, it is dropped from the sample; no attempt is made to recontact it.

Livestock Inventory Surveys (LIS)

In the LIS, estimates are constructed on an individual basis for the very largest producers, which are in the extreme upper tail of the population distribution and are selected with certainty. Because of their uniqueness, it is believed more reliable to estimate data directly for these producers individually, if possible, than to use averages of data reported from other large producers. They are large enough to be well known, and often some information can be obtained from local informed sources within their communities. For other units in the samples, a weighting-class approach is used. The adjustment for unusable reports (including nonresponses) is made by adjusting the inflation rates sample stratum by sample stratum. Table 9 provides some data on the extent of imputation for unit nonresponse of the list sample in the Hog Survey

[1] A link relative is the ratio of a total for a given time period to the total for the same variable in the preceding period for units reporting in both periods. The link relative determines relative change from one time period to the next. The link relative estimator of a total for a given time period is the product of the estimate for the preceding time period and the link relative for the given time period.

TABLE 9

Livestock Inventory Survey—USDA Percentage of List Frame Estimate Attributable to Each Response Type[a]

List sample size	*Hog survey December 1978 (25,587; 14 states)* Percentage of estimated number of operations[b]	*Hog survey December 1978 (25,587; 14 states)* Percentage of estimated inventory	*Cattle survey January 1979 (52,078; 28 states)* Percentage of estimated number of operations	*Cattle survey January 1979 (52,078; 28 states)* Percentage of estimated inventory
Response cases				
Data obtained by				
Mail	22.3	19.6	22.9	23.3
Phone interview	52.1	37.7	50.3	43.6
Personal interview	13.9	21.5	12.8	15.9
Estimated	.1	1.2	.2	1.9
Known zero	1.5	0	1.7	0
Total	89.9	80.0	87.9	84.7
Nonresponse cases				
Refused by				
Mail	.2	.5	.2	.3
Phone	4.4	10.3	5.1	7.2
Interview	.9	5.2	.9	2.3
Total	5.5	16.0	6.2	9.8
Inaccessible	4.6	4.0	5.9	5.5
Total	100.0	100.0	100.0	100.0

[a] The questionnaire item used is total inventory. Unusable responses reflect the percentages of the survey estimate dependent on imputational procedures.

[b] Estimate based on count of operations weighted by the inverse of probabilities of selection. Percentages reflect the proportion of the list frame universe represented by each response type.

and the Cattle Survey. As shown in the table, a higher proportion of large producers than small producers are nonrespondents. The case study comments that "since direct measures of bias cannot be made, effective control is best assured by keeping the percent of unusable reports small."

Several experiments aimed at developing improved imputation approaches for nonrespondents are reported in the case study. One method that shows promise is to try to determine during the data collection process whether a producer actually has any hogs (cattle) and should, therefore, be imputed or not. The method will more precisely identify the class of nonrespondent units for which imputation or adjustment is required. This principle of trying to determine whether a response is applicable to a unit in a survey before imputing a response is important and widely applicable. For example, in Bureau of the

Census income surveys, this principle is applied by type of income, e.g., wage and salary, welfare income.

Survey of Consumer Attitudes (SCA)

The age–income bivariate distribution (seven categories by income, six categories by age) from a recent personal interview sample conducted by the Survey Research Center, which conducts the SCA, is taken as the control for adjusting the current SCA survey. An individual's weight is obtained by dividing the proportion of the control sample in the individual's age–income cell by the proportion of the current SCA sample that falls in that cell. This adjustment is done separately for the two major components of the sample: the random-digit dialing (RDD) telephone sample and the rotating panels of individuals who are recontacted for the survey 6 months after their initial interview. Response rates for the RDD sample are in the range of 70 to 80%; and for the recontact sample, in the range of 50 to 60%. The adjustment method is essentially a poststratification weighting and is intended to adjust for differential coverage of the various population groups by the RRD sample. Other adjustments are also made for other sources of bias and variance.

Office Equipment Survey (OES)

Two stages of ratio estimation were used in constructing the survey estimates from the OES. The first stage, which used first-stage sampling unit totals from the preliminary 1975 county business patterns (CBP) summary tape, did not use the OES survey data and was intended to reduce the impact of variability between the first-stage sampling units on the sampling errors of estimates. The second-stage ratio estimate was a poststratification adjustment made primarily to adjust for establishment nonresponse and to reduce the impact of biases arising from possibly incomplete listing in the survey and also to reduce sampling variability. The adjustment factor was the ratio of employment counts from the published 1975 CBP report by SIC category and employment size class to a corresponding estimate of employment from OES respondents.

Overall, a 15% adjustment was made for all industries combined with individual adjustments in the four SIC categories used that range from 4 to 22%. A later survey round, which included a dependent relisting of the area segments, showed that the initial listings were incomplete, that this fact appeared to account for about 11% of the overall 15% readjustment made in the first survey, and that the poststratification adjustment did indeed help compensate for biases of undercoverage in the sample.

The estimates from the survey, the corresponding figures from *County*

TABLE 10

Calculation of Establishment Adjustment Factors by SIC Group, Office Equipment Survey (OES)

Approximate description of SIC *group*	SIC *codes included in group*	*Total establishment counts from 1975* CBP *(as published)*[a]	*Estimated total establishment from survey (after 1st-stage ratio estimate and adjustment for total nonresponse)*	*Adjustment factor*
1. Manufacturing	20–39	319,380	260,800	1.224
Transportation, communication, utilities	41–42 44–49			
2. Finance, insurance, real estate	60–67	169,502	163,000	1.040
Business, legal services	73, 81, 89			
Educational, social services	82, 84, 86			
3. Retail trade	52–59	369,947	324,300	1.141
Personal services	70–72 75–76 78–80			
4. All industries combined		858,829	747,900	1.148

[a] With unclassifiable establishments distributed among industries.

Business Patterns for 1975, and the ratios of these that served as adjustment factors are shown in Table 10.

Annual Survey of Manufactures (ASM)

The ASM imputes for establishments that do not respond by creating an imputed report on an individual establishment basis. This is done by using designated key items, which may be known from other sources, as in the case of employment or SIC data, and imputing other items based on ratios derived from respondent forms. For the largest establishments, individual methods are used.

Survey of Consumer Finances (SCF)

The SCF uses a combination of weighting-class adjustment for unit nonresponse followed by a poststratification adjustment. Noninterview rates for two survey rounds are shown in Table 11. For unit nonresponse adjustment,

TABLE 11

Incomplete Data for the Surveys of Consumer Finances (1976 and 1977)[a]

	1976		*1977*	
Dwelling units	*Count*	*Rate*	*Count*	*Rate*
Selected	38,257	100.0	17,066	100.0
Vacant	3,687	9.6	1,780	10.4
Occupied	34,570	100.0	15,286	100.0
Nonrespondent	3,232	9.3	1,807	11.8
No contact	2,114	6.1	667	4.4
Refusal	1,118	3.2	1,140	7.5
Respondent[b]	31,335	90.6	13,479	88.2

[a] In years ending with even numbers, the survey is conducted as an April supplement to the monthly Canadian Labour Force Survey. In the years ending with odd numbers, the survey is not a supplement, and is usually conducted between the April and May Canadian Labour Force Surveys.

[b] An occupied dwelling is classified as a respondent, for this table, if labor force data were provided for all eligible members.

each responding sample unit in a poststratum of the SCF is weighted as the product of six factors, the product of the first five being termed the simple survey weight. The five factors are (1) the inverse of the household sampling rate; (2) a rural–urban factor, which compensates for the over- or underrepresentation of population in the selected primary sampling units (PSUs) with respect to the entire non-self-representing area in the province; (3) a nonresponse adjustment, which is the number of dwellings occupied by eligible individuals divided by the number of interviewed households (which is calculated at the subunit level within PSUs in non-self-representing areas); (4) the inverse of the subsampling rate for any subsampling of households that may be applied in clusters experiencing rapid growth; and (5) the inverse of the subsampling rate for any subsampling that may be applied to keep the total sample controlled to a prespecified size. The sum of the simple survey weights for the responding sample units in a poststratum is an estimated number of units for that poststratum. The sixth factor is the ratio of a more accurate estimate of the number of units in the poststratum to the estimated number of units obtained using the simple survey weights.

Readership of Ten Major Magazines (RTMM)

The RTMM study also used an initial weighting-class adjustment followed by a poststratification adjustment. The weighting-class adjustments, which included adjustment for interviews not completed, was carried out separately

TABLE 12

Components of SSU Weights (RTMM)

Stratum	*Component due to disproportionate sample allocation*	*Component due to "nonresponse"*	*Composite (product weight)*
Metropolitan areas			
Under $7000	2.06	1.09	2.24
$7000–12,999	1.01	1.05	1.06
$13,000 and over	.51	.96	.49
Nonmetropolitan areas			
Under $7000	2.05	.80	1.66
$7000 and over	1.08	.98	1.06

for each of five strata and yielded weights referred to as "secondary sampling unit" (SSU) weights. Table 12 shows the five strata and the variation in weights.

The case study explains that the relatively minor adjustment for nonresponse reflected in these weights is at least partly due to the level at which the adjustment was made—that is, at the level of broadly defined income strata. Weighting for differential nonresponse among smaller units, such as SSUs, would have allocated it more specifically to the areas where it occurred but would also have produced greater variation among the weights. The limited extent of nonresponse weighting, along with the limited amount of sample balancing required, contributed to the relatively moderate range of weights used in the study.

The final poststratification adjustment was carried out for 80 cells based on four characteristics: sex (male, female); age (18–24, 25–34, 35–49, 50–64, 65 and over); region (Northeast, North Central, South, West); and residence (metropolitan, nonmetropolitan). The ratio estimate weight for an individual in any given cell was then determined by dividing the U.S. population in the cell by the corresponding sample estimate.

National Longitudinal Survey (NLS)

A weighting-class method was used to adjust the NLS student weights for the approximately 5% questionnaire nonresponse, but not for item nonresponse within completed questionnaires. The NLS case study is an empirical investigation of the nature of missing and faulty data and of the efficacy of alternative item nonresponse adjustment procedures. The study is based on 20 items in the survey that were designed as critical and that, therefore, were subject to telephone follow-up. This provided respondent data for evaluating bias in the survey and for testing imputation procedures.

To adjust for questionnaire nonresponse, five classifier variables were used in defining the classes: race (white, nonwhite); sex (male, female); high school curriculum (general, academic, vocational/technical); and parent's education (less than high school graduate, high school graduate, some beyond high school, college graduate—if available father's education; otherwise, mother's). These variables were each available for 94% or more of the students in the survey, but an "unavailable" category had to be set up for each of the five classifier variables. A total of 540 cells were defined by the variables. A rule that each weighting class must contain at least 20 respondents was used, and a set of rules was established for use in combining "similar" cells that contained fewer than 20 respondents. A total of 87 weighting classes were formed. The sample weights for responding students within a weighting class were adjusted by the ratio of the total sample weights for all students in the cell to the total for respondent students in the cell.

Center for Health Administration Studies (CHAS)

Data from the 1970 Center for Health Administration Studies (CHAS) were adjusted for differential nonresponse separately for each of the four components of the sample. Two approaches were tried: one used the first-stage sampling unit to define weighting classes and the other used categories based on the reason that no interview was completed. A poststratification adjustment was applied after the nonresponse adjustment according to family characteristics of race, residence, size, and income. Sixteen strata were established and, within each, adjustment factors were computed by comparing the CHAS estimates with independent estimates from the Census Bureau's Current Population Survey (CPS) for a similar time period.

The effect of poststratification was examined for two items: mean number of physician visits and mean total hospital expense per admission. The data in Table 13 are taken from a single sample component and are based on poststratification without prior nonresponse adjustment. The original categories are based on family characteristics, and the alternative categories represent a similar adjustment with independent CPS estimates based on individuals rather than families. Poststratification had little effect on estimates for physician visits, but did have effects on the average hospital expenditures per admission, as shown in Table 13.

This case study concludes that without a validation criterion it is difficult to assess the effect of poststratification. However, poststratification is recommended, since it is extremely inexpensive to do and should result in the improvement in the data as long as the units within the adjustment strata are relatively homogeneous.

TABLE 13

Expenditures Per Admission, by Poststratification Adjustment and Demographic Characteristics, CHAS

	Expenditures per admission		
		With poststratification adjustment	
Demographic characteristic	*Without poststratification adjustment* ($)	*Original categories* ($)	*Alternative categories* ($)
Family income			
Nonpoor	649	674	674
Poor	685	715	691
Race			
White	652	684	650
Nonwhite	702	688	730

Health and Nutrition Examination Survey (HANES)

The Health and Nutrition Examination Survey (HANES) case study is an empirical investigation of alternative weighting-class adjustments. Alternative techniques for defining the weighting classes, and alternative variables to use, were both studied. First, a clustering technique (Morgan and Sonquist, 1963) was tried using the variables in Table 14.

TABLE 14

Demographic Variables Available for the Clustering Procedure, HANES

	Demographic variables	*Number of categories*
1	Standard location[a]	35
2	Age	5
3	Race	3
4	Sex	2
5	Family income	13
6	Education of household head	19
7	Degree of urbanization	8
8	SMSA code	3
9	Region	4
10	Household size	17
11	Number of sampled persons in household	6
12	Poverty index	4

[a] Stands or standard locations are counties, cities, or towns in which the examination centers are located. Sample areas from which examinees are drawn for the stands consist of the primary sampling units (PSUs), which may include several counties.

Basically, the clustering procedure used produces a continuing sequence of splits of subgroups of a file into two parts. One dependent variable and a set (or pool) of independent variables is specified for use in the program. At each stage the subgroup that is split is the one that, for the dependent variable, has the largest variance among its members. Once the subgroup to be split is chosen, the particular split to be made is determined by checking all possible splits that can be made based on the values of each one of the independent variables in the pool. The split chosen is the one that maximizes the sum of squares between the means of the two parts.

The particular procedure used for this project would not accept data files larger than 8000 records. Therefore, the HANES file of 14,147 records was subdivided into the following four sets of subgroups with the analysis being applied separately to each set:

(1) The odd- and even-numbered records (that is, alternative records in the file were assigned to two subgroups)

(2) The records contained in each of the four Census Bureau regions: Northeast, North Central, South, and West

(3) The records associated with each of five age groups: 1–5, 6–19, 20–44, 45–64, and 65–75

(4) The records contained in each of the 35 stands[2]

The dependent variable used was the 0–1 response variable: whether or not the person was examined. The percentage of within-subgroup variation "explained" by the weighting classes established with the clustering technique for the first three sets of subgroups varied from 6 to 16% and averaged 10%. For the set of subgroups defined by the 35 stands, the results were found to explain a considerably higher percentage of variation within stands, ranging from 11 to 57% and averaging 37%. However, the latter result was attributed to a large extent to the smaller total group sizes. It was concluded for a variety of technical reasons that the use of the clustering to define specific weighting classes was not feasible although it was useful to identify variables of interest to use. An alternative approach for using variances and covariances to select variables for defining weighting classes was also explored. Based on the joint results it was concluded that the important variables to use were location, age, and sex and that family income, household size, degree of urbanization, and geographical region would be of some use. Six alternative imputation procedures of varying complexity were defined, as shown in Table 15.

It was also decided that the final-stage adjustments to Census Bureau counts carried out in the current procedure were very worthwhile and would be included in all the alternative procedures developed. This decision was based

[2] The 35 stands are listed in Appendix A to the chapter by Chapman in Part II of this volume.

TABLE 15

Variables and Corresponding Number of Classes Used at Each Stage, HANES

Imputation procedure	*Stage 1*	*Stage 2*	*Stage 3*
1	Stand (35) × Age (5) × Sex (2) × Income (2)	Region (4) × SMSA-non-SMSA (2)	Age (5) × Race (2) × Sex (2)
2	Stand (35) × Income (4)	Age (5) × Race (2) × Sex (2)	
3	Stand (35) × Education (4)	Age (5) × Race (2)	
4	Stand (35) × Household Size (3)	Age (5) × Race (2) × Sex (2)	
5	Stand (35)	Age (5) × Race (2) × Sex (2)	
6	Age (5) × Race (2) × Sex (2)		

primarily on the apparent importance of the variables of age and sex in defining weighting classes.

The comparative performance of the alternatives with regard to bias elimination is summarized in Table 13 of the chapter by Chapman in Part II. Based on the evaluation carried out, three general conclusions were drawn. First, all of the proposed procedures show substantially lower estimated biases for the examination variables than does the procedure that makes a single overall weight adjustment (i.e., the unadjusted-weights procedure).

Second, the procedure that involves a single *stage* of weight adjustments, those being to Census Bureau estimates of age–race–sex counts, is approximately as good as the other procedures on the basis of the evaluative results in this study. In fact, all six procedures yielded estimates that did not differ substantially.

Third, for the case in which medical history characteristics were being estimated, and used for evaluation, there are some nontrivial estimates of nonresponse biases remaining after the nonresponse adjustment procedures have been applied. Some of these estimated biases are as large as .5–2.0%. Such biases would presumably be substantially greater than the corresponding sampling errors for these estimates. Because of the similarity of results for the alternative weight-adjustment procedures, it was concluded that this bias cannot be reduced noticeably by using alternative adjustment variables. The bias can perhaps be reduced only by increasing the survey response rate.

Because of the similarity of performance between the alternative procedures, it might be tempting to recommend the easiest procedure to use: the single-stage, weight-up to the 20 age–race–sex Census Bureau counts. However, it was thought useful to include a geographic variable to define weighting classes. Consequently, it was recommended that the stand be included in an adjustment procedure, since nonresponse rates vary substantially by stand, its joint use with other variables is not difficult, its use has an intuitive appeal, and its use may be important in reducing nonresponse bias for some estimates.

In general, it seems reasonable to assume that the use of additional adjustment variables might be helpful in making estimates for some examination measurements not studied here and, in any event, should not introduce biases. It is possible, however, to add to the variances of estimates by using additional variables that create weighting classes that contain too few respondents. But with the use of weight classes such that the minimum weighting class size is 25 respondents, this increase in variance is trivial and can be neglected.

IMPUTATION FOR MISSING OR UNUSABLE DATA (ITEM NONRESPONSE)

Imputation for missing data items refers to techniques for filling in a (plausible) value for a specific data item for which the response is missing, or rejected as unusable, for a unit that responded in the survey. External data techniques use data derived from outside the survey: for example, program entitlements for missing income associated with a welfare program or hospital charges from record sources (SCF, 28A). Among internal data techniques, the hot-deck technique substitutes a reported value from the survey taken from a unit (e.g., a person or an event such as a hospitalization) that has specified characteristics matching those for which the data item is missing (CHAS; Volume 2, Part IV, chapter by Ford; Volume 3, Session VIII; Coder, 1978; Scheuren, 1978). The respondent cell-means technique substitutes the average for the given item based on all respondents in a cell defined by specified characteristics (Volume 3, Session VII). Further constraints may be imposed by raking adjustments, although these may easily lead to overadjustment (Oh and Scheuren, 1978). Much that has been said earlier about the definitions of poststrata and weighting classes also applies to the choice and use of variables for matching in the hot-deck approach (Oh, Scheuren, and Nisselson, 1980; Welniak and Coder, 1980) and to the definition of cells for the cell approach.

Where the objective is to produce statistical estimates of aggregates only, internal data techniques such as the use of cell means may be adequate and will provide estimates with smaller variances than a hot-deck approach (NLS; see, e.g., the chapter by Herzog and Rubin, Volume 2, Part IV; Oh and Scheuren, 1980). However, when an important objective is (also) to provide a microdata base for analysts, hot-deck methods or other methods that try to preserve distributional properties within cells have been found to be more satisfying. An example of an alternative to hot-deck methods is regression on correlated variables known or reported for the unit (Schieber, 1978).

In most of the case studies, if imputation was done for missing or unusable item response, an effort was made to impute based on relations with other items in the survey for which responses were available. This is a form of regression estimation. Unless steps are taken to preserve distributional properties when such methods are used, the effect may be to produce artificially good fits in regressions computed in later analyses of the data. Preserving distributional properties, for example, by adding a hot-deck residual to a regresson value may not be a simple matter, since the nonresponse cases may be atypical in unknown ways.

The choices among alternatives will be based on their computational ease and the costs associated with them. In complex surveys such as the Current Population Survey, for example, the data-processing phase may be very demanding. As pointed out by Bailar and Bailar (Volume 3, Session VII), the data for the survey each month are collected in a 1-week period for approximately 56,000 households and 116,000 persons. These data are edited, processed, and reported within 2 or 3 weeks following data collection. They emphasize that an imputation procedure for such a survey must be both efficient and rapid.

A brief description of item imputation methods illustrated in the case studies follows. Some summary statistics concerning reductions in bias achieved by the methods are presented in the next section.

Employment Cost Index

Total compensation for employment consists of wages and all benefits. For the Employment Cost Index (ECI), the imputation for missing wage data was described earlier; the imputation for benefits is carried out for each of the individual benefits separately. Since respondents (establishments) are given the option of reporting costs for combinations of benefits, a first step is to allocate such combined costs to the individual benefits involved. This is done by allocating the combined costs proportionately to average levels calculated for each benefit based on cases in which they were reported individually. To attempt to decrease the mean square error in the imputation process, an effort is made to use all available data. Since information as to the existence of benefits is usually available, imputations are made benefit-by-benefit. The imputations may involve average levels or quarter-to-quarter movements. Since zero is a valid benefit cost, some movements may be indeterminate, i.e., $0/0$ or $c/0$, where c is some positive number. If the movement is of the form $0/0$, the previous quarter's data are retained unchanged if the benefit is not wage related and changed by the natural change in the wave level if the benefit is wage related. If the movement is of the form $c/0$, a current wage- or non-wage-related level, whichever is appropriate, is used as the imputed value. All movements and levels used in the imputation process are calculated over appropriate cells collapsed to meet the specified criteria.

Livestock Inventory Survey

Individual missing data items that require manual imputation in the Livestock Inventory Survey (LIS) are edited in by the commodity statistician. Refusals and inaccessibles are deleted from the sample and the sample size reduced accordingly. This procedure, in effect, assumes that missing operations are similar to reported operations.

Survey of Consumer Attitudes

The index of consumer attitudes is based on indices computed for each of five questions in the Survey of Consumer Attitudes (SCA)—the percentage of missing data by question is almost always under 10%, the major problem being at the missing unit level. The index for each of the questions is computed based on all the responses to that question, regardless of whether the respondents answered any of the other four index questions. No adjustment is made for nonresponse on the index questions.

Office Equipment Survey

Imputation for missing data items in the Office Equipment Survey (OES) was done only for survey report forms considered usable or complete. To be considered usable, a form must contain complete responses to certain questions relating to the establishment, such as employment size, kind of business (SIC code), and number of pieces of office equipment. The imputation method used for missing data items was a hot-deck procedure. To obtain an imputed value, an establishment with a missing item was clerically matched with another establishment whose major characteristics matched as closely as possible those of the establishment with the missing item. Specific imputation rules and priorities for the matching were specified. It is reported that, with rare exceptions, a questionnaire was used only once for imputation of a particular item. Questionnaires with imputed data were reedited to make sure that the imputation satisfied the editing rules.

Not all missing data items were imputed. The survey sponsor designated only six key items as requiring imputation. For these items the percentage of imputed responses ranged from 1.6 to 7.1%.

Annual Survey of Manufactures

Imputation in the Annual Survey of Manufactures (ASM) is done both for missing data items and for data considered in error on the basis of an elaborate set of data edits. Imputations leans heavily on ratios to designated key items, using numerical values for acceptable ranges developed from respondent reports. The methodology has several interesting features, which the case study in Part II describes in detail. One feature is a technique for making a decision, when a ratio is found to be outside acceptable limits, as to which item report to adjust. When more than one ratio might be used for imputing an item, the alternative imputations are computed, and the "best" possible imputation selected. (The criterion, which is a probability function, is described in the case study.) All imputations must satisfy the edit rules.

Survey of Consumer Finances

Incomplete data for an individual in the Survey of Consumer Finances (SCF) may occur among the labor source items, the work experience items, or income items. If the labor force data are missing as a block, a weighting-class adjustment is made. If the adjustment factor is greater than three, the weighting class is collapsed and the adjustment computed on the basis of the collapsed class. When work experience data are missing, an attempt is made to assign the necessary response on the basis of the response(s) to other item(s): for example, to assign a period of unemployment on the basis of reported unemployment insurance benefits. If such assignment is not possible, a response is imputed by a hot-deck procedure. All data assigned or imputed must pass the edit checks built into the survey. For income data, missing individual income and missing family income are imputed separately by weighting-class adjustments applied to the set of individuals or families for whom income data are reported. A total of 220 poststrata are used for individuals and 200 poststrata for families. Independent figures for the population by poststrata to which the sample is ratio adjusted are obtained from a combination of data from the Labour Force Survey itself and, indirectly, from census sources for the corresponding month. The case study describes an experimental procedure that will make more use of socioeconomic data about the individual or family, paralleling the approach in the Current Population Survey. All imputed data must pass the established edit tests.

National Longitudinal Survey

The National Longitudinal Survey (NLS) case study reports an empirical investigation of hot-deck and weighting-class adjustments for item nonresponse. For discrete items, the weighting-class imputation procedure devised for the study almost uniformly produced estimates whose total error was greater than that of the hot-deck and no-imputation estimates. In general, the hot-deck procedure had some effect in reducing the bias of survey estimates. For items for which nonresponse was more important than inconsistent response, the hot deck generally did better than the no-edit/no-imputation approach. (This phenomenon is discussed later.) However, in a number of instances the bias reduction was accompanied by an associated increase in variance.

Readership of Ten Major Magazines

No imputation for missing data items was done in the Readership of Ten Major Magazines (RTMM).

Center for Health Administration Studies

As noted in the Center for Health Administration Studies (CHAS) case study, it may be preferable to use internal data rather than external data for noninterview adjustments. Differences in definitions and time period and procedural differences between the record source and the survey make it "unlikely that the cost and time spent locating and adapting external data could be justified." Any decision as to the use of external data should depend upon its comparability with the survey and the costs of using them in the procedure case. Nevertheless, the caveat serves as a useful warning against uncritical substitution of external data for missing survey responses.

In the CHAS case study, it was found that the use that could be made of record source information on physician visits was limited and made little difference in the survey results compared with internal data approaches. For estimates of hospital expenditures per admission, however, the effect of item imputation using external data or alternative internal data methods was more striking, as shown by the results in Table 16 selected from the study. For the hot-deck method, admissions were first arranged in geographic order based on the first-stage or primary sampling unit (PSU) in which the interview was located. The A sequence was East to West among SMSAs followed by East to West among non-SMSAs. The B sequence was in reverse PSU order. For each admission with missing expenditure data, a backward search through the file was in effect carried out until the first admission with expenditure data was found in the same classification cell (i.e., the cells used in the cell-mean method). The expenditure associated with that admission was taken as the imputed value.

TABLE 16

Expenditures Per Admission, by Item Imputation Method and Demographic Characteristic

	Expenditures per admission			
		Internal item imputation		
			Hot-deck method	
Demographic characteristic	*External item estimation* ($)	*Cell mean method* ($)	A *sequence* ($)	B *sequence* ($)
Family income				
Nonpoor	674	708	697	684
Poor	715	745	736	730
Race				
White	684	721	715	700
Nonwhite	688	687	629	658

Not only did the alternative affect the averages for various demographic groups but, even more strikingly, it also affected the comparisons of averages between groups. The difference between results obtained by the two hot-deck sequences was thought to reflect a geographic gradient with respect to hospital changes.

Health and Nutrition Examination Survey

The adjustment procedures reported in the Health and Nutrition Examination Survey (HANES) case study were investigated at the item level. The results are summarized in the next section.

BIAS REDUCTION: IMPLICATION FOR NONRESPONSE RATES

The case studies illustrate the fact that the characteristics of nonrespondents are often different from those of respondents and that no method of adjusting for incomplete data short of obtaining the required information is likely to compensate fully for the consequent biases in the survey results. For example, in a simulation study based on data from a hog survey reported in the LIS case study, the results with regard to residual bias after adjustment by alternative

TABLE 17

Percentage Bias in Estimated Total of Expected Farrowings with Alternative Item Imputation Procedures Relative to Estimate with No Nonresponse

	Expected farrowings, relative bias (%)	
Item imputation procedure	*First quarter*	*Second quarter*
Hot-deck imputation	−3	−10
Ratio estimate	−6	−7
Regression estimate	−6	−7
Weighting class[a]	−11	−11

[a] Average for nonrespondents assumed to be the same as for respondents within classes defined on the basis of known characteristics, but not related to the mechanism for selecting the records to be treated as nonrespondents.

imputation procedures are shown in Table 17. In the simulation study, 20% of the responses were randomly (with probability proportional to number of hogs) chosen to be missing; and another 15–20% were chosen to be partially incomplete. The nonresponse adjustment procedure generally removed about half the bias. This is a substantial reduction, but it still left a residual bias that it would not be desirable to ignore.

In the NLS case study (Part II, chapter by Rizvi, Table 4.1), the results shown in Table 18 were found, based on comparing imputed responses with data obtained by follow-up with respondents to resolve missing or unusable

TABLE 18

Average Percentage Bias in Estimate with Alternative Item Imputation Procedures Relative to Estimate with Missing and Inconsistent Data Corrected by Follow-Up with Respondents

Type of item and item procedure	*Domain by race, average relative bias (%)*[a]				
	Total	*Black*	*White*	*Hispanic*[b]	*Other*[b]
Discrete items (51 estimated proportions based on 14 discrete items)					
Hot deck	1.9	3.3	1.9	14.5	3.1
Weighting class	3.1	5.0	2.8	29.7	4.9
No imputation	2.2	3.7	2.2	6.2	3.0
Continuous items (25 estimated averages based on 6 continuous items)					
Hot deck	.8	2.0	.6	7.8	2.0
Weighting class	.9	2.0	.7	8.9	1.4
No imputation	.8	2.0	.7	9.4	1.2

[a] Biases averaged without regard to sign.
[b] Based on small samples.

TABLE 19

Average Percentage Bias in Estimated Mean with Alternative Adjustment Procedures for Individuals not Participating in Medical Examinations or for Whom More Than One Call Was Needed to Obtain Participation in Medical Examination[a]

Alternative nonresponse adjustment procedure	*15 medical history characteristics*	*10 medical examination characteristics*		
		Total	*Dietary calcium*	*Other*
1	.53	.09	.34	.06
2	.52	.11	.27	.09
3	.53	.13	.43	.10
4	.49	.13	.41	.10
5	.55	.10	.35	.10
6	.55	.08	.05	.08
No adjustment	.55	1.20	.64	1.26

Source: Summarized from HANES, Tables 17 and 19.
[a] Biases averaged without regard to sign. Data are from 35 primary sampling units (PSUs) in first survey round. See text for discussion.

data. In this study, for the 14 discrete items analyzed, the rates of item response that were complete and usable (consistent with related items) were 95% or higher, except for 1 item, for which the rate was 87%: the average of missing responses was 0.7%. For the 6 continuous items analyzed, the rates of item response that were complete and consistent were 83% for 2 income items and 93% or better for the other 4, with 13% missing responses for the income items and an average of 2% for the other items. The major reason for responses being found unusable was the failure of respondents to follow the routing pattern (sequence and skip patterns of questions). It may be noted that, in general, the bias of the "no-imputation" (no-edit/no-imputation) estimates is comparable with the bias after hot-deck or weighting-class imputation. A probable explanation lies in the fact that the imputation procedures used forced responses within a routing pattern to agree with the lead-in question. However, a data item may be inconsistent with another data item and yet be correct. In many instances, also, the bias of inconsistent responses and the bias of missing responses had opposite (and offsetting) effects.

In HANES, the results shown in Table 19 were found for six alternative methods (HANES, Section 4.2) of adjustment for missing data arising from person nonresponse. The six alternative procedures were all weighting-class adjustment procedures; they differed in how the weighting classes were defined.

For each of 15 medical history characteristics, the survey mean was computed using each of the six imputation procedures. In these calculations, the set of all sampled persons who supplied medical history data was treated as the entire sample, and the set of examined persons was treated as the respondents.

For each of 10 selected examination characteristics, the set of all examined persons was treated as the entire sample, and the set of respondents was taken to be those examined persons who required only one call to make and keep an examination appointment; those who cooperated on subsequent calls were treated as the nonrespondents.

For the medical history characteristics, the relative biases were small, but not considered negligible. There was little difference in the performance of the alternative procedures explored; none was particularly effective. For the medical examination characteristics, the relative biases after adjustment were considered negligible except for one item, dietary calcium; all of the alternatives were considered to provide important decreases in bias. For dietary calcium, the relative bias in the unadjusted mean was .64%, while the average for the six adjustment procedures was .31%. Again, there was considered to be little difference between the alternative procedures explored.

SAMPLING VARIANCE EFFECTS OF NONRESPONSE ADJUSTMENT

An obvious effect of nonresponse is to reduce the effective sample size and, apart from any bias effect, to increase correspondingly the sampling variance of the statistics affected. To use the illustration of a simple random sample, if responses are obtained from only m units in a sample of n, the sampling variance of a mean estimated from the m respondents will be V/m, ignoring the finite population correction factor. Then, if the variance among respondents in the population is the same as the variance among all units in the population sampled, the sampling variance of the estimated mean of respondents is

$$\frac{V/m}{V/n} = \frac{n}{m} = 1 + \frac{n - m}{m}$$

times the variance obtained if the mean is estimated from the full sample.[3] This factor would also be applicable to weighting-class adjustments.

Adjustments using a hot-deck approach introduces an additional source of nonsampling error, namely, the selection of the units to be duplicated within the equivalent of a weighting-class cell. As used in many surveys, the selection is completely deterministic and depends on the order of the file. The impact of nonsampling error is then in the bias, and the bias will depend on the order of the file (Bailar and Bailar, 1978). If a random ordering is used, the impact of nonsampling error is in the variance as well. Suppose that the response rate is

[3] Various theoretical models for generating nonresponse are possible, as described in Volume 2. For simplicity, the model here assumes that the population is divided into units that would respond if selected for the sample (respondents) and those that would not.

50% or more and that the selection of units to be duplicated is done by simple random sampling without replacement. Hansen, Hurwitz, and Madow (1953, Chapter 5, Section 16, Remark 4) have shown that the effect of the additional stage of sampling is to increase the sampling variance of a mean, compared with a weighting class adjustment, approximately by the factor

$$\left(1 + 3\frac{n-m}{m}\right)\Big/\left(1 + \frac{n-m}{m}\right)^2,$$

assuming that the variance among respondents is the same as among all units in the population.[4] This factor has its maximum value when $(n - m)/n$ is 1/4, i.e., when the nonresponse is 25%, or equivalently $(n - m)/m$ is 1/3; the increase in variance is then 12.5%.

The increase in the sampling variance of an estimated mean, taking account of the reduction in sample size with nonresponse and the hot-deck weighting for nonresponse without replacement (illustrated with simple random sampling), is

$$\frac{V}{n}\left(1 + \frac{n-m}{m}\right)\left[\left(1 + 3\frac{n-m}{m}\right)\Big/\left(1 + \frac{n-m}{m}\right)^2\right]$$
$$= \frac{V}{n}\left(1 + 2\frac{n-m}{n}\right).$$

The increase in the sampling error of a simple random sample mean due to the reduction in sample size, and the additional increase if a hot-deck approach is used, relative to the sampling error with complete response, is illustrated in Table 20 (calculated from the above formula). As can be seen from the table, the percentage increase in sampling error is roughly of the same magnitude as the nonresponse rate; and with increasing nonresponse, the reduction in the effective sample size due to nonresponse accounts for a larger and larger fraction of the increase.

To summarize, if in the computation of sampling errors the responses imputed to nonrespondents are treated as valid responses, the result is to underestimate the actual sampling errors. In terms of Table 20, the relative underestimate is given by the ratio (percentage increase)/(100 + percentage increase). For example, with a 25% nonresponse, the relative underestimate is 22.5/122.5 = .184, or 18.4%. The relative underestimate if a hot-deck approach were not used would be 15.5/115.5 = .134, or 13.4%.

If the selection of responses to be duplicated in a hot-deck approach is done without replacement, any given response will be duplicated only once. If the selection is done with replacement, the number of times a given response may be duplicated is no longer controlled and the effect is to further increase the variance. A sequential hot-deck procedure, in which the sample is put in

[4] This approximation ignores the finiteness of the population. An exact result is given in the reference. Results for other models have been derived by Bailar and Bailar (1978).

TABLE 20

Increase in Sampling Error of an Estimated Mean, Due to Nonresponse and Hot-Deck Duplication, by Nonresponse Rate

Nonresponse rate (%)	*Percentage increase in sampling error compared to case of complete response*		
	Due to nonresponse[a]	*Due to hot-deck duplication*	*Total increase*
5	2.6	2.2	4.9
10	5.4	3.9	9.5
20	11.8	5.8	18.3
25	15.5	6.1	22.5
30	19.5	5.8	26.5
50	41.4	0	41.4

[a] That is, loss of sample size.

some order (e.g., based on geographic location of the units) and for each missing value the previous reported value is duplicated, is of this type (see the chapter by Herzog and Rubin, Volume 2, Part IV). A reported value may be duplicated several times if the ordering of the data file puts several missing values in sequence with no reported value between them. Bailar, Bailey, and Corby (1978) have given a formula for the sampling variance of a mean with a sequential hot-deck procedure assuming that the nonresponses occur randomly through the data file, that the initial value used for imputation is a random selection from the responses in the data file, and that no limit is given for the number of times a particular response is used as an imputed value.

Under their model the effect is to increase the sampling variance of a mean, in comparison with the variance assuming complete response, by the factor

$$1 + 2\left(\frac{n - m}{n}\right)\left(\frac{n^2 + n - n(n - m) - 1}{(m + 1)(m + 2)}\right).$$

This may be written approximately as

$$1 + 2\left(\frac{n - m}{m}\right)$$

if $m + 1$ is large. The increases in sampling error when hot-deck selection is made with or without replacement are computed in Table 21.

Several guidelines follow from these illustrations. The first is that in defining weighting classes or cells for hot-deck imputation, the increase in the contribution of the class/cell to the variance of an estimated mean (or total) rises rapidly with the percentage of cases in the class/cell that represent nonresponse. If hot-deck imputation without replacement is used, the impact on the sampling

TABLE 21

Increase in Sampling Error under Hot-Deck Imputation by Replacement Procedure and Nonresponse Rate

Nonresponse rate (%)	*Percentage increase in sampling error compared to case of complete response when hot-deck sampling is done*	
	Without replacement	*With replacement*
5	4.9	5.1
10	9.5	10.5
20	18.3	22.5
25	22.5	29.1
50	41.5	73.2

variance is relatively small compared with the impact of the loss of effective sample size unless the nonresponse rate itself is low—in which case, the combined impact of both factors is small. This is not so if hot-deck imputation with replacement is used so that the number of times a particular report is used to imputed a response is not controlled. Thus, there is an advantage, which may be a decided advantage, to controlling the number of times an individual report is used.

The second guideline is that weighting classes or cells for hot-deck imputation should not be defined in such a way that the number of cases to be weighted up for or imputed represents a substantial nonresponse rate, unless there is an anticipated compensating reduction in bias so that the mean square error of an estimate of interest is not likely to be increased. With a 10% nonresponse rate in a weighting class, the contribution of the adjustment cell to the sampling variance of an estimate increases by about 11% (1.054 squared) with weighting, by about 20% (1.095 squared) if hot-deck imputation without replacement is used, and by about 22% (1.105 squared) if hot-deck imputation with replacement is used. With a 50% nonresponse rate, the contribution is doubled if either weighting or hot-deck imputation without replacement is used and tripled if hot-deck imputation with replacement is used.

The third guideline relates to methods of estimating the sampling errors of estimates from a survey. Since nonresponse generally (but not always) leads to an increase in sampling error compared with a survey with complete response, it follows that methods for estimating sampling errors that do not reflect the effects of nonresponse in increasing variance will, generally, underestimate the sampling errors. In particular, methods of estimating sampling error that treat imputed data as real observations have this effect. How serious the effect may be depends on the circumstances: an estimate of the relative sampling error of a given statistic as 10% when it actually is 11% might not be considered serious by many users. (However, this error might be considered serious for an estimate of unemployment for which month-to-month differences of a few tenths of a percent are important.) The same estimate (10%) when the sampling error is actually 20% may be disastrous. This suggests that if "exact" methods are not used to estimate sampling error of estimates published from a survey, the user be provided with approximate "multipliers" derived from more exact methods used for a selection of statistics. (An analogy is the publication of design factors—DEFF in Kish's (1965) notation—to convert sampling errors computed on the assumption of simple random sampling when a complex design was actually used.)

Some evidence is presented in the NLS case study of the consequences of using methods for estimating sample errors that do not properly reflect the impact of nonresponse and of adjustments for missing or unusable data. The OES and NLS case studies illustrate how replication may be used to estimate properly sampling errors with poststratification and adjustments for nonresponse.

CHAPTER **5**

Review of Theory

INTRODUCTION

This chapter reviews theoretical aspects of the problems of unit and item nonresponse in sample surveys. The discussion is based principally on Volume 2 of the panel's work, a collection of papers prepared by members of the panel and invited authors. Occasional reference is also made to the symposium papers and related discussions in Volume 3.

Volume 2 is divided into eight parts. Part I gives an overview of the volume and a historical perspective by the late W. G. Cochran. This perspective is particularly valuable since Cochran was an influential member of the team of statisticians who developed the modern theory of probability sampling of finite populations. Parts II and III review data collection methods for reducing and measuring the impact of nonresponse. Part IV presents the theory involving the two methods for the analysis of survey data with nonresponse most commonly adopted by survey analysts—weighting adjustments and hot-deck imputation—and also includes an extension of the hot deck procedure to multiple imputations.

Part V describes a general structural model for survey errors, which include sampling errors, response errors and nonresponse errors, and interactions among those sources of error. The model provides a theoretical framework for comparing the impact of the various sources of error. In practical situations knowledge about response and nonresponse errors must be furnished or estimated using additional data.

Part VI reviews analytical methods based on parametric models for the population items, sometimes known as superpopulation models. The system

INCOMPLETE DATA
IN SAMPLE SURVEYS
Volume 1, Part I

ISBN 0-12-363901-8

of inference in these chapters contrasts strongly with the randomization perspective adopted in the earlier parts. Inferences are based on an assumed stochastic model for the population items: this is in contrast to treating these items as fixed and basing inferences on the distribution governing the selection of responding units from the population. However, the methods derived by this approach are not necessarily distinctive; for example, the weighting methods discussed in Part IV also occur as special cases of the methods based on the linear regression method discussed in Part VI.

Part VII consists of two bibliographies. The first, an annotated bibliography should be sufficient to provide an understanding of the purpose and results of each paper. The second bibliography covers 10 major journals for the period 1970–1979.

Parts I–VI of Volume 2 do not give a unified account of the theoretical aspects of the nonresponse problem, which reflects the current state of the literature on the problem. Rather, they present a wide spectrum of viewpoints, ranging from the somewhat informal and practical chapters on data collection methods in Part II to the general overview of the model presented in Part V and the fairly complex likelihood methodology presented in the second chapter of the superpopulation models in Part VI. In some cases practice appears to wait for theory to catch up; in others, the theory suggests new methods. The panel hopes that the volume will provide a basis for further interactions between the practitioner and the theoretician that will lead to a clearer perspective in future work on the subject.

The rest of this chapter discusses the main issues raised in Volume 2.

SOME DATA COLLECTION METHODS

No analytical methods can correct for nonresponse at the data collection phase. Any method for adjusting for incomplete data is dependent upon assumptions that are unverifiable using only the data at hand. As efforts summarized in the case studies in Part II of this volume and in Volume 3 make clear, there are usually systematic differences between respondents and nonrespondents, and no statistical technique can adjust for all differences. Consequently, the ideal way to handle nonresponse is to obtain complete data. As a practical matter, however, some nonresponse remains after all efforts deemed cost-effective by the survey taker have been completed.

Therefore, it is both desirable to reduce nonresponse to the extent feasible by design and necessary to adjust for the residual incomplete data by analysis. The solutions resulting from such adjustment are nearly always more sensitive to untestable assumptions than one would like. Consequently, in surveys in which nonresponse is anticipated, resources should be devoted to reducing the level of nonresponse at the data collection phase. A variety of strategies are available (see Chapter 4), of which the following require special mention.

Callbacks

Repeated attempts to contact respondents, in the form of callbacks, follow-up letters, or repeated telephone calls, are a widely used method of reducing nonresponse. With trivial exceptions, the subset of a sample who respond on a first contact differ from the whole sample. Classifications of survey results by the number of contacts required to obtain a response usually show systematic differences on the survey variables (see, for example, Jessen, 1978). The role of callbacks in the reduction of nonresponse bias is thus clear.

Rao (Volume 2, Part II) discusses callbacks, including a model due to Deming (1953), that permits assessment of effects of nonresponse when callbacks are made and thus enables optimal strategies in terms of numbers of callbacks to be studied. The population is divided into strata, where μ_j is the mean, σ_j^2 is the variance of the characteristic of interest, and ρ_j is the proportion of the population in stratum j. The probability that a person in stratum j responds at call i is denoted w_{ij}. The mean squared error of the sample mean at call i is derived in terms of these quantities. The model is used in a study by Rao (1966), which also takes into account the cost increase involved in making callbacks. In practice, however, the information available is often too limited for a formal analysis of the optimal strategy, and decisions are based more on practical considerations.

Double Sampling

A variant of the basic technique of callbacks is to restrict them to a subsample of the nonresponding group. The restriction to a smaller sample of nonrespondents for follow-up may allow more intensive efforts to collect information, possibly using different survey instruments; for example, a mail survey may include personal interviews in an attempt to follow up nonrespondents. The overall number of respondents ordinarily will be lower with subsampling of nonrespondents than with a callback strategy. However, if the subsampling fraction is reasonably large and if the response rate in the follow-up subsample is high, then double sampling may nevertheless reduce the risk of nonresponse biases sufficiently to compensate more than enough for the increase in variances.

Discussions of the optimal choice of subsampling fractions of nonrespondents appear in Volume 2, from both a randomization perspective (Part III, chapter by Rao) and a Bayesian perspective (Part III, chapter by Singh). As with callbacks, the information required to apply these theories with precision is often scarce.

Network Sampling

Network sampling applies to household surveys using multiplicity counting rules that do not limit the number of households at which a person is enumerated. It is an alternative to the traditional sampling methods used in surveys that are based on conventional counting rules, such as de jure or de facto residence rules. Conventional rules uniquely link a person to only one household. (Counting rules should not be confused with respondent rules: the former specify which individuals are enumerable at a household; the latter specify which persons are eligible to respond for the enumerable individuals.) Sometimes network sampling can substantially improve the design of household surveys under circumstances for which traditional sampling is inefficient. Circumstances favoring network sampling include small subdomain estimation and certain intractable coverage and nonresponse problems associated with conventional counting rules.

A formal structure for making inference using network sampling and for comparing the design efficiency of network and traditional sampling is stated by Sirken (Volume 2, Part II). He presents an application of network sampling to mortality surveys involving the enumeration of persons who died in a prior reference period and discusses the error effects due to undercoverage and underreporting. The effects of several alternative kinship multiplicities rules are investigated, including those making the decedent enumerable at the households of his surviving children, parents, and siblings.

Substitution Rules

One way to handle nonresponse is to replace the nonresponding unit with a substitute not originally selected for the sample. Objective rules for substitution are often explicitly formulated to attempt to reduce biases likely to occur when the interviewer's judgment controls the choice of the substitute unit. To increase objectivity further, substitute units may be preselected, that is, included as part of the original sample selection.

Two types of substitution rules can be distinguished that are analogous to methods of hot-deck imputation. A *nearest neighbor* (deterministic) substitution rule specifies for each selected unit a hierarchy of alternative units based on proximity to that unit on the listing form. If the selected unit does not respond, then the first alternative is substituted. If that unit also does not respond, then the second alternative is substituted, and so on. A *category* type of substitution rule specifies distinguishing characteristics that the nonresponding units and their substitutes are restricted to have. Within each category, the order in which substitute units are approached may be random or may satisfy a nearest neighbor rule.

Substituted units are sometimes treated as if they were the original selected units. This practice is deceptive and not recommended. Also, as sample units, substitute units have different probabilities of inclusion than the original selected units. Even if nonrespondents and substitutes are carefully matched on observable characteristics, they remain different in that the latter respond and the former do not. A preferable procedure is to treat substitute units as imputed values and not to change the previously calculated nonresponse rates. A difficult question concerns whether substituted units are better proxies for nonresponding units than values imputed from within the sample. The nearest neighbor variant of substitution may give substituted units closer physical proximity to the nonresponding unit than possible by imputation, thus possibly reducing bias while not increasing variance as do some imputation procedures. On the other hand, the ability to match on other characteristics, which may be more important in practice, is somewhat available to both approaches but much more available in imputation.

One negative aspect of substitution is not shared by imputation. The fact that the interviewer has a convenient system for dealing with nonrespondents may reduce the incentive to obtain a response, thus leading to increased nonresponse in the sample. Steps should be taken to ensure that substitution rules do not impede the objective of keeping the nonresponse rates as low as possible. For further details on the substitution method, see Chapman (Volume 2, Part II).

Quota Sampling

Quota sampling stands apart from the other treatments of nonresponse considered here because it does not attempt to achieve a probability sample of the population and avoids the issue of nonresponse by not specifying a well-defined set of units to be included in the sample. The population is divided into subclasses defined by combinations of characteristics, such as age, sex, and income class, for which population totals that can be used to define weights are available. (Quotas are often assigned so that the sample is self-weighting.) Sample sizes are decided for each subclass based on the distribution in the population. Each interviewer is assigned a certain quota of interviews in each subclass, determined so that the required sample size is achieved in aggregate.

Quota sampling avoids nonresponse in the form of noncontact or refusal of specifically selected units, although missing data can occur if interviewers are unable to fill the assigned quotas and missing items may occur in the interviews.

Quota sampling can be combined with probability sampling of higher stage units. For example, census blocks may be selected by a probability sample design and the interviewer assigned a quota of households in the sample block, perhaps distributed among socioeconomic or demographic subclasses. The selection of households is nevertheless made by the interviewer, and the level of response depends on the ability of the interviewer to find enough units on the assigned block or blocks to satisfy the quotas.

When quota sampling is used, it is usually because of its low cost and administrative convenience relative to probability sampling. Its deficiencies are well known. Estimators and estimators of variance are based on unverifiable assumptions about the selection process and may be subject to unknown biases. If quota sampling is used, then the procedures used to arrive at the main sample, the methods used in selecting any supplementary samples, and the assumptions underlying the analysis of the data need to be clearly stated. A more detailed discussion of the method is given by King (Volume 2, Part II).

Randomized Response

Randomized response is an ingenious method, used in personal interview and telephone surveys, designed to reduce nonresponse to sensitive questions for which a respondent may wish his or her answer to remain unknown to the interviewer. A number of alternative questions are included as well as the sensitive item itself. The respondent is asked to answer either the item of interest or an alternative on the basis of a random mechanism for which the probabilities of question selection are known. The interviewer is not told which alternative is being considered. Knowledge of these probabilities allows unbiased estimates of the distribution of answers to the sensitive question to be calculated without knowing which question a specific respondent has answered. The variance, however, is increased.

The efficacy of the technique is related to the specific item of interest, the choice of alternative items, whether the respondents believe in the promised confidentiality, the privacy with which the randomization procedure is performed, and other organizational and methodological issues. For further discussion, see Emrich (Volume 2, Part II).

SYSTEMS OF INFERENCE

The literature on analytical methods reviewed in Volume 2 falls into two categories: papers directed at the specific problem of missing data in sample surveys and papers concerned with the general problem of missing data in statistics.

Most of the work directed at nonresponse in surveys begins with the randomization mode of inference. Population values are treated as fixed, and inference is based on the design distribution, that is, the distribution determined by the sample selection procedure. Then assumptions are made concerning the missing units, usually that they can be treated as though they are random subsamples within (post)strata, but sometimes that there is a stochastic model for the missing variables. Other work on missing data is based largely on assumed stochastic models for all the y variables, with the distribution resulting from sample selection playing a minor role. Both of these approaches offer different insights into nonresponse problems, and both are represented in the panel's work. Both approaches recognize that the treatment of nonresponse relies on implicit or explicit modeling of the values for nonrespondents. The modeler should be aware that the complexity of sample designs in the real world reflects complexity in population values that needs to be adequately modeled to avoid misspecification error. The panel hopes that the juxtaposition of these contrasting systems of inference will lead to a satisfactory synthesis in future work.

Randomized Inference in the Absence of Nonresponse

A useful comparison of the inferential approaches is given in the chapter by Rubin in Volume 2, Part IV. The basic ideas can be illustrated with reference to simple random sampling to estimate the population mean of a single variable. Let y_i denote the value of the variable for the ith unit in the population of N units. Let $y = (y_1, \ldots, y_N)$, and let the population mean,

$$\bar{Y} = \sum_{i=1}^{N} y_i/N,$$

denote the quantity of interest. The vector $d = (d_1, \ldots, d_N)$ indicates sampled and nonsampled units in the survey, where $d_i = 1$ if unit i is selected and $d_i = 0$ otherwise.

In the absence of nonresponse, y_i is observed if and only if $d_i = 1$. The randomization approach bases inferences on the properties of estimators in repeated sampling from the population with the item values y treated as fixed. In other words, inferences are based on the conditional distribution of d given y.

Three steps are used in this inferential system:

(1) Choice of a statistic $S(y_K, d)$, which depends on y only through known sampled values y_K. For example, $S(y_K, d) = \bar{y}$, the sample mean.

(2) Calculation of the first two moments of $S(y_K, d)$ over the design distribution of d given y. This requires that the design distribution be known, as is the case with probability sampling. In particular, under simple random sampling the mean and variance of $S(y_K, d) = \bar{y}$ are

$$E(\bar{y}) = \bar{Y}, \qquad \text{var}(\bar{y}) = n^{-1}(1 - n/N)V,$$

where V is the population variance of the y values.

(3) Estimation of the first two moments of $S(y_K, d)$ from sample values. Usually S is chosen to be approximately unbiased for the quantity of interest, and then the variance of $S(y_K, d)$ is all that is required.

Under simple random sampling, $\bar{y}$ is unbiased for $\bar{Y}$, and its variance is estimated by $n^{-1}(1 - n/N)\hat{V}$, where $\hat{V}$ is the sample variance. Then in large samples,

$$\bar{y} \pm 1.96(\hat{V}/n)(1 - n/N)$$

is an approximate 95% confidence interval for the population mean $\bar{Y}$.

Model-Based Inferences in the Absence of Nonresponse

Model-based inferences require the specification of a distribution for y as well as for d given y. For example, a particularly simple specification is that the y_i are independently normally distributed with common mean μ and variance σ^2. Inferences are based on the predictive distribution of $\bar{Y}$ given the sample values of y. Under simple random sampling, this model leads to the same 95% interval for $\bar{Y}$ as in the randomization approach given earlier. Under the full Bayesian specifications adopted by Rubin (Volume 2, Part IV), the model-based interval is interpreted as a probability interval for $\bar{Y}$, although other interpretations are possible (see, for example, Royall, 1971).

The Response Mechanism

In the presence of nonresponse, randomization and model-based inferences each require a model for the *response mechanism*, that is, for the process that results in nonresponse. Let $r = (r_1, \ldots, r_n)$, where $r_i = 1$ if individual i responds if sampled and $r_i = 0$ otherwise. The response mechanism can be modeled formally by specifying a distribution for r given d and y. In particular, the mean of r_i is the probability that unit i responds if sampled, given the values of d and y.[1]

Some believe that the specification of models for the response probabilities is unjustified. For example, Dalenius (Volume 3, Session VIII) writes,

> some writers postulate "response probabilities" . . . [and] give these probabilities an instrumental role in schemes for coping with the nonresponse problem. I take a dim view of the usefulness of these endeavors . . . it appears utterly unrealistic to postulate fixed "response probabilities" which are independent of the varying circumstances under which an effort is made to elicit a response. Whether an individual selected for a survey

[1] Rubin (Volume 2, Part IV) also specifies a distribution for the recording mechanism; for simplicity we assume here that all respondents are recorded.

> will respond or not may in many instances be determined by factors external to the individual . . . it must be possible to estimate these probabilities. I am at a complete loss to understand how valid estimates can be computed on the basis of the data collected in the course of a real-life survey; it seems unavoidable to introduce assumptions of unknown validity about the probabilities.

The difficulties raised by Dalenius are real, as are the difficulties with other models, but the panel nevertheless believes that any rigorous treatment of nonresponse requires the consideration of various models for the response mechanism. An important step in the assessment of existing methods is to state explicitly the models upon which they are based, or with which they are consistent. Finally, the unverifiable nature of models for the response mechanism means that if nonresponse is a serious problem, sensitivity analyses should be performed for a variety of models for the response mechanism in order to evaluate objectively the potential impact of nonresponse on the survey estimates.

The problems in estimating response probabilities are also real. Errors in estimating individual response probabilities should be taken into account in biases and variance formulas. Also, small estimated probabilities may result in data for a single unit having a very large effect on the estimate and variance even if the estimate is unbiased. This is usually guarded against by limitations placed on weights or by using a ratio estimator [see (4)].

To estimate response probabilities from a given survey does require making assumptions of unknown validity—but assumptions of unknown validity must be made for any other nonresponse procedure unless complete responses from a subsample of nonrespondents is obtained. Thus the sensitivity of the nonresponse procedures to the assumptions on which they are based is of considerable importance; the effects of alternative assumptions should be studied for all nonresponse procedures.

Randomization Inferences in the Presence of Nonresponse

With nonresponse, randomization inferences are based on the distribution of d and r given y. Unlike the design distribution of d, the response mechanism distribution of r is not under the control of the sampler. The three steps of the randomization approach [(1)–(3) above] are duplicated with $S(y_K, d)$ replaced by $S(y_K, d, r)$, a statistic that depends only on the responding values y_i. The distribution of d given y is replaced by the distribution of d and r given y. Rubin (Volume 2) shows the difficulties of this approach when the distribution of the response mechanisms depends on the values of Y, but these difficulties do not occur when probability sample designs are used.

In practice, methods for dealing with nonresponse from the randomization perspective have been stated as though the distribution of r corresponds to another stage of (possibly stratified) random sampling, and the probability of response has been assumed constant within specified subgroups of the population. Thus no new tools are required to draw inferences when confronted with nonresponse. Oh and Scheuren (Volume 2, Part IV) aptly coin the term *quasi-randomization* to describe response models that define a response distribution to be a design distribution. The model described by Platek and Gray (Volume 2, Part V) does not place restrictions on the form of the response probabilities, but the methods with which they illustrate their theory assume constant probabilities of response in subclasses for the response distribution. Their general model is used to evaluate the biases and variances of estimates under several imputation procedures in an example with varying response probabilities that are correlated with a characteristic of the population.

Model-Based Inferences in the Presence of Nonresponse

In the model-based approach, inferences are based on a model for the joint distribution of y, d, and r. Given a joint distribution of y, d, and r, no additional tools are required to draw inferences when confronted with nonresponse with either the randomization or model-based approaches. The principal additional requirement is the specification of the model for the response mechanism and possible computational problem in the calculation of estimates.

A key concept in the superpopulation approach to sample surveys is *ignorability* (Volume 2, Part IV, chapter by Rubin). The joint distribution of y, d, and r specified by the full model can be factored in the form

$$f(y, d, r) = f(y)f(d|y)f(r|d, y),$$

where f denotes the probability density function for continuous variables or the distribution function for discrete variables. The density function $f(y)$ is specified by the model for the data, $f(d|y)$ is specified by the model for the sampling mechanism, and $f(r|d, y)$ is specified by the model for the response mechanism. The sampling mechanism is called ignorable[2] if, for example, the distribution $f(d|y)$ does not depend on y or depends on y only through values recorded for all units in the population. The response mechanism is called ignorable if, for example, the response distribution of sampled items does not depend on y or depends only on items that are recorded for all units of the population. Thus, the probability of response can depend on design variables and fully recorded item variables, but cannot depend on variables that are subject to nonresponse.

The importance of ignorability in the model-based view of inference rests on the fact that if the sampling or response mechanism is ignorable, model-

[2] Ignorable for inference based on the likelihood; detailed conditions are given in Little (Volume 2, Part VI).

based inferences can be made without having to specify the distribution of that component of the model. Thus, for data without nonresponse, if the sampling mechanism is ignorable, inferences can be based on the distribution $f(y)$, and the distribution of sample selection $f(d \mid y)$ plays no part in the analysis (in contrast to the randomization approach in which this distribution is central to inferences). If a model is assumed, the sampling mechanism is ignorable for probability sample designs, provided the design variables (known for all units of the population) are recorded for analysis. Probability sampling is thus desirable under the model-based approach to ensure that the distribution of d given y does not have to be modeled. With quota sampling, however, the probabilities of selection may depend on unrecorded variables in unknown ways, and these relations should be included in the model to provide correct inferences.

The response mechanism is not under the control of the sampler, and hence it is not ordinarily ignorable. If the probability of response depends on the values of nonresponding items, the form of this association needs to be modeled. Since data are only available for respondents, this aspect of the model is not capable of verification, and so the results may be sensitive to untestable model assumptions. Thus, a variety of model specifications, including when appropriate different models for various items, is advisable to assess the sensitivity of results.

Little (Volume 2, Part VI) gives examples of two forms of nonignorable models. In the first form the distribution of r given d and y is modeled explicitly as a form of stochastic censoring. In the second form the distribution of y given r is specified, and the distribution of y for nonrespondents ($r_i = 0$) is linked to the distribution of y for respondents ($r_i = 1$) by a Bayesian prior distribution. Such models are of very recent origin and provide a possible theoretical basis for future practice rather than a theory of current practice. (In Volume 2, Part VI, Little reviews the more established literature of models for item nonresponse where the response mechanism is assumed ignorable.)

Much of the literature reviewed in Part VI of Volume 2 is based on models that assume the item values are independent, identically distributed random variables with normal or multinomial distributions. Many people who base inference on the distribution of sample selection are skeptical of the use of such models for making inferences, since they can give poor results if they do not adequately reflect heterogeneity in the population sampled. If models are to be used more extensively as a basis for inferences, they must be accurately specified so as not to be contradicted by the data. Little (Volume 2, Part VI) indicates ways in which stratification and clustering can be included in the model specification (also see, for example, Scott and Smith, 1969; Fay and Herriot, 1979). However, more work is needed to bridge the crediability gap in these approaches.

Although in the absence of nonresponse, people who use randomization avoid using models as a basis for inferences, they do use models of the population to aid the selection of efficient sampling designs and associated estimators.

Furthermore, everyone agrees that models are required to impute values for missing data. These models may be implicit, as in poststratification and weighting-class methods, or explicit, as in methods based on a distribution for the population values.

WEIGHTING

In probability sample designs, the appropriate procedure for compensating for unequal selection probabilities is to assign a weight

$$w_i = 1/\pi_i \tag{1}$$

to sampled units i, where

$$\pi_i = E(d_i|y) \tag{2}$$

is the mean of the selection indicator for unit i, that is, the probability that unit i is selected. These weights are used to estimate population quantities. For example, the population mean $\bar{Y}$ is estimated by

$$\bar{y}_w = \sum_{i=i}^{N} w_i d_i y_i / \sum_{i=i}^{N} w_i d_i, \tag{3}$$

where only the sampled units (for which $d_i = 1$) contribute to the summations. The numerator of $\bar{y}_w$ is the Horvitz–Thompson estimate of the population total (Horvitz and Thompson, 1952). The reason for including weights is that $\bar{y}_w$ is then approximately unbiased for $\bar{Y}$ for a large class of sample designs, since from (2)

$$E(\bar{y}_w|y) \simeq \sum_i w_i y_i E(d_i|y)/\sum_{ii} w_i E(d_i|y) = \bar{Y}.$$

In the presence of nonresponse, a natural extension of this idea is to assign to units that are sampled and respond a weight

$$w_i^* = 1/\pi_i \varphi_i,$$

where $\varphi_i = E(r_i|y,d)$ is the probability of response unit i. The analogous estimator to (3) is then

$$\bar{y}_w^* = \sum_{i=i}^{N} w_i^* d_i r_i y_i / \sum_{i=i}^{N} w_i^* d_i r_i. \tag{4}$$

Unfortunately the φ_i are unknown. In practice the sample is usually divided into subclasses $C_1, \ldots, C_k$, and φ_i is estimated as the proportion of sampled units that respond in the class to which unit i belongs. This leads to the estimator

$$\tilde{y}_s = \sum_{i=i}^{N} d_i r_i y_i/(\pi_i \varphi_i) / \sum_{i=i}^{N} d_i r_i/(\pi_i \varphi_i). \tag{5}$$

Oh and Scheuren (Volume 2, Part IV) discuss this estimator for the case in which the selection probabilities π_i are equal. Platek and Gray (Volume 2, Part V) discuss the numerator of (5) as an estimator of the population total. They call the subclasses $C_1, \ldots, C_k$ *adjustment* cells.

Platek and Gray distinguish two types of subclasses according to whether they are based on variables used for the sample design, such as geographic strata, or on variables that are measured without nonresponse in the data collection phase of the survey. They call the former *balancing areas*, or design-dependent balancing areas, and the latter *weighting classes*. Oh and Scheuren call $\tilde{y}_s$ a weighting-class estimator, regardless of which type of subclass is used. When the proportion of the population in each subclass is known, an alternative estimator to (5) can be formed. Oh and Scheuren call this a *poststratified* estimator since the poststrata population weights are known. On the other hand, Schaible (Volume 3, Session III) refers to the weighting-class approach as a form of poststratification.

The units within adjustment cells usually have unequal and unknown probabilities of response. Estimates of the response probability within adjustment cells are usually calculated as average response probabilities, i.e., as though the nonresponse mechanism specified that the response probabilities are equal within adjustment cells. The fact that the real nonresponse mechanism differs from the hypothetical nonresponse mechanism leads to biases that may be large if the response probabilities within adjustment classes are highly variable and especially if, in addition, the y_i values in the subclasses have large variances and are highly correlated with the response probabilities. This fact emphasizes the importance of defining the adjustment subclasses so that the φ_i are relatively constant within subclasses.

These considerations lead to a desire for relatively homogeneous classes. However, to avoid possibly large variances it is desirable to define classes so that the number of respondents within the classes are not so small that the weight assigned to a class is highly variable due to sampling variability (in which case the weighting class estimator may have an unnecessarily large variance). At the extreme, if there are no respondents in a cell, the denominator of the weight is zero and a bias may result. There is still a problem with only one or a few cases in many of the cells. Then the variance of the within-class weights may be large, and the estimate may suffer from the same drawback as the design-weighted estimate (3) when the weights are highly variable (i.e., in ways that differ widely from the weights appropriate to optimal allocation of the sample). The effort to reduce biases by creating small cells may then be at the expense of a large variance. This leads to a desire to keep adjustment cells large enough so that the coefficient of variation of the number of respondents in a cell is not too large. In simple random sampling a commonly used rule is that the number of respondents in a class should not be less than about 20. With such a rule, variance increases will be minor or negligible. Moreover, with such a rule, the remaining nonresponse biases after nonresponse adjustments may be only trivially larger than the remaining biases using more and smaller cells. Considerable experience suggests that the use of highly detailed weight or poststratification classes ordinarily is not substantially more effective in reducing biases or variances than are moderately detailed classes.

The fact that the response probabilities are unknown and have to be estimated adds a component to the variance of the weighting class estimator (4). Assuming a constant response probability φ_h in class h, the estimated probability of response $\hat{\varphi}_h$ has an estimated variance of $\hat{\varphi}_h(1 - \hat{\varphi}_h)/m_h$ under a quasi-randomization model, where m_h is the number of sampled elements in C_h. If the m_h are too small, this quantity can represent a nontrivial component of the variance. This component is kept relatively small or trivial by increasing the size of the weighting class, such as by the rule that $m_h \geq 20$. Note that under simple random sampling the usual stratified sampling variance estimator appropriately includes the variance component from the nonresponse adjustment, conditional on the observed response rates.

These considerations suggest that even with this relatively simple technique, the choice of weighting class is a problem that deserves careful attention. As Oh and Scheuren state (Volume 2, Part IV),

> These elementary considerations suggest that practitioners employ more than one set of nonresponse weighting adjustments with the actual number of alternatives being a function of the resources available, the amount of nonresponse, the degree of uncertainty about the response mechanism, and the extent to which different purposes are to be served by the same data set. It is further recommended that, as is consistent with much existing practice, when employing these alternative adjustments, steps to be taken to reduce the adverse variance impacts that can occur if the subpopulations chosen have small m_h or large n_h/m_h.

Moments of the poststratified and weighting-class estimator under simple random sampling are derived, and estimators compared for an artificial generated data set, by Oh and Scheuren. Their conclusions are worth quoting, since they are relevant to other methods of handling nonresponse:

> (1) All too often weighting adjustments for nonresponse are chosen for their convenience rather than their appropriateness (Bailar, Bailey, and Corby, 1978). Part of the reason for this is that, despite protestations to the contrary, we may be grossly underestimating the seriousness of the nonresponse bias in surveys. It is essential that at the design stage plans be made for the nonresponse adjustment including plans to collect good predictor variables (if possible) for use in the adjustment process.

(2) Inadequate attention is being given to validation studies which measure the nonresponse bias and, until recently, to improving our techniques for handling nonresponse or for thinking through the analysis implications of the residual errors which remain in the survey data even after adjustment. (The need for validation studies has special force in large-scale surveys where the biases due to response and nonresponse errors tend to dominate the mean square error.)

(3) Practitioners should explicitly state the response mechanism assumed by the adjustment procedure they employ. Evidence, if available, should be cited on why such a mechanism may be plausible.

(4) The impact of alternative estimators should be examined so that questions of sensitivity to alternative assumptions can be brought out concretely.

(5) It is desirable that practitioners provide sets of survey weights rather than a single estimator. . . . This may be especially important for secondary users whose analysis goals were not incorporated in developing the adjustments.

(6) The value of replication cannot be stressed too much as a way of introducing robustness into the variance estimation. Replication also seems an essential aid in interpreting differences introduced in our results by alternative estimators.

(7) Since response models almost never hold exactly, the only truly robust approach to the problem of bias is to keep nonresponse to a minimum. There is no adequate substitute for complete response, as the simulations in the chapter concretely illustrate.

It should be noted in summary (and to put this adjustment procedure in perspective) that a weight-adjustment procedure based on units is simple to apply; it handles unit nonresponse for all items simultaneously; and when applied with reasonable care and with low nonresponse rates, it may yield reasonably acceptable results. However, while nonresponse biases may be reduced, on the average, they may still be substantial after adjustments for particular items. In addition to overall adjustment procedures, analysts may prefer to deal separately with individual items that are of importance for a particular purposes by introducing special imputation or weighting procedures for them.

IMPUTATION

This section reviews procedures for handling nonresponse by replacing missing data by estimates of the missing items. Point estimates consisting of linear combinations of the data can be calculated from survey data compiled in this way, although the variance of these estimates, where necessary, needs to be adjusted to allow for nonresponse.

A useful procedure for evaluating survey estimates in the absence of nonresponse is to see what values they imply for nonsampled units. For example, an estimate of the population total may be written in the form

$$\sum_{i \text{ sampled}} y_i + \sum_{i \text{ not sampled}} \hat{y}_i,$$

where $\hat{y}_i$ is the implied prediction of y_i for nonsampled units. Estimators that do not lead to intuitively sensible values of $\hat{y}_i$ are called into question. This predictive view of estimation is closely related to the Bayesian view of inference for sample surveys.

The analogous procedure for estimates of the total $N\bar{Y}$ in the presence of nonresponse is to decompose estimates in the form

$$\sum_{\substack{i \text{ sampled} \\ \text{and responding}}} y_i + \sum_{\substack{i \text{ sampled} \\ \text{not responding}}} \hat{y}_i + \sum_{\substack{i \text{ not} \\ \text{sampled}}} \hat{y}_i,$$

where the middle summation is over imputed values for units that do not respond. Imputation can thus be viewed as an extension of the predictive method of estimation in the case of nonresponse. Some but not all weighting estimates can be viewed in this way.

Imputation is a general and flexible method of tackling the missing data problem. However, it is not without pitfalls. In the words of Dempster and Rubin (Volume 2, Part I),

> The idea of imputation is both seductive and dangerous. It is seductive because it can lull the user into the pleasurable state of believing that the data are complete after all, and it is dangerous because it lumps together situations where the problem is sufficiently minor that it can legitimately be handled this way and situations where standard estimators applied to real and imputed data have substantial biases.

Most imputation methods involve two steps, the classification of the unit into a subclass of the population where the y values are relatively homogeneous and the imputation of values of y for that subclass. There are a variety of methods for the choice of imputed value for each missing unit:

(1) The value of one of the responding items in the subclass, selected at random or in some deterministic way with or without replacement or based on a measure of distance from the nonrespondent unit (hot-deck imputation) (Volume 2, Part IV, chapter by Ford).

(2) The value from a previous survey or other information (cold deck imputation).

(3) The average of the responding units in the subclass (mean imputation).

This form of imputation is equivalent to weighting respondents in each subclass by the ratio of sampled to responding units (discussed earlier). Other estimators of the center of the distribution, such as the median, can also be considered.

(4) The predicted mean of the subclass estimated by a regression model with y as the regressand; this is a generalization of mean imputation (see chapter by Little, Volume 2, Part VI).

(5) The value of a unit substituted at the data collection phase (see chapter by Chapman, Volume 2, Part II).

(6) The predicted mean of the subclass plus a residual generated from a model error distribution using pseudorandom numbers.

All these methods are based on the same underlying model for the response mechanism. If one interprets the resulting nonresponse adjustments as unbiased, one must assume that, on the average, over the chosen subclasses, the probability of response is uncorrelated with y. Methods (1), (5), and (6) incorporate variation between individuals in the subclass in the estimates in an attempt to preserve the distribution of y. For methods (1) and (6) the estimates have a higher variance for estimates of means than analogous estimates for which means are imputed. However, the failure of mean imputation methods to reflect the variability of nonresponding values in a subclass can have undesirable consequences. For example, if values of y_i are grouped into levels, possibly with other data cross-classified by those levels, all nonrespondents in a subclass have the same imputed value of y_i and hence fall in the same category. Thus, estimates of proportions (other than those based on aggregates for the weighting classes) may have greatly increased variances and biases. Methods (1), (5), and (6) for imputation distribute nonrespondents over the categories and thus avoid such unsatisfactory heaping unless items for one or more respondents are imputed too often.

Another situation in which hot-deck imputation appears to have an advantage concerns the estimation of variance of sample estimates based on imputed data. It is well known that if means are imputed for missing values and the variance of the sample mean is estimated from the completed data, ignoring the effects of imputation, the resulting estimated variance is too low. If the fact that imputation has been used is ignored, the variance of individual y_i is underestimated and the sample size is overestimated. However, it is perhaps less well known that hot-deck procedures also lead to underestimates of the estimated variance. The question of variance computation in the presence of nonresponse has not received much attention in the statistics literature, and it appears to be a topic on which more research is needed. (See discussion of multiple imputations in the next section.)

Hot-deck procedures are particularly suited to data processing on the computer, and they have developed in sophistication, sometimes with little attention to underlying theory. To quote Ford (Volume 2, Part IV):

> Because hot-deck procedures originated in survey practice with little theory to direct their development, the statistical literature provides few definitions or results about these procedures. Widespread practice in the absence of well-developed theory clouds the subject with ambiguities and inconsistences. For example, there is no general agreement on the exact definition of a hot-deck procedure. This chapter reflects the present state of development not only by pointing to the lack of well-developed theory but also by using, for the sake of discussion, definitions and principles which are still being debated.

Even within the framework of selection from well-defined subclasses, a variety of hot-deck procedures can be defined according to the proportion of responding items and whether selection is made with replacement, without replacement, or by a sequential procedure that loops through responding items if they are needed more than once. Ford (Volume 2, Part III) and Herzog and Rubin (Volume 2, Part IV) provide expressions for the variance of these methods under simple random sampling and a quasi-random selection mechanism. Ford also discusses hot-deck methods in which the responding variables are ordered according to other items before sequential selection, a form of implicit stratification analogous to systematic sampling. Other forms of the hot deck are described in Sande (Volume 3, Session VIII). The relative strengths and weaknesses of variants of hot-deck procedures are not clear, although they may emerge with future theoretical and empirical work. Ford cites studies (Ford, 1976; Cox and Folsom, 1978; Ernst, 1978; Pare, 1978; Schieber, 1978) that compare the hot-deck method with weighting and regression imputation. He concludes (Volume 2, Part III),

> Because each of these empirical studies is confined to the investigation of one particular survey, broad generalizations of their results are difficult to make. These studies do, however, support the theoretical conclusion that the standard errors of hot-deck estimators are underestimated by computations which assume that the data set was originally complete. They also indicate that there *may* not be much improvement in the mean square error of a hot-deck estimator in comparison to an estimator which omits the missing values unless there is auxiliary information which is highly correlated with the survey data. They do *not* consistently show that one missing data procedure is better than another.
>
> These empirical studies should encourage other survey organizations to carry out experiments which detect the effects of missing data and missing data procedures in their own surveys. The objective of such studies should be not only to find procedures yielding low mean square errors of estimates but also to ensure that the multivariate structure of the data set after imputation is representative of the population.

MULTIPLE IMPUTATION

A major step toward the establishment of a theoretical basis for hot-deck methods is taken in the multiple imputation theory of Rubin, represented in Volume 2 by the chapter by Rubin and Herzog (Part IV). These authors emphasize *interval* rather than *point* estimation: that is, methods are assessed not only by the estimates of population quantities that they provide but also by the estimated precisions, in the form of estimated variances and associated probability or confidence intervals for the population quantity. This emphasis reflects a healthy trend toward the presentation of sampling errors, including effects of treatment of nonresponse for survey estimates, a trend that can be expected to continue in the future.

The inferential properties of elaborate imputation procedures for complex surveys are hard to evaluate. However, Rubin and Herzog show that for the simple case of nonresponse for a single simple random sample and a random response mechanism, where the correct inference is known, methods based on filling in the missing values and treating the data as complete never give the right answers.

Specifically, suppose there are n sampled units, of which a random subset of m units respond and $n - m$ units do not respond. The correct 95% interval for the population mean of a variable Y (assuming m is large and ignoring finite population corrections) is

$$\bar{y} \pm 2\sqrt{\hat{V}/m}, \tag{5}$$

where y and $\hat{V}$ are the sample mean and variance of responding items. Imputing the sample mean for missing units and acting as if the data were complete leads to the 95% interval

$$\bar{y} \pm 2(m/n)\sqrt{\hat{V}/m}, \tag{6}$$

which has the correct center ($\bar{y}$) but underestimates the correct interval width by a factor of m/n. If, instead of imputing the sample mean, values are imputed by random selection from responding units with replacement, then the resulting interval is

$$\bar{y}_i^* \pm 2\sqrt{\hat{V}_1^*/n}, \tag{7}$$

where $\bar{y}_1^*$ and $\hat{V}_1^*$ are the sample mean and variance of responding and imputed items. Taking expectations over repetitions of the hot-deck selection (with the same set of responding units) gives

$$E(\bar{y}_1^*) = \bar{y}, \qquad E(\hat{V}_1^*) = \hat{V}.$$

Hence, the center of the interval is incorrect, but the expected center is correct if a random response mechanism is assumed; the width of the interval is underestimated by a factor of $\sqrt{m/n}$ in expectation. An adjustment for this factor is sometimes made in practical survey work. It is a minor adjustment if item responses are high, and it is often not made for items for which the proportion responding is reasonably high (say, 95% or more). The adjustment should be made if item responses are low.

A way of improving the hot-deck interval is to repeat the hot-deck imputation I times. Let $\bar{y}_i^*$ and $\hat{V}_i^*$ be the sample mean and variance of responding and imputed values for the ith set of imputations. Calculate the averages of the I means and variances across imputations,

$$\bar{y}^* = \sum_{i=1}^{I} \bar{y}_i^*/I, \qquad \bar{u}^* = (\sum_{i=1}^{I} \hat{V}_I^*)/I,$$

and the variance of the I sample means across imputations,

$$w = \sum_{i=1}^{I} (\bar{y}_i^* - \bar{y}^*)^2/(I - 1).$$

Herzog and Rubin suggest the 95% interval

$$\bar{y}^* \pm 2(w + \bar{u}^*/n)^{1/2} \tag{8}$$

for the population mean. The center of the interval is the average of the centers from each imputed data set. The squared width of the interval is the average of the squared widths from each imputed data set ($\bar{u}^*/n$) plus an additional component of variance (w), which allows for the randomization introduced by the hot deck. Herzog and Rubin show that $\bar{y}^*$ is a more efficient estimate of the population mean than $\bar{y}_1^*$ and the efficiency increases with the number of sets of imputations I; and that the interval (8) is closer to the correct interval (5) than intervals (6) or (7).

The width of the interval (8) is in fact still underestimated, even if the number of imputations I is large. Herzog and Rubin provide a theoretical explanation of this result, based on Bayesian principles. Correct imputation of missing items can be viewed as selecting values from their posterior distribution given the data and model. In the procedure described, items are selected from the true distribution of values in the population, as represented by the responding items. An additional random quantity should be added to the imputations to account for the disparity between the true and posterior distributions, and the variance of this quantity should appear in (8). This final adjustment is not easy to derive in complex situations. However, if the amount of data is large, the effects of omitting this step are small.

Of course in this example there is no reason to perform multiple imputations or hot-deck procedures, since the correct interval (5) can be obtained directly. The value of multiple imputations is in more complex situations, in which the correct answer is not easy to derive. The principle of forming a set of imputed data sets, performing a complete data analysis on each set, and then combining the results in a manner analogous to that of (8) can be applied quite generally.

For example, survey practitioners often prefer not to accept standard errors based on a superpopulation model that may not be correctly specified. Sometimes they calculate design-based standard errors using balanced repeated replication, the Jacknife or Taylor series expansions. With multiple imputation, a model can be used to generate imputed values and a set of imputed data sets formed. Design-based estimates and standard errors can then be formed for each data set and combined as in (8) to form an estimate and standard error that allows for nonresponse. The model-dependent part of the process is then combined with the imputations, and, hence, sensitivity to the model is reduced.

Imputed values in this example are generated by a quasi-random (or ignorable) response mechanism. Nonignorable mechanisms can easily be modeled making scale and location changes to the imputed values. Thus multiple imputations may provide a fairly simple method of assessing sensitivity to nonignorable models. Practical applications of the method are still at an early stage of development, but the technique certainly holds promise for future research.

A MODEL FOR TOTAL SURVEY ERROR

The work reviewed thus far assumes that the value of the variable of interest (say, y_i) is recorded without error for each responding unit. In practice, however, a variety of factors may lead to response errors—incorrect information from the respondent, recording errors from the interview, coding errors, and punching errors. There is a relation between response error and nonresponse. A particularly large response error might lead to a value being dropped from the data in the editing phase of a survey. Conversely, adjustment, imputation, or accepting information from other sources in effect replaces nonresponse by a response that is almost always subject to error.

Models for response errors have been developed (Hansen, Hurwitz, and Bershad, 1961; Fellegi, 1964) that distinguish components of response bias, response variance, and response covariance accumulated at different stages of the survey process. The model described by Platek and Gray (Volume 2, Part V) extends this work to include nonresponse in the form of nonresponse or imputation error.

The Platek–Gray model is based on the randomization of inference. Inferences are based on modification of the Horvitz–Thompson estimator of the population total,

$$\sum_{i=1}^{N} d_i y_i / \pi_i,$$

where d_i, y_i, and π_i are as defined in the section on systems of inference. Imputation leads to the modified estimator

$$Y_{\mathrm{T}} = \sum_{i=1}^{N} d_i \hat{y}_i / \pi_i,$$

where y_i is the observed or imputed value for unit i. Platek and Gray consider four methods of imputation: weighting within adjustment cells (defined in the section on weighting); duplication in adjustment cells (a form of hot-deck procedure); substitution of historical or external source data; and zero substitution.

The objective is to derive expressions for the bias and variance of each estimator. This is done by taking expectations with respect to three distributions:

(1) The distribution F_1 of the sampling mechanism d given the set of the item values y; expectation is denoted by E_1.

(2) The distribution F_2 of the response mechanism r given d and y; expectation is denoted by E_2.

(3) The distribution F_3 of the response given d, y, and r; expectation is denoted by E_3.

The first two distributions are familiar from above sections. The third distribution relates to the model for response errors. Specifically, a respondent's recorded value y_i is related to the "true value" Y_i by the expression

$$y_i = Y_i + {}_{\mathrm{R}}\varepsilon_i,$$

where ${}_{\mathrm{R}}\varepsilon_i$ is the response error. Taking expectations with respect to F_3 gives

$$E_3(y_i | d, r, Y) = Y_i + {}_{\mathrm{R}}b_i,$$

where ${}_{\mathrm{R}}b_i$ is the *response bias* for the unit i, and

$$E_3(y_i^2) - [E_3(y_i)]^2 = {}_{\mathrm{R}}\sigma_i^2$$

is the *response variance* for unit i. The response bias and variance are assumed to be unaffected by the sampling and response mechanisms.

A nonrespondent's imputed value can be written as

$$y_i = Y_i + {}_{\mathrm{NR}}\varepsilon_i,$$

where ${}_{\mathrm{NR}}\varepsilon_i$ is the imputation error. If y_i is obtained from external records, then ${}_{\mathrm{NR}}\varepsilon_i$ has a structure similar to response error and can be assumed to have an imputation bias and variance unaffected by the sampling and response mechanism. If y_i is obtained from the survey data, then the imputation error has a complex structure that depends on the sampling and response patterns d and r and on the response errors in completed items.

The main focus is on the bias and variance of estimators rather than on the bias and variance of specific values. Biases are obtained by taking expectations with respect to F_3, F_2, and F_1, in turn. The bias is developed into two components: response bias and imputation or nonresponse bias. Variances are obtained by expansions of the unconditional variances into sums of expected conditional variances and variances of conditional expectations. These variances are derived as the sums of five quantities: the sampling variance (SV); the simple response variance (SRV); the correlated response variance (CRV); the variance contributed by the events of responding or nonresponding within units (VRR); and the covariance component contributed by the covariance between the event of being sampled or not and the event of responding or not among pairs of units (CVRR). For a given population, the components of bias and variance can be calculated and compared for different estimates. Platek and Gray describe an example of this procedure for a hypothetical population (Volume 2, Part V).

SUPERPOPULATION MODELS FOR NONRESPONSE

Ignorable Models

The inferential basis of the modeling approach to survey analysis with incomplete data has been reviewed. This final section reviews the specific models and associated methods that are described in Part VI of Volume 2.

The first chapter of Part VI is based on models where the sampling mechanism and the response mechanism are ignorable; hence, inferences can be based solely on a model for the distribution of the item values y. Models are specified in a parametric form. For example, the model

$$y_i \underset{\text{nd}}{\sim} N(\mu, \sigma^2), \tag{9}$$

which denotes that the y_i are independent and have a common normal distribution with parameters μ (mean) and σ^2 (variance); or the model

$$y_i \underset{\text{ind}}{\sim} N(\mu_h, \sigma_h^2), \tag{10}$$

where y_i belongs to stratum (or weighting class) h and μ_h, σ_h^2 are parameters specific to that stratum; or the normal linear regression model

$$y_i \underset{\text{ind}}{\sim} N\left(\sum_{j=1}^{k} \beta_j x_{ij}, \sigma^2\right), \tag{11}$$

where $x_{i1}, \ldots, x_{ik}$ are the values of other item variables or variables indicating strata, and β_j is the regression coefficient of y_i on x_{ij}.

The parameters in these models are not the same as population quantities, which, in principle, can be measured by a complete census of the population. Rather they represent the simplified structure of the population as formulated by the model. This structure is of primary interest in analytic surveys, in which the parameters themselves may be the focus of interest. In descriptive surveys, however, in which population quantities are of primary interest, the parameters can be viewed as intermediate quantities for predicting the nonsampled and nonresponding values. In practice, model-based interval inferences about population quantities (such as the population mean $\bar{Y}$) and related parameters [the superpopulation mean μ in model (9)] often differ only by finite population correction factors. If these are small, then parameters can be equated to population quantities for the purpose of estimation. For the most part the focus in Little's chapters is on inferences about parameters.

Models such as (9), (10), or (11) lead to likelihood functions for the parameters, given the observed data. Point estimation for the parameters can be based on the method of maximum likelihood. Standard errors of estimates are based on the observed or expected information matrix, obtained from the second derivatives of the log-likelihood with respect to the parameters. Other principles of model-based inference, such as the method of moments, are of course possible. However, the large-sample likelihood methods are efficient and are often justified in survey work in which the amount of data is considerable.

The methods obtained by this approach can be briefly illustrated with reference to the simple models described above. For (9), the maximum likelihood (ML) estimate of μ is $\bar{y}$, the mean of the responding items. Its estimated variance is the sample variance v of responding units divided by the number m of responding units. These inferences correspond to design-based inferences assuming a random response mechanism and simple random sampling [see (5)].

For (10), the ML estimate of μ_h is $\bar{y}_h$, the mean of the responding units in stratum h. If the number N of population units in stratum h is known, the population mean is estimated by

$$\sum_h N_h \bar{y}_h / \sum_h N_h,$$

which is the poststratified estimator discussed by Oh and Scheuren (Volume 2, Part IV), If N_h is unknown, it can be estimated by Nn_h/n, where n_h and n are the sample sizes in stratum h and in the whole sample, respectively. This leads to the weighting-class estimator discussed by Oh and Scheuren and by Platek and Gray. Thus both of these methods can be derived with appropriate model assumptions about the Y values.

Models (9) and (10) lead to imputing means for the missing values; model (10) leads to a regression estimate

$$\hat{y}_i = \sum_{j=1}^{k} \hat{\beta}_j x_{ij},$$

where $\hat{\beta}_j$ is the ML estimate of β_j. It is noteworthy that (10) is a special case of (11) with the x_{ij} defined as dummy variables indicating the strata. Thus, mean imputation (or weighting) is in fact a special case of regression imputation. The regression model leads to a very flexible set of methods for introducing covariate information into the imputed value for a missing item.

The models specified so far treat nonresponse for a single-item variable, with other variables completely observed in the survey. The maximum likelihood method also provides estimators when more than one variable is subject to nonresponse. Little distinguishes methods for monotone, or nested, data patterns, where the variables can be ordered so that variables are successively more complete, and methods for arbitrary patterns, where the data do not have a monotone response pattern. In the former case, estimation is often straightforward, and standard errors can be easily derived. In the latter case, iterative methods are usually needed to find ML estimates. Methods for continuous variables are based on multivariate normal or multivariate linear regression models. Methods for categorical variables are derived from log-linear models based on the multinomial distribution.

Nonignorable Models

The ignorable models just outlined have a fairly extensive literature, although the literature is not specifically related to survey data. In contrast, nonignorable models are much more recent in origin and have a rather fragmented literature. They indicate rare attempts to model the distribution of the response mechanism. The technicalities of maximum-likelihood estimation for these models are trying for a reader not familiar with likelihood theory. Thus we limit discussion to one example.

Suppose that two variables y_i (income) and x_i (age) are recorded and that y_i is subject to nonresponse. Suppose that the probability that unit i responds is

$$\Phi(\alpha + \beta x_i + \gamma y_i), \tag{12}$$

where Φ is the cumulative standard normal density function. If $\gamma = 0$, the probability of response depends only on values that are observed (x_i), and the response mechanism is ignorable. Methods that correctly use the covariate x_i to impute for y_i are then not subject to bias. But if $\gamma \neq 0$, the probability that y_i is missing depends on missing y_j values, the response mechanism is nonignorable, and all the methods discussed up to this section lead to bias. Little (Volume 2, Part VI) describes methods that correct for this selection bias and thus appear to solve the nonresponse problem; however, the solution is highly sensitive to the specification of the form of the response mechanism (12), and this form cannot be directly verified by the data, since they are restricted to values where a response has occurred, i.e., $r_i = 1$. Thus the future value of these models probably lies in sensitivity analyses, based on the multiple imputation methods discussed above or on ML estimation for a variety of model specifications.

Biographical Sketches of Panel Members and Staff

Ingram Olkin (Chair) is professor of statistics and education and former chairman of the Department of Statistics, Stanford University. He has particular interests in multivariate analysis and in models in the social sciences. He has been editor of the *Annals of Statistics* and on the editorial boards of statistical, educational, and mathematical journals and has served frequently on governmental agencies and committees. He is a fellow of the American Statistical Association, the Institute of Mathematical Statistics, the International Statistical Institute, and the Royal Statistical Society. He received a B.S. degree from the College of the City of New York, an M.S. degree from Columbia University, and a Ph.D. degree from the University of North Carolina.

Barbara A. Bailar is the associate director for Statistical Standards and Methodology, Bureau of the Census. Her principal interests are measurement errors in surveys and censuses, the application of statistical methods to survey sampling, and the estimation of data for small areas. She is a fellow of the American Statistical Association and a member of the International Statistical Institute. She received a B.A. degree from the State University of New York at Albany, an M.S. degree from the Virginia Polytechnic Institute, and a Ph.D. degree from the American University.

Barbara A. Boyes was assistant commissioner for survey design, Bureau of Labor Statistics, U.S. Department of Labor. That office was responsible for the statistical design and validity of such major economic indicators as the consumer price index, the producer price index, the employment cost index, and the wage surveys of the Bureau of Labor Statistics. She was formerly with the Bureau of the Census. She dealt extensively with the problems of nonresponse both in demographic surveys and in complex economic surveys. She was a fellow of the American Statistical Association and a member of the International Statistical Institute. She received a B.A. degree from the University of Minnesota and has also studied at the American University.

Arthur P. Dempster is a professor of theoretical statistics at Harvard University. During 1978–1979, when the Panel was meeting actively, he was ASA fellow at the Bureau of the Census. His interests and research activities range over design, data analysis, modeling, and posterior inference as applied to problems in social and biomedical statistical analysis. He received B.A. and M.S. degrees from the University of Toronto and a Ph.D. degree from Princeton University.

Robert M. Elashoff is professor, Department of Biomathematics, University of California, Los Angeles. His main interests are epidemiology, discriminatory analysis, and missing data in multivariate statistics. He received a B.S. degree from Suffolk College, an A.M. degree from Boston University, and a Ph.D. degree from Harvard University.

Robert L. Freie is a statistician with the U.S. Department of Agriculture. He has recently assumed responsibility for livestock, dairy, and poultry statistics, but most of his work has been devoted to current agriculture statistics, in the general area of survey design, sampling, and estimation. He teaches in the USDA Graduate School, and he is a member of the American Statistical Association. He received a B.S. degree from Iowa State University and studied at North Carolina State University.

Louis Gordon is a statistician with the Energy Information Administration, U.S. Department of Energy. Previously he was assistant professor of statistics at Stanford University. He is particularly interested in nonparametric inference, large-sample theory, and survey design. He received a B.S. degree from Michigan State University and M. S. and Ph.D. degrees from Stanford University.

Robert M. Groves is an associate research scientist in the Survey Research Center and an assistant professor in the Department of Sociology, University of Michigan. His principal interests are survey design and the measurement of nonsampling errors in surveys. His current work uses telephone surveys and computer-assisted telephone interviewing to study those issues. He received an A.B. degree from Dartmouth College and M.A. and Ph.D. degrees from the University of Michigan.

Morris H. Hansen is senior vice-president of Westat, Inc., a social science and statistical research and consulting organization. He has served as an international consultant on sample surveys. Previously he was at the Bureau of the Census, where he served as associate director responsible for the research and development program concerned with census and sample survey methods, quality control, computers and their applications, and geographic methods. He is a member of the National Academy of Sciences, a fellow of the American

Statistical Association and the Institute of Mathematical Statistics, and a member of the International Statistical Institute. He received a B.S. degree from the University of Wyoming and an M.S. degree from American University.

Harold Nisselson is senior statistical advisor at Westat, Inc. Previously he was associate director for statistical standards and methodology at the Bureau of the Census. His principal interests are in the application of quantitative methods to sample design and the collection and analysis of data in sample surveys in a variety of field of application. He is a fellow of the American Statistical Association and a member of the International Statistical Institute, the International Association of Survey Statisticians, and the Institute of Mathematical Statistics. He received a B.S. degree from the City College of New York.

Richard Platek is director of the Census and Household Survey Methods Division, Statistics Canada. His work has involved survey designs for large national surveys and censuses, both in Canada and in third world countries, development work in nonresponse, imputation, and estimation. He is a fellow of the American Statistical Association and a member of the International Statistical Institute. He received B.A., B.Sc., and M.A. degrees in mathematics and mathematical statistics from London and Carleton universities.

M. Haseeb Rizvi, who served as staff associate to the panel, is associated with Sysorex International Inc., Cupertino, California. Previously he had teaching assignments at Stanford University and Ohio State University. He is primarily interested in ranking and selection procedures, nonparametric inference, and multivariate analysis. He is an elected member of the International Statistical Institute. He received B.Sc. and M.Sc. degrees from the University of Lucknow, India, and a Ph.D. degree from the University of Minnesota.

Donald B. Rubin is professor of statistics at the University of Chicago. Previously he was chairman of the statistics division at Educational Testing Service (ETS), Princeton, New Jersey; during the latter stages of the Panel's work he was on leave from ETS, serving as president of Datametrics Research, Inc., and as coordinating and applications editor of *Journal of the American Statistical Association.* His principal research interests in statistics include estimation from incomplete data, causal inference in observational studies, model-based estimation in sample surveys, and applications of Bayesian and empirical Bayesian techniques. He is a fellow of the American Statistical Association, the Institute of Mathematical Statistics, and the John Simon Guggenheim Foundation. He received an A.B. degree from Princeton University and M.S. and Ph.D. degrees from Harvard University.

Frederick J. Scheuren is director, Statistics of Income Division, Internal Revenue Service. Previously he was acting chief mathematical statistician,

Social Security Administration. His research has included a number of large-scale studies of the relations between income and wealth measures reported on records and obtained in surveys; a prominent feature of that research was assessing alternative adjustments for survey nonresponse. He has participated actively in the American Statistical Association and the Conference on Income and Wealth. He received a B.A. degree from Tufts College and M.A. and Ph.D. degrees from the George Washington University.

Joseph H. Sedransk is professor, Department of Mathematics, State University of New York, Albany. His main research activities are the theory of sample surveys, especially the design and analysis of analytical surveys, and Bayesian statistical inference. He received a B.S. degree from the University of Pennsylvania and a Ph.D. degree from Harvard University.

Monroe G. Sirken is associate director, Office of Mathematical Statistics, National Center for Health Statistics. His principal interests relate to total survey design with particular emphasis on the nonsampling error effects of survey designs. His work involves developing statistical models for analyzing the cost-and-error effects of survey designs and conducting survey experiments to estimate the cost-and-error parameters specified by models. He received B.A. and M.A. degrees from the University of California, Los Angeles, and a Ph.D. degree from the University of Washington.

William G. Madow was study director for the panel and consultant, Committee on National Statistics. Previously he was staff scientist at Stanford Research Institute and consulting professor at Stanford University. His main interests are in survey design and analysis and statistical inference. He is a fellow of the American Statistical Association and the Institute of Mathematical Statistics and an elected member of the International Statistical Institute. He received A.B., M.A., and Ph.D. degrees from Columbia University.

References to Part I

Bailar, J. C., and Bailar, B. A. (1978). Comparisons of two procedures for imputing missing survey values. *Proceedings of the Section on Survey Research Methods*, American Statistical Association, pp. 462–467.

Bailar, B. A., and Lamphier, C. M. (1978). *Development of Survey Methods to Assess Survey Practices*. Washington, D.C.: American Statistical Association.

Bailar, B. A., Bailey, L., and Corby, C. (1978). A comparison of some adjustment and weighting procedures for survey data. In N. Krishnan Namboodiri (ed.), *Survey Sampling and Measurement*. New York: Academic Press.

CASRO Task Force on Completion Rates (1982). *On the Definition of Response Rates*. Port Jefferson, N.Y.: The Council of American Survey Research Organizations.

Chapman, D. W. (1976). A survey of nonresponse imputation procedures. *Proceedings of the Social Statistics Section*, American Statistical Association, pp. 245–251.

Cochran, W. G. (1977). *Sampling Techniques* (3rd ed.). New York: Wiley.

Coder, J. (1978). Income data collection and processing for the March income supplement to the Current Population Survey. *Proceedings of the Data Processing Workshop: Survey of Income and Program Participation*. Washington, D.C.: U.S. Department of Health, Education and Welfare.

Cox, B. G., and Folsom, R. E. (1978). An empirical investigation of alternative item nonresponse adjustments. *Proceedings of the Section on Survey Research Methods*, American Statistical Association, pp. 219–221.

Deming, W. E. (1953). On a probability mechanism to attain an economic balance between the resultant error of response and the bias of non-response. *Journal of the American Statistical Association* 48:743–772.

Drew, J. H., and Fuller, W. A. (1980). Modeling nonresponse surveys with callbacks. In *1980 Proceedings of the Section on Survey Research Methods*, American Statistical Association, pp. 639–642.

Ernst, L. R. (1978). Weighting to adjust for partial nonresponse. *Proceedings of the Section on Survey Research Methods*, American Statistical Association, pp. 468–473.

Fay, R. E., and Herriot, R. A. (1979). Estimates of income for small places: An application of James–Stein procedures to census data. *Journal of the American Statistical Association* 74:269–278.

Fellegi, I. P. (1964). Response variance and its estimation. *Journal of the American Statistical Association* 59:1016–1041.

Ford, B. L. (1976). Missing data procedure: A comparative study. *Proceedings of the Social Statistics Section*, American Statistical Association, pp. 324–329.

Hansen, M. H., Hurwitz, W. N., and Bershad, M. A. (1961). Measurement errors in censuses and surveys. *Bulletin of the International Statistical Institute* 38(2): 359–374.

Hansen, M. H., Hurwitz, W. N., and Madow, W. G. (1953). *Sample Survey Methods and Theory*, Vol. I: *Methods and Applications*. New York: Wiley.

Hanson, R. H. (1978). *The Current Population Survey: Design and Methodology*. Bureau of the Census Technical Paper 40, U.S. Department of Commerce.

Horvitz, D. G., and Thompson, D. J. (1952). A generalization of sampling without replacement from a finite universe. *Journal of the American Statistical Association* 47:663–685.

Jessen, R. J. (1978). *Statistical Survey Techniques*. New York: Wiley.

Kish, L. (1965). *Survey Sampling*. New York: Wiley.

Morgan, J. N., and Sonquist, J. A. (1963). Problems in the analysis of survey data, and a proposal. *Journal of the American Statistical Association* 58: 514–35.

National Research Council (1979). *Privacy and Confidentiality as Factors in Survey Response*. Panel on Privacy and Confidentiality as Factors in Survey Response. Committee on National Statistics, National Academy of Sciences, Washington, D.C.

Office of Federal Statistical Policy and Standards (1978). Glossary of Nonsampling Error Terms: An Illustration of a Semantic Problem in Statistics. Statistical Policy Working Paper 4, Subcommittee on Nonsampling Errors, Federal Committee on Statistical Methodology. U.S. Office of Management and Budget, Washington, D.C.

Oh, H. L., and Scheuren, F. J. (1978). Multivariate rankiag ratio estimation in the 1973 (exact match study). *Proceedings on the Survey Research Methods*, American Statistical Association, pp. 716–722.

Oh, H. L., and Scheuren, F. J. (1980). Estimating the variance impact of missing CPS income data. *Proceedings of the Section on Survey Research Methods*, American Statistical Association, pp. 408–415.

Oh, H. L., Scheuren, F. J., and Nisselson, H. (1980). Differential bias impacts of alternative Census Bureau hot-deck procedures for imputing missing CPS income data. *Proceedings of the Section on Survey Research Methods*, American Statistical Association, pp. 416–420.

Pare, R. M. (1978). Evaluation of 1975 Methodology: Simulation Study of the Imputation System Developed by BSMD. Business Survey Methods Division, Statistics Canada, Toronto.

Rao, P. S. R. S. (1966). A Study of Call-Back Policies in Sample Surveys. Technical Report No. 10, Statistics Department, Harvard University.

Royall, R. M. (1971). Linear regression models in finite population sampling theory. In V. P. Godambe and D. A. Sprott (eds.), *Foundations of Statistical Inference*. Toronto: Holt.

Scheuren, F. J. (1978). Discussion of the paper entitled "Income data collection and processing for the March income supplement to the Current Population Survey," by John Coder. *Proceedings of the Data Processing Workshop: Survey of Income and Program Participation*, U.S. Department of Health, Education and Welfare.

Schieber, S. J. (1978). A comparison of three alternative techniques for allocating unreported Social Security income on the survey of low income aged and disabled. *Proceedings of the Section on Survey Research Methods*, American Statistical Association, pp. 212–218.

Scott, A. J., and Smith, A. M. F. (1969). Estimation in multistage surveys. *Journal of the American Statistical Association* 64:830–840.

Welniak, E. J., and Coder, J. F. (1980). A measure of the bias in the March CPS earnings imputation system and results of a sample bias adjustment procedure. *Proceedings of the Section on Survey Research Methods*, American Statistical Association, pp. 421–428.

Wiseman, F., and McDonald, P. (1980). Towards the Development of Industry Standards for Response and Nonresponse Rates. Marketing Science Institute, Cambridge, Mass.

PART II

Case Studies

CHAPTER **1**

Overview

Harold Nisselson

1. INTRODUCTION

Data collected in censuses and surveys are, almost without exception, incomplete—typically from nonresponse, but also from other causes. It is a widespread, if not general, practice for those conducting surveys to make adjustments for incomplete data. The argument is made that this approach provides at least as valid a base for data analysis as not making adjustments, while preserving the maximum information on the individual data items and simplifying the analyst's task. The argument can also be made, if this approach is adopted, that the data producer—having access to the full range of the individual data and knowledge of the idiosyncrasies of the particular data collection—can fill in missing data better than individual data users, as well as avoid the additional effort that would be required if each data user were to attempt to repeat the same work.

In this volume, statisticians associated with 10 large-scale and significant survey efforts have undertaken to describe in some detail how they have treated their survey data to try to compensate for incompleteness and to enhance its value to data users. Because the phenomenon of incomplete data is so pervasive and because it is unusual to find an adequate description in data publications of the methods by which the problems of incomplete data are treated, these chapters are a valuable contribution to the literature of survey methodology. In addition to describing what was done, several chapters present an analysis of alternative methods. Two of them (Chapters 9 and 11) are almost unique among ongoing surveys in presenting an evaluation of alternative adjustment methods for incomplete data by comparison with external sources

INCOMPLETE DATA
IN SAMPLE SURVEYS
Volume 1, Part II

ISBN 0-12-363901-8

TABLE 1

List of Case Studies and Abbreviations for Text Citations

Case study	Abbreviation
1. The Employment Cost Index: A Case Study of Incomplete Data	ECI
2. USDA Livestock Inventory Surveys	LIS
3. The Survey Research Center's Surveys of Consumer Attitudes	SCA
4. Treatment of Missing Data in an Office Equipment Survey	OES
5. Annual Survey of Manufactures	ASM
6. Incomplete Data in the Survey of Consumer Finances	SCF
7. An Empirical Investigation of Some Item Nonresponse Adjustment Procedures—National Longitudinal Survey	NLS
8. Readership of Ten Major Magazines	RTMM
9. Total Survey Error—Center for Health Administration Studies	CHAS
10. An Investigation of Nonresponse Imputation Procedures for the Health and Nutrition Examination Survey	HANES

TABLE 2

Summary of Sample and Survey Designs of the Case Studies

Case study	*Sample design*[a]	*Data collection unit*	*Primary data collection method*
ECI	Two-phase stratified sample with controlled selection in second phase	Sample occupation within sample "establishment"	Mail[b]
LIS	Combination of list and multistage area sample	Producer of cattle, hogs	List sample—mail Area sample—personal interview
SCA	Random-digit dialing	Randomly selected adult household member[c]	Telephone interview
OES	Multistage area sample	Establishment	Personal interview
ASM	Single-stage Poisson sampling with variable probabilities	Each establishment of a sample company	Mail
SCF	Multistage area sample	Each individual in sample household	Personal interview
NLS[c]	Stratified sample of secondary schools, sample of seniors within school[d]	Sample senior student in sample school	Mail[e]
RTMM	Multistage area sample	Sample individual in sample household	Personal interview
CHAS	Multistage area sample	Individuals in sample households	Personal interview
HANES	Multistage area sample	Sample person selected from screening interviews of household sample	Physical examination

[a] All the case studies are based on probability sample designs.
[b] Personal visit in base period when survey was initiated.
[c] For the index of consumer sentiment, unit is head of household or spouse.
[d] Sample design for base period when survey was initiated.
[e] Third follow-up.

of data for individual missing responses. For the convenience of the reader, a list of the case studies and the abbreviations that will be used to refer to them are given in Table 1.

Additionally, the 10 case studies represent examples of the design and execution of complex sample surveys at the hands of skillful and experienced practitioners of survey methodology. They give considerable information about types of problems to be expected in the conduct of even well-planned surveys. Thus the studies have aspects of the most general interest. A summary of the sample and survey designs used in the case study surveys is given in Table 2.

The objective of this chapter is to provide an introduction and road map for the ten case studies of the treatment of incomplete data in surveys that are the subject of this volume. Sections 2–4 provide some background on the problem; Section 5 presents an overview of the case studies and some case-by-case comment.

2. BACKGROUND

Nonresponse in a survey may be of one of two kinds:

1. *Item nonresponse*, in which the survey instrument is obtained for a unit but responses for one or more data items are not obtained; and
2. *Unit nonresponse*, in which the survey instrument is not obtained at all for a unit in the census or survey sample.

Incomplete data that is caused by unit nonresponse occurs for a variety of reasons; for example, inaccessibility of an area for personal interview because of weather, the failure of an interviewer to contact a respondent or to obtain a response within the limits of time established for data collection or the survey resources, or the refusal of a respondent to provide any of the information requested. In addition, it may occur *de facto* through loss or delay of completed responses in the mails. A response may be received but be so incomplete that the instrument is not considered usable and is treated as a unit nonresponse. The case studies in this volume provide some data on these reasons for nonresponse.

Conceptually, incomplete data also occur when a unit in the target population of the survey is not included in the survey frame. This is generally referred to as a problem of *undercoverage*, rather than nonresponse. However, although the problems of undercoverage are not within the scope of this volume, the method of poststratification and adjustment to independent controls described in the case studies are sometimes suitable for adjusting survey data to try to compensate for undercoverage; and, in their application, they may have that effect (SCA, OES, SCF, RTMM, CHAS, HANES).

The reverse problem of overcoverage of the target population can also occur because of nonresponse. This may happen when the survey frame is impure, i.e., includes both units that are within the scope of the survey and those out of scope, so that it is uncertain as to whether or not a nonresponse should be considered to represent incomplete data (LIS, ECI). Intensive follow-up of a subsample of nonrespondents may be used to estimate the proportion of cases in that category for which adjustment should be made.

Incomplete data from item nonresponse also occurs for a variety of reasons: for example, the respondent's lack of knowledge of the information requested, the respondent's refusal to provide data such as income that is considered sensitive, or the interviewer or respondent missing an item in a mail survey because of a confusing skip pattern or other problem in the instrument design. In actual survey practice, it also occurs when an item response is obtained but is considered to be in error on the basis of consistency checks or other editing criteria. If the response rejected by these criteria is not accepted ("gold-plated") by an analyst reviewing the case and if the respondent is not then followed up to confirm or modify the response, a "missing" data item is created by the rejection of the response, and the missing data may be replaced by imputed data. Again, although the problems of "bad" data and editing are not within the scope of this volume, the methods described herein are relevant to those problems. For example, the ASM study describes a method, when the ratio of two variables in an editing check is found to be out of tolerance, of deciding which of the two data items to accept and which to adjust.

Problems of potential underimpution or overimpution similar to those of under- or overcoverage for unit nonresponse may be posed by item nonresponse when it is uncertain as to whether the absence of a response represents missing data. Responses to related items may help to resolve such uncertainties and improve data imputation (SCF, ECI).

3. CONTROL OF INCOMPLETENESS IN DATA

The case studies demonstrate that the characteristics of nonrespondents are different from those of respondents and that no method of adjusting for incomplete data short of followup to obtain the required information can fully compensate for the consequent biases in the survey results. Thus, the safest way to treat the problem of incomplete data is to avoid it. Fortunately, a review of the case studies shows that, if an effort is made, response rates can be maintained at relatively high rates.

Nevertheless, a question remains as to what minimum targets for response rates should be set and whether data with nonresponse rates above a given level should be published by a survey organization. One suggestion, where the survey data may be used for purposes of official policy or decision making, is the standard set in the President's Reporting Burden Reduction Program initiated under President Ford for surveys sponsored by the federal government (McIntire, 1978); i.e., that surveys with expected response rates under 75% would require special justification for approval by the Office of Management and Budget (OMB), and that no survey with an expected response rate under 50% would be approved. In principle, a more refined answer is that nonresponse should be controlled below the point where the potential bias in the survey data could obscure the measurement of finding sought. It seems doubtful that the OMB standard just quoted is sufficient to meet this standard—and many, if not most, federal government surveys have substantially higher response rates. Various models for assessing the potential bias are described by Cochran (1977, Chapter 13). In practice it may be very difficult to apply this principle because large-scale surveys generally have multiple objectives and the data are likely to be used by a number of analysts for a variety of objectives.

SIC code	Type of yarn	First quarter	Second quarter
	Yarn consumed	356,447	335,849
02281 02	Cotton	32,569	30,419
02281 10	Carded	[r]25,665	[a]23,915
02281 20	Combed	6,904	[a]6,504
02281 37	Rayon and acetate	95,723	86,305
02281 30	Spun, 100 percent	32,957	[a]27,833
02823 01	Filament, 100 percent	62,766	58,472
02823 11	Acetate	43,737	[a]40,993
02823 21	Rayon	19,029	[a]17,479
02824 11	Nylon filament	15,321	14,713
03229 33	Glass filament	33,299	32,668
02280 13	All other yarns, including blends and mixtures	179,535	171,744
02281 43	Polyester, including content in yarn blends and mixtures	80,203	78,866
02281 04	Cotton content of blends and mixtures	[r]32,553	[a]32,016
02281 35	Rayon and acetate content of blends and mixtures	39,049	[a]36,658
02283 11	Wool, alpaca, and mohair yarn, and content in blends and mixtures	3,430	2,614
02281 45	Acrylic fibers	5,183	5,220
02281 46	Paper	([1])	([1])
02281 42	Silk	666	[b]611
02281 44	Saran and olefin	12,412	9,508
02241 55	Rubber, elastic, lastex, etc.	167	[b]153
02281 41	Nylon	2,469	[b]3,011
02280 11	All other fiber yarn and fiber content of blends and mixtures	[1]3,403	[1]3,087

[a]10 to 25 percent of this item is estimated. (See "Description of Survey" for discussion of estimation of missing reports.) [b]26 to 50 percent of this item is estimated. [r]Revised.

[1]Paper yarns are included with all other yarns.

Source: Current Industrial Reports, Series MQ-22T, Manmade Fiber Broadwoven Gray Goods, second quarter 1970.

Fig. 1. *Yarn consumed (in thousands of pounds) by manmade fiber weaving mills.* [*From U.S. Bureau of the Census* (*1974, Appendix II, Exhibit 6*).]

Two conclusions may be drawn from these considerations. First, the safest course is to try to control nonresponse to relatively low levels. Second, information about the adjustment of data for nonresponse should be published to the fullest extent possible for the guidance of data users. This is particularly important when the impact of the adjustment varies widely from item to item. Figure 1, taken from the U.S. Bureau of the Census, Technical Paper No. 32 (1974), illustrates this point. It also illustrates the fact that data that might otherwise be suppressed because of a large imputation rate may be published for reasons of convenience.

Figure 1 also illustrates that fact that a relatively low response rate at the unit level may fail to warn of relatively high nonresponse rates for individual items, especially for items that apply to relatively few units in the population.

4. MEASUREMENT OF RESPONSE RATE

The objectives of controlling nonresponse and informing data users require that a definition of the response rate or nonresponse rate be established and that records be maintained to provide the necessary information. As demonstrated in the case studies of federal surveys done by Bailar and Lanphier (1978), there is no general agreement among survey practitioners as to a definition of nonresponse, and records of nonresponse may not have even been maintained.

For simple random sampling and where qualitative data are of interest, a definition of the response rate at the unit level in terms of unweighted counts of the number of units in the sample and the number from which a (usable) response was obtained provides a useful guide. The proportion of units responding in the sample is an estimator of the proportion of units expected to respond in samples of the given size. If the population sampled consists of units that would always be either respondents or nonrespondents, the proportion of units responding in the sample is an estimator of the proportion of respondents in the population—a measure of the coverage of the units in the population by responses in the survey. The SCA illustrates, however, that in random-digit dialing (RDD) telephone surveys[1] the determination of the number of units in sample—which involves a determination as to whether or not a telephone number at which no respondent contact is made represents a unit within the scope of the survey—may not be a simple matter. A similar problem was noted earlier in connection with the discussion of impure survey frames. These

[1] These are surveys that use the set of all possible four-digit numbers within existing telephone exchanges as the sampling frame.

problems aside, for complex sample design and/or with subsampling applied in follow-up for nonresponse, this definition is less informative. As illustrated in the case studies, where the probabilities of selection (inclusion in the basic sample and/or in follow-ups) vary, the counts should be weighted inversely by the probabilities of selection to obtain an unbiased estimate of the survey of units in the population.

Where quantitative data are of interest, unit counts, whether weighted or not, are less informative than rates based on a relevant measure of size of the units. For example, if s_i denotes the measure of the size of the ith unit (e.g., sales or receipts for business establishments, value of shipments for industrial establishments) and p_i the probability of selection, a nonresponse rate defined as

$$\sum_{\mathrm{NR}} w_i s_i \Big/ \sum w_i s_i$$

where $w_i = 1/p_i$, the numerator is the weighted sum for the nonrespondents (NR), and the denominator the weighted sum for all units in the sample, provides a guide as to the potential for nonresponse bias. Note that if the p_i are roughly proportional to the s_i, this ratio is approximated by the unweighted count. As illustrated by the LIS, if response rates vary by size of unit, nonresponse rates based on counts of establishments and on a measure of size will differ. For example, in the December 1978 Hog Survey (see chapter by Freie in this volume), the nonresponse rates based on the different definitions were

unweighted count of farms	14.3%
weighted count of farms	10.0%
percentage of estimate attributable to nonrespondents	20.0%

Related to the question of weighted or unweighted measures of nonresponse is the impact on the potential bias in survey estimates of incomplete data associated with units of differing size when the units have been selected with varying probabilities. It is common to find the feeling that incomplete data from nonresponse, for example, is a matter of greater concern with regard to "large" units than "small" units, and greater effort is devoted to obtaining responses from large units. Since the contribution of the ith unit to the estimate of a total, say X, is the weighted value $w_i x_i$ rather than the unit report x_i, it is the weighted value that should be used as the criterion. With optimum selection probabilities, it will frequently be the case that, apart from units large enough to be selected with certainty, the potential impact of nonresponse by small units is about as great as that of large units. In these circumstances, the allocation of greater effort to obtaining responses from large units must be justified by considerations such as the cost of obtaining responses from units of different size or the accuracy of adjustments (imputations) for nonresponse that can be made for units of different size.

5. OVERVIEW OF THE CASE STUDIES

Table 3 gives a general overview of the methods described in the case studies for treating incomplete data.

The term "adjustment for unit nonresponse" refers to techniques of weighting respondents to sample totals including nonrespondents [see the chapter by Oh and Scheuren in Volume 2 and the chapter by Bailar and Bailar in Volume 3 of this treatise; see also Bailar, Bailey, and Corby (1978)]. These techniques implicitly assume that, on average, within weighting classes, nonrespondents are similar to respondents. The term "poststratification adjustment" refers to techniques of stratifying respondent units, making an estimate for the poststratum from the respondent units on the basis of their sampling weights, and then inflating the sampling weight of each respondent unit by the ratio of the control figure for the poststratum to the estimate based on the respondents [see Cochran (1977, Section 5.A.2) and Hansen, Hurwitz, and Madow (1953, Ch. 9, Sec. 23)]. Typically, in a population survey, the poststrata will be defined on the basis of characteristics such as age, sex, and race for which control totals in terms of number of persons will be available from the Bureau of the Census (RTMM) or an independent survey (SCA). In the establishment surveys, the preference for an item to be used for a poststratification adjustment is one that reflects economic activity, such as employment or total sales, but may be the number of establishments, with the poststrata incorporating a size stratification [see OES, Tupek and Richardson (1978), and Robison and Richardson (1978)]. Poststratification adjustments may lead to reductions in the sampling errors of estimates, but the principal effect of such adjustments may be more important for reducing biases arising from differential coverage of units from poststratum to poststratum (see the chapter by Thomsen and Siring in Volume 3). In multistage samples, an initial weighting-class adjustment based on primary sampling unit characteristics may be used, followed by a poststratified adjustment based on the people, farms, or other final units in the sample (Hanson, 1978).

The term "imputation for missing data items" refers to techniques for filling in a value for a specific data item missing for a unit that responded in the survey. The hot-deck technique substitutes a reported value from the sample taken from another unit having specified characteristics matching those of the unit for which the data item is missing [see the chapters by Ford and by Oh and Scheuren in Volume 2 and the chapters by Bailar and Bailar and by Sande in Volume 3; see also Bailar, Bailey, and Corby (1978)]. The respondent cell-means technique substitutes the average for the given item based on all respondents in a cell defined by specified characteristics. The external data technique uses data derived from outside the survey, e.g., program entitlements for missing income associated with a welfare program (SCF, CHAS).

Some highlights of the case studies are summarized now.

5.1. Employment Cost Index

There are two points of special interest in the Employment Cost Index (ECI) survey. The first relates to the treatment of nonresponse in the first-phase survey. This phase was intended to obtain measures of employment size by occupation class to be used in the sampling for the second-phase survey. Nonrespondents were assigned a nonzero measure of size for each occupation class. This avoided bias in the second phase although it increased the sampling variances of the estimates. The second relates to the very detailed use of all available information about sample units for imputing for missing data items, especially in the case of missing data on the value of employment benefits other than wages and salaries. Since the survey is a repetitive one, it was sometimes possible to impute a current estimate from an earlier report for a unit. The sample design represents an ingenious approach to satisfying constraints imposed by the survey sponsors.

5.2. Livestock Inventory Surveys

A point of interest in the Livestock Inventory Surveys (LIS) is the special attention paid to the nonresponse of large producers. For the largest producers, an effort was made to create a report using information from a variety of sources. The LIS is a good example of a multiple-frame sample design.

5.3. Survey of Consumer Attitudes

The Survey of Consumer Attitudes (SCA) is the only case study based on a RDD telephone sample. In such surveys there is always a problem in determining whether a selected telephone number for which no contact is made represents a nonrespondent household or a unit not within the scope of the survey. Considerable effort in the SCA is devoted to this problem. Also of interest are the efforts to increase response rates by attempting to convert initial cases of refusal to be interviewed and, since the survey is a continuing one, to obtain current interviews with units that were nonrespondents in earlier survey rounds for reasons other than refusal.

5.4. Office Equipment Survey

The Office Equipment Survey (OES) illustrates a multistage establishment sample design. Apart from the careful documentation of the survey outcomes, the ingenious poststratification adjustment of the sample may be noted. It is of interest that a later survey round, which included an independent relisting of the

TABLE 3

General Overview of Methods Used in Case Studies for Treating Incomplete Data

Case study	*Method of adjustment for unit nonresponse*	*Method of imputation for item nonresponse*				*Comments*
		Other items reported	*Hot deck*	*Respondent cell means*	*External data*	
ECI	Phase I—Respondent means			× (phase I)		
	Phase II—Weighting class			× (phase II)		
LIS	Large producers—Individual imputation	× (individual imputation)				
	Other producers—Weighting class					
SCA	Weighting class		Not applicable			
	Poststratification weighting to independent controls					
OES	Weighting class		× (key items)			Effects of unit nonresponse adjustments are reflected in variance estimates
	Poststratification weighting to independent controls					

ASM	Weighting class	×		×		
SCF	Weighting class Poststratification weighting to independent controls	×		×	× (selected items)	Groups for item imputation defined by AID analysis
NLS	Weighting class		No adjustment made			Experimental comparison of hot deck, weighting class, and no imputation for item nonresponse Evaluation of variance impacts of adjustment
RTMM	Weighting class Poststratification weighting to independent controls		Not applicable			
CHAS	Weighting class Poststratification weighting to independent controls			×		Experimental comparison of external data, hot deck, and cell means Evaluation of poststratified weighting
HANES	Weighting class Poststratification weighting to independent controls		Not applicable			Experimental evaluation of weighting-class definition Experimental comparison of six alternative weighting procedures for unit nonresponse

area segments, showed that the initial listings were incomplete and that the poststratification adjustment did indeed contribute to compensating for biases of undercoverage in the sample. Finally, the variances of survey estimates were computed by the BRR method. For each of 24 balanced half-samples, the nonresponse and two-stage ratio–estimation adjustments were computed separately and independently. Estimates of variance were made based on these 24 sets of estimates, which appropriately reflect both the imputation and estimation procedures.

5.5. Annual Survey of Manufactures

The case study from the Annual Survey of Manufactures (ASM) has two points of special interest. The first is the experimental comparison of responses to a short and long survey instrument. Although the unit response rates did not differ, the establishments asked to respond on the short instrument showed lower item nonresponse rates than experienced with the corresponding items for establishments asked to respond on the long instrument. The second point of interest relates to the fact that extensive editing is based on ratios computed from the responses on various items; e.g., unit values, ratios of subclasses to totals. An interesting technique is presented for making a decision as to which item report to adjust when the ratio is found to be outside acceptable limits.

5.6. Survey of Consumer Finance

The case study from the Canadian Survey of Consumer Finance (SCF) illustrates how extensive and ingenious use can be made of data reported for an individual to improve the imputation for missing data. The imputation techniques include both hot deck, with extensive matching of characteristics, and external data. Since income is consistently subject to relatively high nonresponse rates, compared with other items such as labor force status, the techniques are of particular interest.

5.7. National Longitudinal Survey

The third follow-up of the National Longitudinal Survey (NLS) describes a rare case in which response data became available in later survey rounds for nonresponses in the initial survey and thus provided a basis for evaluating alternative imputation strategies. The survey instrument had a complex skip pattern, and this was the most serious source of error in the estimates. Discarding inconsistent responses and replacing them with blanks was not found to be desirable for reducing bias. From this point of view, a hot-deck technique, which replaced the inconsistent responses with data from an individual who responded in a consistent manner, did reduce the overall bias in estimates. Because the hot-deck technique leads to an increase in the variance of estimates (see the chapter by Ford in Volume 2 and the chapter by Bailar and Bailar in Volume 3), it produced larger mean square errors than no imputation. This suggests that a desirable solution would be to use the hot-deck technique to minimize bias and to increase the initial sample size to control the sampling error of estimates.

The evaluation of variance estimation techniques showed that standard statistical packages, which treat imputations as responses, tend to underestimate sampling errors. The effect, as expected, decreases with decreasing nonresponse rates. The use of balanced repeated replications (BRR) (McCarthy, 1966) to provide estimates of sampling error reflecting the impact of imputation is illustrated.

5.8. Readership of Ten Major Magazines Survey

The case study from the Readership of Ten Major Magazines (RTMM) Survey presents a wealth of detail on the careful administration of a field interview survey and attention to controlling nonresponse.

5.9. Center for Health Administration Studies

The survey of health services use and expenditures conducted by NORC for the Center for Health Administration Studies (CHAS) provides a comparative study of external-data, hot-deck, and cell-means techniques for item imputation. Although check data were not available for this study, the comparative analysis led to a recommendation for the use of hot-deck techniques among these, compared with no imputation. Poststratification adjustments are also found to be desirable.

5.10. Health and Nutrition Examination Survey

The case study from the Health and Nutrition Examination Survey (HANES) gives a detailed description of the application of the AID technique to the problem of defining weighting classes. Six nonresponse adjustment procedures are compared with regard to selected medical-history data available for half the medical examination nonrespondents and with regard to ten medical examination characteristics for those persons examined, where those persons who required only one call to make and keep an examination appointment were treated as respondents and all other examinees were treated as nonrespondents. The conclusion is reached that a simple poststratification adjustment by age–sex–race groups is as effective as the more complex procedures examined. Finally, the bias remaining after adjustment was generally relatively large compared with the bias reduction achieved.

6. SOME SUMMARY COMMENTS

The choice of an imputation method should depend on the objectives of the survey. Where the survey is intended only to provide statistical estimates of aggregates (ECI), techniques such as the use of cell means may be adequate. However, where an important objective is (also) to provide a micro data base for analysts, hot-deck methods are likely to be more satisfying [see the chapter by Sande in Volume 3; also see Tupek and Richardson (1978)].

Poststratification adjustments, which are widely used in survey practice, are often helpful in reducing bias as well as sampling variance. The bias improvement, of course, is affected by any bias in the control totals themselves. There is a body of theory and experience to guide the use of poststratification.

More use should be made in survey practice of variance estimation techniques such as BRR or jackknife, which can incorporate the impact of adjustments for missing data.

The various techniques have different impacts on variance and bias. These are best examined in the framework of a total survey-error model (see the chapters by Platek and Gray and by Lessler in Volume 3 of this treatise).

REFERENCES

Bailar, Barbara A., Bailey, Leroy, and Corby, Carol (1978). A comparison of some adjustment and weighting procedures for survey data. In N. Krishnan Namboodiri (ed.), *Survey Sampling and Measurement*. New York: Academic Press.

Bailar, Barbara A., and Lanphier, C. Michael (1978). Development of survey methods to assess survey practices. *1978 Proceedings of American Statistical Association*, San Diego, Calif.

Cochran, W. G. (1977). *Sampling Techniques* (3rd ed.). New York: Wiley.

Hanson, Robert H. (1978). The Current Population Survey: Design and Methodology, Technical Paper 40, Bureau of the Census, U.S. Department of Commerce, Washington, D.C.

Hansen, M. H., Hurwitz, W. N., and Madow, W. G. (1953). *Sample Survey Methods and Theory*, Vol. I: *Methods and Applications*. New York: Wiley.

McCarthy, Philip J. (1966). Replication: An Approach to the Analysis of Data from Complex Surveys, NCHS Series 2, No. 14, Washington, D.C.

McIntire, James T. (1978). Guidelines for reducing the burden of public reports to federal agencies. In President's Reporting Burden Reduction Program, Fiscal Year 1978. Memorandum from Office of Management and Budget, February 17, 1978.

Robison, Edwin, L., and Richardson, W. Joel (1978). Editing and imputation of the 1977 truck inventory and use survey. *Proceedings of the Section on Survey Research Methods*, American Statistical Association, pp. 203–208.

Tupek, Alan R., and Richardson, W. Joel (1978). Use of ratio estimates to compensate for non-response bias in certain economic surveys. *Proceedings of the Section on Survey Research Methods*, American Statistical Association, pp. 197–202.

U.S. Bureau of the Census (1974). Standards for Discussion and Presentation of Errors in Data, Technical Paper 32. Washington, D.C.: U.S. Government Printing Office.

CHAPTER **2**

The Employment Cost Index: A Case Study of Incomplete Data

Barbara A. Boyes*
Margaret E. Conlon

1. INTRODUCTION

In the fall of 1975, the Bureau of Labor Statistics (BLS) introduced a new statistical series, called the Employment Cost Index (ECI), designed to meet the need for a timely, unambiguous, and comprehensive measure of changes in the price of labor. Although data are collected for both wages and benefits, at present only wage data are published in the index [See Sheifer (1975) and Hoy (1978)].

As in any statistical series, the problem of how to handle the missing data has to be solved. Because of the complexity of the survey design, both unit nonresponse and item nonresponse existed at several different levels. An attempt was made to use nonresponse adjustments that optimized both cost effectiveness and statistical validity in all cases.

The ECI was selected as a case study dealing with incomplete data for several reasons:

(1) It illustrates the complexities that a statistician encounters with the large-scale repetitive surveys used in government work and the techniques used to obtain the maximum amount of data while minimizing the burden on respondents.

(2) It also illustrates the use of omission and imputation as methods to adjust for incomplete data in a probability survey of establishments.

* Deceased.

INCOMPLETE DATA
IN SAMPLE SURVEYS
Volume 1, Part II

ISBN 0-12-363901-8

(3) It uses administrative matching of reporting units from two different surveys to reduce respondent burden.

(4) It gives evidence of a correlation between response rate and establishment size.

This paper describes the 1979 ECI; design and estimation changes have been implemented since then.

2. SURVEY BACKGROUND

The ECI is a probability sample survey designed to measure changes in the price of a standardized unit of labor service. The ECI's objective is to provide periodic publications of national Laspeyres-type indexes for wages, benefits, and total compensation. In addition, subindexes for five major industries (MID), nine major occupational groups (MOG), four geographic regions, and union/nonunion and metropolitan/nonmetropolitan areas are calculated.

Currently, the population for the ECI consists of establishments with one or more employees in the private nonfarm economy (excluding households) in the United States. Efforts are under way to expand the scope of the survey to federal, state, and local governments and, later, to household and farm economies.

For the base period, establishments were initiated into the voluntary survey by personal visits. After the selected occupations for the establishment had been matched and employment collected, data were collected for wages. The respondents to this survey were generally the persons most familiar with the occupational wage structure of the firm. In small firms, the respondents were generally the owners; in larger firms, accountants or personnel officers. Each respondent was asked to report wages for the selected occupations, either as an hourly rate or as hours and straight-time pay. After initiation, a quarterly mail update was requested, in which the respondent reported changes in the wages reported for the selected occupations in the previous quarter. When the survey was expanded to include benefits, another personal visit took place. Each respondent was asked to provide information on the existence of benefits and if they were wage-related or non-wage-related. Those respondents who seemed reluctant to cooperate were not pressed for benefits data if it seemed that their cooperation for wage data would be jeopardized. Benefits were divided into two categories: those legally required with ceilings and all others. Each respondent was permitted to supply cost data for the benefits in one of the following ways: an expenditure, a rate (e.g., 5% of gross earnings), or practice and usage (e.g., two weeks' vacation for employees in a certain tenure category, with 30 employees in that category). Participants were also given the option of reporting the cost figures that represent costs for more than one benefit item ("combined benefits"). For example, a respondent may be able to report only one cost figure for health and life insurance, if the respondent receives coverage for both plans from the same insurance company and is charged at one rate for both.

Cost data were then converted to cents-per-hour figures for calculation of the index. Quarterly updates of benefits are solicited and the establishments are asked to report "no change" or the amount of change in the benefit plan as reported the previous quarter.

Since the ECI survey was to produce both major industry and major occupation subindexes, adequate samples of units in both these groups were necessary. For the universe of occupations, the 1970 Census of Population tabulations by occupation within the nine major occupation groups cross-classified by two-digit 1967 Standard Industrial Classification (SIC) were used. These formed a suitable universe in that they provided complete coverage of occupational groups and weights for sample selection of occupations. For purposes of ECI sampling, consolidations were made in some cases so that the data used in sampling were for 62 industry groups.

For industries, the 1971 BLS list of all reporting units from the unemployment insurance system (obtained through the cooperation of the individual state agencies) provided the basic universe of establishments in the sampling frame. These reporting units, although referred to as establishments, do not necessarily conform to the establishment definition (an economic unit, generally at a single physical location, where business is conducted or where services or industrial operations are performed). The units are defined by the employers as they see fit to report their tax liability under the unemployment insurance laws subject to the rules the state may choose to impose. These rules vary from state to state; for example, in some cases, a reporting unit may be a single establishment at a single location; in others, it may be all company units in a county. For several industries in which the unemployment insurance listing might have been incomplete, supplementation from more reliable sources was used to reinforce the sampling frame.

The development of the sampling survey design required knowledge of the likely overall budget, considerations of available data resources for sample design use, the development of estimates of unit costs and variance components, the identification of basic index objectives and the precision goals, the probable character of the survey's data collection procedures, and the development of rough approximations to various sampling variance components. These estimates of variance together with estimates of unit costs were used in the examination of several alternative sampling designs.

Along with these factors, it was necessary to consider the need for information on the occupational structure within each of the reporting units from the unemployment insurance system. This information was not then available, yet the unit of sampling for the ECI had to be an occupation within an establishment. Another question was how to achieve a measure of control on the extent of respondent-reporting burden once a unit was selected for the sample. Requesting a large volume of information at a reporting unit might lead to refusal to cooperate in a voluntary survey. On the other hand, if only one or two occupations were to be reported per period, there would be a waste in initiation as well as in subsequent data collection activities. The interaction of all the results and considerations led to a recommendation for a single-stage double sampling of reporting units. The first phase would establish the occupational mix for a sample of occupations in the second phase. Double sampling would provide the occupational distributions necessary for selecting the sample for the second phase, would be reasonable from the standpoint of cost-variance effects, would facilitate the achievement of the precision goals for each of the broad occupation groups, and would provide a starting point for achieving bounds on the respondent-reporting burden.

For Phase I of the double sample, the detailed occupations were stratified into certainty and noncertainty occupations within each of the 62 industry groups. Data suggested that within each industry group the five detailed occupations with the largest employment as identified from the 1970 census would be designated as "certainty" occupations. The remaining detailed occupations would be stratified into nine major occupational groups (MOG) for each industry group. Eighteen occupations (two from each of the nine MOGs) would be selected, based on a systematic probability-proportionate-to-size (PPS) approach (relative proportion of employees in detailed occupation to total employees in noncertainty occupations in the MOG in each industry group) for a total of 23 occupations.

All reporting units in the unemployment insurance system were considered eligible establishments for Phase I. For Phase II, establishments recorded as "out of business" or "out of scope" (not classified in the correct two digit SIC code) were not included in the sampling frame.

The first phase of the double sample of establishments consisted of about 10,000 units selected across the 62 industry groups. The sampling interval was determined and "certainty" establishments were designated. The selection of the remaining sample units was by a PPS (to total employment) procedure. These 10,000 units were surveyed by mail to obtain occupational mix information concerning the respective 23 Phase I sample occupations in the appropriate one of the 62 industry groups described above.

The second phase of the double sampling led to the final sample of about 2000 establishments and the identification of the final sample of occupations to be priced within each establishment. The second-phase sample was selected so that there would be a joint probability selection of establishments for all 23 detailed occupations in a given subindustry, such that the number of selected sample occupations in an establishment would be consistent with the expected number of occupations and conditional on the relative weighted employment obtained from the first-phase sample units and such that the number of occupations selected within each sample establishment would be within reasonable upper and lower respondent burden boundaries of reporting occupational compensation rates.

The occupational mix data obtained from the Phase I sample units were used to establish measures of size for the second-phase sample. The Phase II selection was an adaption of a PPS procedure whose principles were first suggested by Lahiri (1951). Within each establishment, relative measures of size for each sampled occupation were derived as the ratio of the establishment occupational employment to the total occupational employment across establishments. Then the maximum relative measure of size for each establishment was identified and used as a basis for a PPS selection of the Phase II sample establishments. This was done to ensure that those establishments having a large percentage of the total employment for a sample occupation would have a high probability of selection into the Phase II sample. Once the sample establishments for Phase II were selected, the relative measures of size were standardized using the maximum relative measure of size for the establishment. These standardized measures were used in conjunction with a two-way controlled selection process to arrive at the subsample of occupations to be priced for each Phase II establishment. Presurvey planning had established desirable levels for respondent burden (tentatively set at 10 quotes per establishment) and the required number of quotes per occupation to attain the precision required for the overall index and all subindexes. These constraints were used to establish a pattern of selection for the sample occupation quotes. A selection of sample occupations was performed using a system of patterns to arrive at the final Phase II sample of 2000 establishments and their corresponding sample occupation quotes. For those establishments in which the system of patterns produced a multiple selection of an occupation, the weight was adjusted upward. Supplements to this sample were later required for the construction industry, the trade industry, and the finance, insurance, and real estate industry because of nonresponse, an outdated frame, and insufficient quotes for some occupations. Credit for much of this basic survey design belongs to Joseph Steinberg (1975), formerly of BLS.

2.1. Estimation

Estimates for the overall index and the subindexes are published quarterly. They include both quarter-to-quarter and year-to-year relatives, although the actual indexes are calculated but not published at present. Reference to the estimates as "indexes" means that the estimates are ratios that relate the current quarter to the base period. The current indexes are not pure Laspeyres measures relative to the base period in the sense that constant weights are not associated with each quote over time.

The general form of our estimate of the national index at time T involves the product of a series of quarter-to-quarter relatives

$$R_T = \prod_{t=1}^{T} R_{t,t-1},$$

where

$$R_{t,t-1} = \sum_s \sum_k C_{sk}\hat{\hat{X}}_{sk,t} / \sum_s \sum_k C_{sk}\hat{\hat{X}}_{sk,t-1},$$

where C_{sk} is the employment represented by the kth sample occupation in the sth two-digit SIC (this measure of employment is based on the 1970 Census of Population) and $\hat{\hat{X}}_{sk,t}$ is the estimated total average hourly compensation for the sth SIC, kth occupation at time t.
It may also be noted that $\hat{\hat{X}}_{s,k,t}$ is itself a ratio,

$$\hat{\hat{X}}_{sk,t} = \frac{\sum_{i(t)\in N} \delta_{ski} w_{ski,t} E_{ski} \bar{X}_{ski,t}}{\sum_{i(t)\in N} \delta_{ski} w_{ski,t} E_{ski}},$$

where

$\bar{X}_{ski,t}$ is the reported total average hourly compensation for the ith establishment, kth occupation, and sth SIC at time t. It comprises two parts, $\bar{X}_{1ski,t}$, the average hourly wage, and $\bar{X}_{2ski,t}$, the average hourly benefit cost. These values may be reported or imputed.

E_{ski} is the reported employment for the ith establishment, kth occupation, and sth SIC at the time of initiation.

$w_{ski,t}$ is the "sample" weight associated with the skith unit of observation at time t. The weight of a unit may vary over time because of sample supplementation.

δ_{ski} is the random variable reflecting the number of times the skith unit is selected at every stage of sampling except the Phase I occupational sample.

$i(t) \in N$ denotes that the set of establishments being summed over is a function of time t. In particular, the summation is only over the set of establishments having data (either real or imputed) in both time t and $t - 1$.

This estimate is a product of ratio estimates. In addition, the numerator and denominator of each ratio taken individually involve sums of ratios themselves. The sample weights can vary over time, but for the relative $R_{t,t-1}$ relating time t to $t - 1$, the weights are the same. Also, the declining sample may result in some of the C_{sk} being zeroed out for some of the relatives because of lack of sample.

One effect of the current form of the index is that some of the subindexes may appear to be inconsistent with the national index. Because of this consideration, there are now plans to modify the current estimation procedure, scheduled for implementation in the future. Further details concerning the estimation are available upon request (Wright and Kaufman, 1978).

No variances are presently produced for the ECI, since it is still in the developmental stages. Procedures being implemented for calculating estimates of the variance of the published ECI wage indexes are described in a paper by Marks and Frevert (1978).

3. INCOMPLETE DATA

Incomplete data have occurred on many different levels in the ECI. One of the most logical divisions for the purposes of discussion of methods is that of the Phase I and Phase II segments of the survey. In addition, because of the complex imputation procedures, the discussion will be divided further into wages and benefits for the Phase II survey.

Of the approximately 10,000 Phase I sample units selected, about 1000 were units that also participated in the BLS Occupational Employment Survey (OES). To reduce time and cost, as well as respondent burden, data for these 1000 units were extracted from information provided by these reporters for the OES. Because some of the occupation definitions were different for the two surveys, not all of the information needed for the ECI Phase I survey was available from OES; therefore, some imputation occurred for nonmatched ECI occupations in OES establishments.

In addition, for the 9000 units solicited directly for ECI data, not all were able to provide information on every occupation. Some establishments were either unable or unwilling to report any data at all for Phase I, whereas others were out of business, out of scope, or unreachable by the address furnished.

Table 1 shows an approximate distribution of the sample-by-size class and several response variables of interest. The nonresponse rate, which is defined as the number of nonresponding establishments (refusals, post office returns, unusables, and no responses) divided by the number of responding and nonresponding establishments, for this main survey was 28.5%. Because of split

TABLE 1

Percentage Distribution of Responses by Size Class for Phase I Occupational Survey (ECI)

		Size class				
Response category	*Total*	*1–19*	*20–99*	*100–499*	*500–3999*	*4000+*
OES data	10.3	.5	3.6	12.6	21.8	28.7
Response	58.8	66.6	67.0	56.5	47.8	38.0
Refusal	1.1	.5	.5	.8	2.1	4.5
Out of business	3.0	4.8	3.6	3.2	1.2	.2
Post office return	3.3	7.4	3.8	2.6	.2	.0
Out of scope	.3	.3	.4	.3	.2	.0
Unusable	.2	.0	.1	.2	.2	.3
No response	23.0	19.9	21.0	23.8	26.7	28.3
Total sample units in size class	10,140	2201	2790	2427	2141	581
Nonresponse rate	28.5	29.3	26.5	28.4	29.6	33.2

and merged schedules, it is not certain that these numbers are exact. Response categories in Table 1 are defined as follows:

Response	Employment data for some or all of the 23 sample occupations were provided on the questionnaire and returned to the regional office
Refusal	Employment data for all occupations were refused by the respondent and the questionnaire returned to the regional office.
Establishment is out of business	
Post Office return	The post office was unable to deliver the mail questionnaire to the respondent with the address provided by BLS. A one-third subsample of this category was selected for the Phase II survey
Out of scope	Classified in wrong two-digit SIC
Unusable	Data for some or all occupations were provided but were unusable for the Phase I survey because the schedule was either not readable or a duplicate of a schedule already received.
No response	No response, either with or without data, was ever received. (This differs from a "refusal" where the questionnaire was returned even though it did not contain any data.)
OES data	Data for the Phase I survey were obtained from respondent input for the OES survey.

As respondent's schedules were returned to the BLS, Washington, they were reviewed by the staff for completeness and consistency. There were no alternatives for refusals; likewise, no subsampling of nonrespondents occurred at this time.

Schedules passing review and edit were then keypunched on cards. Data were again reviewed for clarity and completeness. This second review consisted of a multitude of computer edits producing flagged data, which were further reviewed and refined to correct inconsistent responses. If all the information from an establishment was inconsistent, the reply was treated as a nonresponse and the data were imputed. If only part of the information was inconsistent, this information was corrected by reviewing the actual schedule. After schedule review and edit on both levels was completed and the survey closed, the imputation process for Phase I began.

Because the occupational distribution information collected in Phase I was to be used to select the sample of occupations at each establishment in Phase II, it was necessary to impute values to all occupations at all establishments sampled in Phase I. Thus each occupation would have a probably of selection in Phase II. These values were imputed from a class of similar establishments.

The imputation process for Phase I was as follows:

First, for those establishments reporting data for some but not all occupations, it was assumed that there was no employment in the establishment for the unreported occupations, and zeros were imputed.
Second, nonresponse adjustment cells were classified by SIC and by size class. Because of the large number of cells, it was sometimes necessary to collapse cells until an acceptable response rate was reached. Because these estimates of occupational size were to be used to select the sample for Phase II, it was necessary to achieve an acceptable response rate to have as reliable an estimate of occupational employment as possible.

Usually, cells were collapsed across three-digit SIC and size class, so that values would be imputed for the most homogeneous groups possible. Because data were required on an occupation-by-establishment basis, imputation was also done on this level. For each reported occupation and establishment in the imputation cell, it was determined whether the occupation was certainty, multiple-hit probability, or probability. Then a weighted occupation employment was calculated according to an algorithm specific to one of the three categories. An average employment was then calculated as follows:

(1) If the occupation employment was a missing value from an establishment whose data were derived from the OES survey, the average employment of the non-OES responding units in the imputation cell was imputed.

(2) If the occupation employment was a missing value from an establishment that reported data for some, but not all, occupations, zero employment was imputed.

(3) If the occupation employment was a missing value from an establishment that did not report data for any occupation (i.e., refusal, nonresponse, etc.), the ratio of average occupational employment to average total employment of all "respondents" (i.e., response solicited or response compiled from OES) in the cell was applied to each individual establishment's total employment to impute a distribution of occupational employment for the establishment.

Of course, units out of business and out of scope were excluded from the imputation process.

Once the Phase II sample had been selected and personal visit initiation of establishments begun, response rates were monitored in the Washington office. Oversampling had been provided in the designated sample to account for expected out of business, out of scope, and refusals. For the base period, overall rates for out of business and out of scope and for refusal were about 10 and 15%, respectively. The overall out-of-business and out-of-scope rates were higher than the Phase I out-of-business and out-of-scope rates (3.3%) because Phase II was conducted by personal interview, which made it easier to identify out-of-scope cases than in the mail questionnaire returned to Washington in

Phase I. The Phase II refusal rate was also higher than the Phase I rate (1.1%); the greater difficulty and time involved in providing both employment and wage data for Phase II than in supplying employment data alone made refusal more likely. However, this nonresponse in Phase II created a sample shortage in the construction industry and the salesworker occupation group. It was therefore decided that a supplemental sample would be required for these groups to produce the necessary indexes.

3.1. Incomplete Data for Wages

For the September 1975 base period for wages, there were about 2400 establishments solicited for initiation by personal visit. Of these, about 1600 were found to be on the unemployment insurance list, 95 were permanent losses, about 130 were out of scope, and about 165 were out of business. Approximately 55 schedules were merged into 40, and another 20 schedules were split. In addition, approximately 350 (14.6%) establishments refused to cooperate. All these measurements were unweighted.

If a respondent refused to cooperate in the base period, no alternate was assigned. There was no subsampling of nonrespondents for further field follow-up.

As in Phase I, a regional staff review of schedules for clarity and internal consistency was performed. Those schedules requiring additional information were updated by telephone, if possible, or by another personal visit. Schedules passing regional review were keypunched and transmitted to Washington.

Response to the base-period solicitation was classified and coded. The schedules were then machine-edited as usual for invalid codes and consistency. Flagged data were reviewed by economists, and where necessary, the regional office recontacted the respondent for clarification.

No explicit adjustment for nonresponse was made for the base period at that time. It was believed that a study of the data would help in developing a meaningful nonresponse cell and nonresponse adjustment. Because no adjustment was made, an implicit assumption was made that respondents and nonrespondents in an estimation cell had similar characteristics and that nonresponse rates were uniform throughout for all size classes within an estimating cell.

To determine how to make adjustments for the base period and to analyze other variables of interest for the ongoing survey, a study of nonresponse was conducted in the fall of 1977. Table 2 was a result of part of this study.

Rates in Table 2 were computed by dividing the number of nonresponses (temporary, seasonal, strike, refusal) by the number of in-scope sample units. Rates were classified by four variables of interest: number of jobs in the establishment, establishment employment, region, and major industry division.

The results of this and other studies have led to the conclusion that there is a high correlation among establishment size, response rate, and reported level of

TABLE 2

Quarterly Nonresponse[a] Rates for Selected Subclasses of the Sample (in Percents), Phase II

	Quarter ending					
Subclass	*Dec. 1975*	*Mar. 1976*	*June 1976*	*Sept. 1976*	*Dec. 1976*	*Mar. 1977*
Region						
Northeast	23.8	24.6	24.1	23.7	25.2	26.6
North central	20.4	21.5	22.3	21.2	22.6	22.8
South	25.1	27.0	28.0	27.2	30.0	26.7
West	21.5	19.1	22.9	20.9	21.9	27.5
Industry						
Mining	6.9	10.3	13.8	10.3	13.8	10.3
Construction	19.7	20.0	21.6	20.9	22.0	21.6
Manufacturing	26.1	26.7	26.2	26.5	28.9	28.8
Transportation	19.2	18.2	20.1	18.2	21.4	21.5
Wholesale/retail trade	20.9	22.1	23.3	22.4	24.0	25.7
Finance, insurance and real estate	21.4	21.6	20.9	23.1	25.4	23.3
Services	24.4	24.6	24.3	24.2	23.9	25.2
Overall	22.3	23.5	24.5	23.5	25.3	25.6

[a] Nonresponse includes refusals and otherwise unattainable responses but excludes out of scope and out of business.

quarter-to-quarter change. In particular, establishments with small employment have a very low response rate and the largest number of no changes quarter to quarter. These smaller establishments are probably more likely to refuse because they lack the staff and resources to respond to questionnaires. Establishments from larger-size classes have response rates that are generally in the 90–95% range. A nonresponse adjustment for the base period making use of this fact is envisioned for implementation at some future time.

Tables 3 and 4 provide some proportions of response or nonresponse by industry and status for the base-period wages for both establishments and occupations. Regional and major industry divisions did not show the marked correlation with nonresponse rate that size class did.

Once the establishment has been initiated and a quarterly mail update collected, a temporary nonresponse or seasonal nonresponse may occur, or the schedule may still be pending at the close of collection. In these situations, the previous quarter's data (reported or imputed) are moved by the average wage change of a matched sample of establishment/occupations at some collapsed cell level. Wages generally form a monotonically increasing function, and establishments that experience temporary or seasonal closures, such as amusement parks, do not reopen with wages at the level at the time of closing but with

TABLE 3

Proportion of Industry Total by Status Code for Base Period, September 1975, for Establishment[a]

Status code	Industry							
	Mining	*Construction*	*Manufacturing*	*Transportation*	*Wholesale/ retail trade*	*Finance insurance, and real estate*	*Services*	*Total*
107 Reporting	77.1	57.3	67.9	75.6	70.5	76.2	68.4	69.1
214 Temporary nonresponse	.0	.0	.0	.0	.0	.0	.0	.0
321 Seasonal nonresponse	.0	.0	.1	.0	.0	.0	.3	.1
428 Strike nonresponse	.0	.5	.6	.4	.2	1.4	.6	.5
535 Refusal	5.7	10.3	17.7	9.8	12.5	13.6	15.4	14.4
642 Out of business	2.9	17.8	6.6	4.4	6.3	4.1	5.3	6.8
749 Out of scope	2.9	8.6	5.2	2.7	6.4	2.7	4.2	5.2
856 Other permanent loss[b]	11.4	5.5	1.9	7.1	4.1	2.0	5.8	3.9
Total number of sample establishments in industry	35	185	896	225	590	147	358	2436

[a] Does not include supplements to sample.
[b] Not possible to locate establishment or two reporting units were reported on one schedule.

TABLE 4

Proportion of Industry Total by Status Code for Base Period, September 1975, for Occupations in Reporting Establishments[a]

Status code	Industry							
	Mining	*Construction*	*Manufacturing*	*Transportation*	*Wholesale/ retail trade*	*Finance insurance, and real estate*	*Services*	*Total*
107 Reporting	62.5	33.4	56.3	55.4	40.4	48.4	54.7	50.2
214 Temporary nonresponse	.0	.0	.1	.1	.0	.0	.1	.0
321 Seasonal nonresponse	.4	.2	.1	.0	.1	.0	.5	.1
428 Strike nonresponse	.4	1.3	.5	.2	.1	.3	.2	.4
535 Refusal	.0	.2	1.7	.7	.6	.3	.5	1.0
642 No job match[b]	36.8	65.0	41.3	43.4	58.8	51.0	44.0	48.3

[a] Does not include supplements to sample.
[b] Was not possible to locate any employees in establishment whose job matched the selection occupation. The imputation process in Phase II created this situation.

wages having increased in accordance with the wage movement during the period they were closed. Thus temporary and seasonal nonresponses are moved forward by the average rather than the lateral wage movement. The cells for imputation are predetermined and follow a prescribed sequence based on satisfying certain criteria. The criteria for determining the collapse level is that cells are collapsed (starting with the most detailed cell) until the weighted employment of the respondents reaches 60% of the weighted employment of all "active" schedules in the cell. An establishment/occupation can receive imputed data for no more than four consecutive quarters before being reviewed.

If an establishment did not respond for four consecutive quarters, it was dropped from the sample. No attempt was made to recontact it. This allows the flexibility of using respondents' data even if they are occasionally nonrespondents for one reason or another.

3.2. Incomplete Data for Benefits

After benefits were initiated, a shuttle form (the same form is returned to the respondent each quarter for the addition of the current quarter's data to that of the previous quarter's) was used to collect quarterly benefit updates. The regional offices were given the responsibility of obtaining new data elements reflecting benefit changes and updated existence information if necessary. Data were coded and transmitted to Washington where new cent/hour benefit costs are calculated, if necessary, and a series of edit checks performed.

For benefits, owing to the variety of reported data, there are a multitude of edit checks. There are edits for valid status codes and checks on valid ranges of hours-related data. There is a validation of new benefit entries and a review for internal consistency among benefits (e.g., social security and federal railroad retirement cannot both apply to the same occupation). In addition to the consistency edits, there are also screening programs, which are scheduled to be implemented.

Once the benefit costs are available for all responding establishments, the next step is to make sure the cost figures are all in a similar format. This means that each cost figure must represent the cost of one and only one benefit within an establishment/occupation. This is not automatically true because the respondent is given the option of reporting cost figures that represent costs for more than one benefit item ("combined benefits"). This is an important option because the respondent does not always have cost figures available for each benefit separately. Because the occupational practices of individual establishments change from time to time (for example, an establishment may change from combined benefits to individual benefits or from offering two combined benefits to three combined benefits), it was operationally more efficient to impute benefits individually. The following problems arise from "combined benefits":

(1) Since imputations for missing data are done benefit by benefit, these "combined benefits" cannot be used in any of the imputation calculations, at least not in their combined form. Since there are a significant number of "combined benefits" (a crude approximation is 10% of all benefit data), not using the combined data in the computation of the index could add bias to the ECI.

(2) The respondent always has the option of reporting "data not available" for a given quarter. If two benefits that are combined together in a prior quarter both receive a status code of "data not available" for the current quarter, imputational difficulties can develop unless the computer system can keep track of the varying pattern of benefit combinations from one quarter to another by cross-checking "combined benefits" for all possible situations. This procedure, however, seems to be too complex to implement efficiently. Besides, it does not solve the first problem. The allocation of "combined benefits" is relatively simple and solves both problems.

The main idea behind the allocation procedure is the calculation of an average level for each benefit. Only benefit costs that are not involved in any combining (i.e., benefit costs that are reported individually) can be used in these calculations. The combined data can then be allocated proportionately to these levels. Since the distribution of benefits will vary for each benefit, the basic estimation cells will have to be combined until enough data are available to calculate reliable estimates.

The basic assumption made with this procedure is that the actual cost proportions of individual benefits involved in the combining are the same as the cost proportions of individual benefits not involved in any combining. If this is true, this process should be relatively unbiased. In any event, the allocation does increase the number of cost figures used in the imputation, thereby reducing the variance (and, it is hoped, the total error).

After the "combined benefits" have been allocated, the imputation process begins. All imputations are made to fill in any gaps in a respondent's benefit data (i.e., a nonresponse adjustment). There are no problems with respondents who supply all necessary wage and benefit data. A number of respondents, however, will only partially respond. Furthermore, since wages represent approximately 75% of total compensation, a decision was made that no wage data would be excluded from the total compensation indexes because of a lack of benefit data. This is done to maximize the use of available data and, it is hoped, to reduce the mean square error of the index. Therefore, to obtain total compensation figures for all wage respondents, imputations must be made (when necessary) to obtain a total benefits figure. Adding wages to the total benefit figures yields total compensation. The total compensation figures can then be used to calculate total compensation indexes.

To decrease the mean square error in the imputation process, all available data must be used. Since the benefit existence information is usually available, imputations can be made benefit by benefit.

3.3. Base-Period Imputations

As mentioned earlier, all wage respondents will have a total benefit cost figure. This benefit cost will be (1) totally supplied by respondent, (2) partially supplied by respondent and partially imputed for, or (3) totally imputed for. During the base period, all imputations will be average benefit levels. There are three possible situations where levels are imputed for a given benefit.

(1) Where a wage-related benefit plan exists and data are not available, the average cost, based on only those plans in the imputation cell for which data are wage-related and available, is imputed.

(2) Where a nonwage-related benefit plan exists, but data are not available, the average cost, based on only those plans in the imputation cell for which data are non-wage-related and available, is imputed.

(3) Where it is not known whether or not the benefit practice exists, the average cost, based not only on all plans in the imputation cell for which the benefit practice exists but also on those cases for which no plan exists (zero cost), is imputed.

The distribution of each benefit is different. This implies that calculating imputation levels across fixed cells will yield reliable estimates for some benefits and poor estimates for others. To compensate for the distributional difference, the estimation cells will be collapsed until a reliable estimate can be calculated. However, no collapsing was performed beyond a major occupational group by major industry.

3.4. Ongoing Benefit Imputations

During subsequent quarters, the nonresponse imputations require five computations:

(1) wage-related quarter-to-quarter benefit changes (movements),
(2) non-wage-related quarter-to-quarter average movements,
(3) wage-related average levels when the benefit practice is known to exist,
(4) non-wage-related average levels when the benefit practice is known to exist,
(5) overall average levels when it is not known whether the benefit practice exists or not.

The wage- and non-wage-related average movements are calculated over available wage- and non-wage-related benefit plans, respectively, based on only those benefit plans that exist for both quarters. The average levels are calculated in a manner similar to the base-period average levels.

When the existence information is known and the movement is of the form c_2/c_1 ($c_1 \neq 0$), this ratio applied to the prior quarter's benefit cost, and the resulting quantity is used as the current quarter's imputed value. This assumes that the prior quarter's cost is not zero. If the prior quarter's cost is zero, then the appropriate wage- or non-wage-related current quarter's level is used as the imputed value. A problem arises if the movement is of the form $0/0$ or $c/0$, which is possible since zero is a valid benefit cost.

When the movement is of the form $0/0$, then the prior quarter's data are moved either laterally, if the benefit is non-wage-related, or by the national quarter-to-quarter wage relative, if the benefit is wage-related. When the movement is of the form $c/0$, then a wage- or non-wage-related level, whichever is appropriate, is the imputed value for a "data not available" benefit.

When the existence information is not known or when the existence information is known but data have not been available for a specified number of consecutive quarters, then the imputation is the same as the base-period imputation where the existence information is not known.

All movements and levels are calculated over appropriate collapsed cells. The cells for benefits are slightly different from those for wages and are described in the next section. The collapse criteria in terms of an acceptable "weighted response rate" are also somewhat different. Collapsing may introduce some bias, but the mean square error is, it is hoped, reduced.

3.5. Collapse Cells

It is sometimes advisable to collapse a cell having a low response rate with a cell that is similar with respect to the variable being measured but has a high response rate. In this way the mean square error is reduced by making the decrease in variance greater than the corresponding increase in bias. For this reason, collapse cells and criteria for collapsing have been established for the ECI. The collapse cells are based on grouping cells that are homogeneous with respect to the variable of interest. A sequence is also established in which the cells that are collapsed together are progressively less homogeneous. The sequence of collapses for the wage data is SIC/occupation, SIC/major occupational group (MOG), cluster of related SICs/MOG, subindustry (e.g., durable goods)/MOG, and major industry division (MID)/MOG. If the cells have been collapsed to the MID/MOG level and the criteria for an acceptable level of response has still not been satisfied, then, for certain MOGs, there are additional collapses: professionals with managers; nontransport operatives with transport operatives; laborers with service workers; salespersons with clerks; and craftsmen with nontransport operatives. (The last two collapses are employed only if salespersons and craftsmen, respectively, do not meet the collapse criteria—not vice versa). Once these have been made, then the industry collapse from SIC to cluster, cluster to subindustry, and subindustry to MID is repeated. (The collapse across MOGs has rarely been used and will not be one of the allowable collapses in the future.) The collapse pattern for benefits is similar to that for wages except that (1) certain establishment/occupation benefits are identified as being wage-related and others as being non-wage-related, and imputation is always done separately for these two categories and (2) the collapsing of MOGs is not allowed. In forming the collapse cells and sequence, an attempt was made to group cells with similar wage and benefit levels and changes.

Since collapse cells are formed without regard to region, union/nonunion status, and metropolitan/nonmetropolitan status, lack of sufficient sample results in collapsing across these (publication) characteristics.

In practice, however, only approximately 6% of the "active" establishment/occupation responses are imputed for on a quarterly basis; therefore, the effect on the published data is probably not large.

Since the ECI is a quarterly survey and the nonresponse adjustment for quarter-to-quarter change is applied to individuals (establishment/occupation), validation of quarterly imputation data is feasible.

Because of lack of resources, however, BLS has been unable to make an assessment at this time. We plan to initiate an evaluation program as soon as the resources become available. In addition, preliminary plans for an ECI Quality measurement survey have begun. This survey will probably begin after the wages and benefits systems are in place.

REFERENCES

Hoy, Easley (1978). General survey design aspects of the employment cost index. *Proceedings of the Section on Survey Research Methods*, American Statistical Association, pp. 695–699.

Lahiri, D. B. (1951). A method for sample selection providing unbiased ratio estimates. *Bulletin of International Statistical Institute*, 33, Part III, pp. 133–140.

Marks, Harry, and Frevert, David (1978). ECI variance estimation. *Proceedings of the Section on Survey Research Methods*, American Statistical Association, pp. 706–710.

Sheifer, Victor J. (1975). Employment cost index: A measure of changes in the "price of labor." *Monthly Labor Review* July: 3–12.

Steinberg, Joseph (1975). Sampling aspects of the employment cost index. Unpublished paper, U.S. Bureau of Labor Statistics.

Wright, Douglas A., and Kaufman, Steven F. (1978). Employment cost index: Estimation procedures. *Proceedings of the Section on Survey Research Methods*, American Statistical Association, pp. 700–705.

CHAPTER 3

USDA Livestock Inventory Surveys

Robert L. Freie

1. SURVEY DESCRIPTION

1.1. General Considerations

The Livestock Inventory surveys are recurring surveys used in the estimation of hog and cattle inventories for the United States and for major producing states. These surveys are conducted quarterly for hogs in 14 states and semi-annually for cattle in 28 states. These states account for 85 to 90% of U.S. inventories.

Reports published from these surveys provide key information on current and future supplies of meat and other livestock products. These data are used widely within both the private economy and government, with benefit to producers, consumers, and many others interested in the producing and marketing of livestock and products.

The survey population of interest is that comprising all producers of cattle and hogs. For purposes of establishing reporting units, livestock is associated with land "operated," not through ownership. Therefore, the target population must include all operations of land with livestock. Usually, these are farm or ranch operating units including land owned, leased, and managed.

The questionnaires are designed first to define the operation in terms of land operated and then to obtain livestock numbers regardless of ownership. The livestock questions include total inventory, a breakdown by important subclasses, and information on births and deaths. The number of data items usually totals about 25 for the hog questionnaires and 20 for the cattle questionnaires. In addition, farm and ranch names and names of other individuals associated with partnership operations are requested so potential duplication

INCOMPLETE DATA
IN SAMPLE SURVEYS
Volume 1, Part II

ISBN 0-12-363901-8

can be identified and appropriately handled. Examples of questionnaires are provided in Appendix A.

1.2. Sampling

Surveys utilize two sampling frames, a list frame of farm and ranch operators or operations and an area frame for sampling small land area units or segments. The list frames are developed and maintained independently by state. A variety of available list sources are used in this effort and include annual state farm censuses, assessor's records, USDA lists, brand lists, and lists maintained by state governments for inspection or control purposes. Smaller specialized lists are often merged with a large, general farm list to improve coverage and control information. Control data, usually in the form of some historic livestock count, are maintained and used in stratification for sampling. The list-frame coverage accounts for about 80% of the livestock population.

The area sampling frame provides the complete coverage frame and is constructed with stratification by land use. Major stratification is based on intensity of cultivation for crops. In addition, city or built-up areas are identified as well as large nonagricultural areas such as parks, recreational areas, and military bases. Segments vary in size according to land use but are typically about 1 sq. mi. in intensively cultivated strata, several square miles and larger in the more open areas used primarily for livestock grazing, and about 0.1 sq. mi. in city and residential areas.

As indicated, list frames are stratified by size of operation prior to sample selection. Eight to ten strata are most commonly used, although this will vary according to the quality of the control or stratification variable. New list samples are selected at least annually and usually correspond to list updating. These samples are generally reused for interim surveys with some minor variations in sample rotating and resampling. List sample sizes average about 1800 per survey and state. For the most part, samples are optimally allocated to strata within states based on recent survey variances.

The area segment samples are selected for general-purpose agricultural surveys used primarily for estimating crop acreages and livestock inventories. A compromise sample allocation is used to serve the multipurpose needs. Typically, a state has between 300 and 350 segments, and about 20% are rotated out each year and replaced with new segments. For field enumeration, segments are located on county highway maps, with boundaries identified on large-scale aerial photography.

There are several procedures available for associating agricultural data with area segments. For the Livestock Inventory surveys, a "weighted segment" procedure is used. This procedure requires inventory numbers to be collected for an entire farm operating unit. Where this unit includes land operated outside the segment boundaries, the data must be prorated to the segment in proportion to the fraction of farm acreage that is inside the segment. For purposes of estimating livestock, the weighted segment procedure offers substantial gains in sampling efficiency over alternative methods.

The area sample is completely enumerated around June 1 each year. The names of all land operators inside segment boundaries are identified at that time. Area-frame operator names are matched against the list frames. Operations identified with the list belong to the area "overlap" domain. Those not found on the list belong to the "nonoverlap" domain and become the basis for estimating list-frame incompleteness.

In December, another area-frame survey is conducted, based on a subsample of area tracts enumerated in June. A tract is identified with the reporting unit and consists of the land inside segment boundaries under one operating or management arrangement. The December sample is selected with emphasis on those tracts with livestock reported in June. The overlap/nonoverlap classification is reviewed to account for changes in area-frame operators and list-frame updates.

For the March and September hog surveys, the classification made during the previous survey is used in lieu of full area-frame surveys. Current data are collected only from area tracts belonging to the nonoverlap domain as determined for the previous survey (either December or June). By necessity, no list-frame changes are permitted during these intervals.

1.3. Data Collection and Processing

The data collection for each survey covers a period of about 2 weeks. Methods of collection include mail, telephone, and personal interviews. For the list-frame sample, a mail questionnaire is used. After allowing a brief time for mail response, follow-up interviews are made by telephone or personal visit, with the intention that all mail nonrespondents be contacted to obtain the desired information.

Data from area-frame respondents are collected primarily through personal interviews. Personal interviews are required for the June and December surveys, whereas telephone interviews are permitted and used on a limited basis for the March and September hog surveys.

Interviewing is done primarily by part-time employees. Men and women hired for this work usually have a knowledge of agriculture, time available to work periodically throughout the year, and, of course, the ability to conduct interviews. They receive formal training for livestock surveys before the June and before the December survey periods each year. Responsibility for data collection lies with field offices located in the individual states, which follow survey plans, instructions, and materials prepared in the central Washington, D.C., office. In addition, regional training schools are held for field-office personnel prior to their training of interviewers.

Both editing and summarizing of the data are handled primarily by the field offices, using common computer edit and summary systems developed for network processing. Questionnaires receive a check for completeness and a brief manual review for major inconsistencies, interviewer notes, and respondent comments upon receipt in the field office. Then the data are keyed for processing. The computer edit is composed primarily of consistency checks between related data cells and, in addition, assures that all sample units have been accounted for. All data failing the consistency checks are flagged for manual review and resolution through verification or correction. Edit checks are classed as "critical" or "noncritical," depending on the nature and seriousness of the inconsistency. Data associated with noncritical checks can be accepted or corrected by the reviewer. Data failing the critical checks must be corrected. If needed, data are verified or corrected by recontacting the respondent by telephone. Interviewers are trained to include comments on the questionnaire to explain unusual situations.

An individual missing data cell would require manual imputation by a reviewer familiar with the usual relationships among data cells, either prior to keying the data or during the computer editing process. If not imputed earlier, cells with data missing would be identified with the consistency checks of the computer edit and, therefore, would be flagged. Questionnaires that are unusable because of respondent refusal or inaccessibility are coded accordingly for processing. Unusable questionnaires are keyed to account for all sample units and to provide a summary by type of response.

The summary programs provide stratum-level expansions and sampling errors for all survey items. Once state-level summaries are completed, the data files are accessed on a computer network and regional- and U.S.-level summaries are obtained.

Simple, unbiased estimators are used in generating survey expansions from the livestock surveys. The basic formulations for estimates and variances correspond to sample design and are presented in Appendix B.

The formulation provided for the area-frame portion of the estimate is appropriate in June when the area sample is completely enumerated. During other survey periods, a stratified subsample of tracts (not primary sampling units) is used for the area-frame sample. The post- or substratification technique utilizes and takes advantage of information obtained for each tract in June. This introduces an additional tract-level probability-of-selection factor in the estimation formula. The variance formulation becomes considerably more complex and was developed around the product formula for variances, using the survey-estimated population number of tracts for each substrata and the current average tract value within substrata.

In multiple-frame survey designs, estimators that weight together the separate frame estimates of overlapping domains can be used with some gain in precision. Although not used, this would be possible with the livestock surveys in June and December, when overlapping area-frame data are collected. The small reduction in sampling errors is judged as not sufficient justification for increasing survey complexity. If used, the area-frame overlapping data would receive only a small weight because of the relative inefficiency of the area frame. The data are used, however, in separate area-frame expansions of livestock inventories.

Survey results, as described, are not published directly but are used as the principal information source for preparing official United States Department of Agriculture (USDA) reports. The total estimation process utilizes other available information in a form that provides additional checks and balances on livestock inventories. Survey estimates of births and deaths, together with previous inventory estimates and reliable information on imports, exports, and slaughter, provide an indication of current inventories. Records of federally inspected slaughter provide the best in available checklist information. These data can be used in revising previously published estimates if necessary.

For any report, the survey results and other available information are first reviewed and analyzed in the state field offices. A commodity statistician sets preliminary state estimates. In the central office, a group of eight to ten commodity statisticians, including field representation, is designated to review data at U.S. and regional levels. Participants in this group, officially named the Crop Reporting Board, work individually but essentially arrive at a concensus and set U.S. and regional estimates. Advantage is taken of the higher degree of relative precision of the survey and other data at the regional and U.S. levels. For example, survey estimates of livestock inventories from an area sampling frame are quite useful at these levels but have more limited value in setting state estimates because of higher relative sampling errors. Once U.S. and regional estimates are set, the preliminary state estimates are accepted or adjusted to conform as needed. Security precautions are employed for all phases of data analysis and report preparation to prevent premature disclosure of information. Publication dates and release times are established well in advance to provide all users equal opportunities for accessing the data. Sample pages of a livestock inventory report are provided in Appendix C.

2. WHY THIS SURVEY WAS SUGGESTED AND SELECTED AS A CASE STUDY

Current estimates of livestock inventories are critically followed by the livestock industry. Development of reliable probability-based surveys for these estimates has been one of the primary challenges for the current USDA agricultural data series. Area sampling frames are the only complete frames available but are not particularly efficient for livestock estimation. Stratified list frames of producers, although difficult to develop, cumbersome to manage, and impossible to keep up to date, permit cheaper data collection methods and improved sampling precision. Sample methodology, using both sampling frames, acquires advantages from both.

The principal area of incomplete data for these surveys is associated with total questionnaire nonresponse, primarily because some livestock producers refuse to provide the information. Respondent cooperation is voluntary. Missing data for individual cells are rare because the questionnaires are relatively short and include no personally sensitive questions. Total nonresponse is not viewed as critical but is of constant concern to USDA. Also, nonresponse seems to be increasing and is approaching 15% of the list-frame sample. Area-frame nonresponse has a much smaller impact on survey reliability and has a smaller nonresponse rate. This is due partly to data collection by the personal interview method and also to the fact that usable data can sometimes be taken by observation. For comparison, only about 4% of the area-frame questionnaires are coded nonusable (refusal). The main discussion of this chapter is directed to problems of list-frame nonresponse.

In any survey, the first effort should be to minimize nonresponse by whatever data collection methods are available. For remaining nonrespondents, however, some assumptions have to be made. Unfortunately, in most survey applications, the validity of adopted assumptions cannot be tested or resulting biases measured. If adequate information is known about individual nonrespondents, this should be utilized in the imputation process.

Because of importance and uniqueness, many aspects of the multiple-frame livestock surveys have been included in statistical research activities of USDA. These studies include investigations of alternative imputation procedures and an evaluation of current problems associated with nonrespondents in the list-frame sample.

The first effort, "Missing Data Procedures: A Comparative Study" by Ford (1976), discusses alternative missing-data procedures (see Section D.1 in Appendix D). Using a simulation experiment, six procedures were evaluated for bias and variance estimation, using list-frame data from a USDA livestock survey. The author concluded that double sampling ratio or regression procedures were preferable to any of the hot-deck procedures investigated.

The second study (Part 2) by Ford (1978) compared three alternative missing-data procedures—ratio, regression, and hot-deck—and the current operational procedure for estimating the total number of hogs from a USDA survey (see Section D.2 in Appendix D). No significant differences were found among procedures. The lack of any differences was believed attributable to the low correlation between the control variable and the reported data. The control variable is a measure of size used for stratifying the list universe. The current procedure of simply adjusting for reduced sample sizes resulting from missing units has the advantage of operational simplicity. Adjustments are made at the stratum level, so to further reduce nonresponse biases within-stratum correlations have to be improved. However, improved control data will also support more extensive stratification for greater sampling efficiency. Therefore, given that better control data can be obtained, it is uncertain whether alternatives provide significant advantages over the current procedures.

Another study, "The Effect of Refusals and Inaccessibles on List Frame Estimates" by Gleason and Bosecker (1978), looked at using information reported in a previous survey for imputation. (A summary of conclusions and recommendations from this report is included in Appendix D.) Since cattle surveys are semiannual and hog surveys quarterly and many sample units overlap between surveys, there is some potential for taking advantage of the survey-to-survey correlations in imputations when a nonrespondent has reported in a previous survey. Also, a special effort was made through personal enumeration, and 40–50% of previous refusals were converted to respondents. Survey-to-survey correlations are generally sufficient to be of assistance in imputation. However, this would be useful for only part of the nonrespondents and would be of no value for surveys following a frame update and new sample selection.

An evaluation of response by stratum demonstrates that nonresponse is more frequent among large producers. Also, there is evidence that a higher proportion of nonrespondents than respondents does have livestock, particularly among operators refusing to cooperate in surveys. Although the first consideration is controlled, at least partially, by adjusting for nonresponse at the stratum level, the second suggests that some downward biases do exist with present procedures. In the study "The Use of Partial Information to Adjust for Nonresponse" by Keith N. Crank (1979), effort was made to obtain information about nonrespondents as to whether they do or do not have the livestock species of interest (see Section D.4 in Appendix D). An adjustment could then be made, assuming the average of nonrespondents with livestock is the same as the average of respondents with livestock. This is believed to be a reasonable assumption for use in handling missing data and is to be evaluated by the USDA in upcoming surveys. However, there are still uncertainties as to the ability to collect sufficient supplemental information during the time constraints of an operational survey. Also, there will always be a class of nonrespondents for which this information is not available, and additional research is needed to determine the best assumptions for handling these unknowns. For example, this group would include sample units normally coded as inaccessibles, which often include a high proportion of out-of-business operations that can no longer be located. Table 1, from the report by Crank (1979), demonstrates the importance of various assumptions, particularly for unknown nonrespondents, on the list-frame expansions for selected states.

TABLE 1

Ratio of current estimates to estimates based on varying assumptions,[a] along with the estimated coefficient of variation (C.V.) for each estimate[b]

		Current procedures		*Assumption a2, b, C1*		*Assumption a2, b, C2*		*Assumption a2, b, C3*	
State	*Species*	*Est.*	*C.V. (%)*	*Est.*	*C.V. (%)*	*Est.*	*C.V. (%)*	*Est.*	*C.V. (%)*
Illinois	Hogs	1.00	3.9	1.02	3.9	1.05	4.1	1.02	3.9
Iowa	Hogs	1.00	2.9	1.03	2.8	1.05	2.8	1.03	2.8
Nebraska	Hogs	1.00	3.4	1.03	3.3	1.06	3.3	1.03	3.3
Illinois	Cattle	1.00	4.7	1.02	4.7	1.06	5.0	1.02	4.7
Nebraska	Cattle	1.00	2.4	1.04	2.4	1.06	2.5	1.04	2.4
Total	Hogs	1.00	2.0	1.03	2.0	1.05	2.0	1.03	2.0
Total	Cattle	1.00	2.2	1.03	2.2	1.06	2.3	1.03	2.2

[a] The following assumptions were made about the use of supplemental information:

Assumption a2: The percentage of positive nonrespondents is different than the percentage of positive respondents.

Assumption b: The average of positive respondents is the same as the average of positive nonrespondents.

Assumption C1: The proportion of positive unknowns is the same as the combined proportion of both positive respondents and positive known nonrespondents.

Assumption C2: The proportion of positive unknowns is the same as the proportion of positive known nonrespondents.

Assumption C3: The proportion of positive unknowns is the same as the proportion of positive respondents.

[b] Data from Keith N. Crank (1979).

3. INCOMPLETE DATA

3.1. Measurements

Measures of incomplete data for the list-frame component of recent hog and cattle surveys are provided in Tables 2–4. Data provided in these tables correspond to the response codes used in survey processing. Five codes identify usable questionnaires; four codes identify unusable questionnaires. Most of these breakouts are self-explanatory; however, some additional comments may help explain the estimated, known-zero, and inaccessible categories.

The estimated group is usually limited to the very largest producers. These producers are in the extreme upper tail of the sample distribution and are selected with certainty. Because of their uniqueness, it is believed that estimating data for these operations is more reliable than is using an average of data reported from other large operations. They are large enough to be well known, and often some information can be obtained from local informed sources within their communities.

The "known-zero" category is for sample units that have indicated in a prior survey that they are no longer farming (out of scope) or are at least indefinitely

TABLE 2

Livestock Inventory Surveys (USDA):
List-Frame Sample Response[a]

	Hog survey December 1978 (14 states)	*Cattle survey January 1979 (28 states)*
List sample size	25,587	52,078
Percent usable		
Data obtained by		
Mail	22.1	22.9
Phone interview	46.2	44.9
Personal interview	15.8	16.0
Estimated	.6	.9
Known zero	1.0	1.8
Total	85.7	86.5
Percent unusable		
Refused by		
Mail	.4	.2
Phone	6.6	5.6
Interview	2.7	1.9
Total	9.7	7.7
Percent inaccessible	4.6	5.8
Total	100.0	100.0

[a] Based on unweighted sample counts.

out of business as livestock producers. "Inaccessibles" is a category for which the operator or other informed person cannot be located within the time frame of the data collection period.

Probably Table 4 provides the most meaningful measure of incomplete data. The potential impact of biases resulting from missing data and the imputation process can be evaluated in terms of the survey estimates. Since direct measures of biases cannot be made, effective control is best assured by keeping the percentage of unusable reports small. It is apparent from comparisons of the three tables that a higher proportion of the larger producers refuse to cooperate on the livestock surveys.

3.2. Data Collection Efforts

The intent of survey procedures is to obtain usable data for the entire sample. No substitutions for nonrespondents are permitted and subsampling of

TABLE 3

Livestock Inventory Surveys (USDA):
List-Frame Sample Weighted Response[a]

	Hog survey December 1978 (14 states)	*Cattle survey January 1979 (28 states)*
List sample size	25,587	52,078
Percent usable		
Data obtained by		
Mail	22.3	22.9
Phone interview	52.1	50.3
Personal interview	13.9	12.8
Estimated	.1	.2
Known zero	1.5	1.7
Total	89.9	87.9
Percent usable		
Refused by		
Mail	.2	.2
Phone	4.4	5.1
Interview	.9	.9
Total	5.5	6.2
Percent inaccessible	4.6	5.9
Total	100.0	100.0

[a] Weighted by the inverse of probabilities of selection. Percentages reflect the proportion of the list-frame universe represented by each response type.

nonrespondents for more intensive effort is not used. Within the time frame allowed, enumerators are expected to carry out as many call backs as necessary by phone or personal visit to obtain the desired information. Refusals are a more serious, but separate, problem. Procedures used in dealing with refusals are best described in the accompanying sheet from the Supervising and Editing Manual prepared for these surveys.

3.3. Treatment in Processing

Treatment of incomplete data in processing is handled primarily with coding for response type. Questionnaires for which usable data are unattainable are coded with one of the four nonresponse codes. This coding circumvents any further processing but does allow for the accountability of all sample units prior to summarization. All data associated with usable questionnaires must satisfy

TABLE 4

Livestock Inventory Survey (USDA):
Percentage of List-Frame Estimate Attributable to
Each Response Type[a]

	Hog survey December 1978 (14 states)	*Cattle survey January 1979 (28 states)*
List sample size	25,587	52,078
Percent usable		
Data obtained by		
Mail	19.6	23.3
Phone interview	37.7	43.6
Personal interview	21.5	15.9
Estimated	1.2	1.9
Known zero	.0	.0
Total	80.0	84.7
Percent unusable		
Refused by		
Mail	.5	.3
Phone	10.3	7.2
Interview	5.2	2.3
Total	16.0	9.8
Percent inaccessible	4.0	5.5
Total	100.0	100.0

[a] Total inventory is questionnaire item used. Unusable responses reflect the percentages of the survey estimate dependent on imputational procedures.

the computer edit process described previously. Missing cell data would be identified with the various computer checks for consistency between related questionnaire items. However, usually missing data for key items would have been identified during the initial check-in process. In either case, data must be obtained through a callback or manually imputed based on expected relationships with other reported data items.

No precise record of the number of cell imputations or their impact is made. However, their frequency is low and known to be of minor concern in relation to the importance of unit nonresponse. Counts of the times that data fail individual consistency checks are obtained, but missing data are not identified separately and would be only a small part of these counts.

At summary, data are assumed complete for all records coded as usable responses. Counts of usable records within strata (n_h) are made in summary. Population counts for each stratum (N_h) are prepared as external parameters to summary. This procedure for variance estimation follows the assumption that

8.14 Treatment of Refusals

Refusals must receive special treatment with the final objective being to obtain an accurate report. The following suggested procedure serves as a guide.

A. Refusals from Previous Surveys

(a) Mail refusal - Interview mail refusals during the current survey. This is done by telephone or personal contact depending on the size of the operation.

(b) Telephone refusal - Try personal contact by an experienced enumerator, supervisor or SSO staff depending on importance of operation.

(c) Personal interview refusal - A special pre-survey letter from the State Statistician might be helpful in explaining the need for and use of data and emphasizing confidentiality. Relating our estimates to a source familiar to the operator might also be helpful.

It sometimes helps to rotate enumerators or use a supervisory enumerator or an SSO staff member to contact a refusal on a prior survey. Where practical, scheduling personal contacts between surveys by the SSO staff may be helpful.

B. Refusals from Current Survey

If a respondent refuses, it is not advisable to contact that respondent again during that survey unless it was a phone or mail refusal and some question was raised about the survey. In the event of a personal interview refusal, enumerators should be instructed as to the type of information required and to what extent they should secure that information. For every refusal, the enumerator should be instructed to write down the reason given by the respondent for refusing. Any information provided by the respondents or observed during the visit should be noted.

Enumerators should be made aware that if the operator is a sampled list questionnaire or a non ag tract, the above information should be sufficient since no data are required to be estimated. However, if the refusal is a farmer with land in an area tract or a preselected EO, then a complete report must be estimated. In these instances the enumerator should be instructed to contact local sources of information such as government agencies, agribusiness, etc, to obtain as much information as possible.

nonresponse is random and assumes respondents and nonrespondents are from the same population (refer to Appendix B for formulation).

The variances are computed for all data items and assume simple random sampling with reduced sample sizes for nonresponse. This formulation is consistent with current assumptions made about nonrespondents. The approach has decided advantages over variances computed with imputed data: It provides nearly a true measure of sample variability (not bias) and eliminates the potential dangers of false security from reduced variances for most imputation methods.

3.4. Publication

Missing data problems and their magnitude are not directly addressed in USDA reports. A brief statement on data reliability (see page 2 of Appendix C) is provided in each report, suggesting some of the limitations in the survey data used for the report and alluding to "omissions" as one of the nonsampling errors to which survey estimates are subject. Since they provide a continuous series of estimates, USDA livestock reports are most critically followed by users, who become very familiar with their content. These users are interested principally in the point estimates and frequently have limited statistical backgrounds. It is extremely difficult to prepare satisfying statements on data limitations that will be meaningful to most users. Also, those estimating procedures used by the USDA that incorporate additional sources of data in addition to the survey results complicate this issue. Nevertheless, data users could be made more aware of levels and changes in survey response rates.

APPENDIX A. LIVESTOCK SURVEY QUESTIONNAIRES

Crop Reporting Board
Economics, Statistics, & Cooperatives Service
U.S. Department of Agriculture

Hog and Pig Survey

DECEMBER 1, 1978

Form Approved
O.M.B. Number 40-R3774
Approval Expires 3-31-81

C.E. 11-0087

M

Stratum	ID - Segment	Tract	Subtract
00		01	01

Survey	Resp.	Office	Office
813 1	910	911	920

Dear Reporter

Your **HELP** is needed to **MAKE HOG** and **PIG ESTIMATES** as **ACCURATE** as possible.

Your name was selected in a small sample of farmers in the State and a report is needed even if you have no hogs and pigs or only a few. Questions refer to hogs and pigs on all the land you operate. Facts about your operation will be kept confidential and used only in combination with similar reports from other producers.

Response to this survey is voluntary and not required by law. However, your cooperation is very important to insure timely and accurate estimates.

Please help reduce survey costs by completing this inquiry and returning it as soon as possible. Should your report be delayed in reaching us, one of our interviewers may request your assistance by phone or in person. The enclosed envelope requires no stamp. Thank you.

Respectfully,

Bruce M. Graham

Bruce M. Graham, Chairman
Crop Reporting Board

Please make corrections in name, address and Zip Code, if necessary.

Is your operation known by another name, than printed above?

☐ NO

☐ YES *Enter name* ____________________

LAND OPERATED NOW

The following questions refer to the hogs and pigs on all the land you operate. Therefore, we first must determine the total acres you operate. Please make any necessary corrections when acres operated are entered. Include cropland, pastureland, woodland and wasteland.

1. How many **ACRES** are now in **YOUR ENTIRE FARM** or **RANCH?** .... [900 ____ .0]

*(**Include** all land owned, rented or managed, but **exclude** land rented to or managed by others.)*

(Please turn to page 2.)

- 2 -

HOG AND PIG INVENTORY

Please report below all **HOGS** *and* **PIGS** *on the land you operate regardless of ownership. Include hogs and pigs purchased and still on hand.*

3. **HOGS and PIGS for BREEDING**

   a. **Sows, gilts** and **young gilts** bred and to be bred. 301

   b. **Boars** and **young males** for breeding. 302

   c. **Sows** and **boars** no longer used for breeding 303

4. **HOGS and PIGS FOR MARKET and HOME USE**
   (Exclude breeding hogs already reported.)

   a. Under 60 lbs. *(Include pigs not yet weaned.)*. 311

   b. 60 — 119 lbs. 312

   c. 120 — 179 lbs 313

   d. 180 lbs. and over *(Exclude hogs no longer used for breeding.)* 314

5. **TOTAL number of HOGS and PIGS** – *[add 3a through 4d]* 300

EXPECTED FARROWINGS

6. **SOWS and GILTS** *(reported in Item 3a)* **EXPECTED TO FARROW:**

   a. From now through December 1978, January and February 1979? 331

   b. During March, April and May, 1979? 332

PREVIOUS THREE MONTHS FARROWINGS

9. **SOWS** and **GILTS FARROWED** during September, October and November 1978 until now. 326

   10. **PIGS** from these *(Item 9)* litters:

   a. Now on hand. 327

   b. Already sold 328

PURCHASES

11. Hogs and Pigs purchased since **June 1, 1978** . . . .
   a. Now on hand 317
   b. Already sold or slaughtered . . 318

If NO Hogs and Pigs were purchased in last six months skip to item 13.

M - 3 -

12. **FEEDER PIGS** purchased during November 1978. .... | 340 |

a. Average **PRICE PER HEAD**. .... Dollars and Cents | 341 . ___ ___ |

b. Average **WEIGHT PER HEAD**. .... Pounds | 342 |

DEATHS AFTER WEANING

13. **DEATHS** of **WEANED PIGS** and **OLDER HOGS** during September, October and November. .... | 335 |

OPERATION DESCRIPTION OF LAND

Additional information is needed on your operation to assist in detecting possible duplication in reporting. (Please make any necessary corrections when operation description information has been entered below.)

18. Do you *(the individual or operation listed on the face page)* operate **AGRICULTURAL LAND** in a partnership or joint operating arrangement? *(Exclude landlord-tenant, cash rent or share crop arrangements.* *(Check One)* ☐ YES - *continue* ☐ NO - *turn to page 4.*

19. Who are the persons in this partnership or joint land arrangement with you?

a. Name ________________________________ Telephone No. ____________
(Last) *(First)* *(Middle)*

b. Address ____
(Route or Street) *(City)* *(State)* *(Zip)*

c. Partnership or Operation Name ______________________________

a. Name ________________________________ Telephone No. ____________
(Last) *(First)* *(Middle)*

b. Address ____
(Route or Street) *(City)* *(State)* *(Zip)*

c. Partnership or Operation Name ______________________________

20. How many acres of land are in this partnership or joint operating arrangement?. .... Acres | |

a. How many of these acres were included in Item 1, page 1?. .... Acres | |

21. How many hogs and pigs are now on the Item 20 acres .... Number | |

a. How many of these hogs and pigs were included in Item 5, page 2? .... Number | |

(Please turn to page 4.)

M - 4 -

22. The results of this survey will be released December 21, 1978.
Would you like to receive a copy?

☐ YES = 1

☐ NO = 2

001

COMMENTS

Please comment on any unusual death loss, average gains, or production problems affecting your answers.

Any comments on problems or factors affecting hog production in your area will be appreciated.

That completes the survey. Another hog survey will be conducted in about three months and we may need to contact you again. Thank you for your help.

Reported by ______________________ **Date** ______________

Telephone Number __________ (Area Code) ______________ (Number)

912

GPO 933 605

OFFICE USE

Economics, Statistics, & Cooperatives Service

U.S. Department of Agriculture

Cattle & Calf Inquiry

January 1, 1979

Form Approved
O.M.B. Number 40-R3774
Approval Expires 3-31-81

C.E. 12-0066

M

Stratum	ID - Segment	Tract	Subtract
00 _ _	_ _ _ _ _ _ _ _ _	0 1	0 1

Survey	Resp.	Office	Office
812 1	910 __	911 __	920 __

Dear Reporter:

Your **HELP** is needed to **MAKE CATTLE** and **CALF ESTIMATES** as **ACCURATE** as possible.

Your name was selected in a small sample of farmers in the State and a report is needed even if you have no cattle and calves or only a few. Questions refer to cattle and calves on all the land you operate. Facts about your operation will be kept **confidential** and used only in combination with similar reports from other producers.

Response to this survey is voluntary and not required by law. However, your cooperation is very important to insure timely and accurate estimates.

Please help reduce survey costs by completing this inquiry and returning it as soon as possible. Should your report be delayed in reaching us, one of our interviewers may request your assistance by phone or in person. The enclosed envelope requires no stamp. Thank you.

Respectfully,

Bruce M. Graham, Chairman
Crop Reporting Board

Please make corrections in name, address and Zip Code, when necessary

Is your operation known by another name than printed above?

☐ NO

☐ YES → *Enter Name* __________________

LAND OPERATED NOW

Cattle and Calf questions refer to the numbers now on the total acres operated. Therefore, information is needed for the total acres you operate. If you have completed a cattle questionnaire in the past year, your reported acres may have been entered below. Please make corrections or complete.

Include cropland, pastureland, woodland and wasteland.

1. How many **ACRES** are now in **YOUR ENTIRE FARM** or **RANCH?** .... [900 | .0]
   (Include all land owned, rented or managed, but exclude land rented to or managed by others.)

(Please turn to page 2.)

-2-

CATTLE AND CALVES INVENTORY

Please report below all CATTLE *and* CALVES, *(Including those on feed), regardless of ownership, on the land you operate. (Also include those owned by this farm or ranch that are now on public or grazing association land.)*

How many are:

3. **BEEF COWS** *(Include heifers that have calved)* .... [351]

4. **MILK COWS**, whether dry or in milk *(Include milk heifers that have calved)* .... [352]

   4b. How much milk was produced by these (Item 4) milk cows yesterday? ...... Pounds [501] or Gallons [ ]
   (One gallon = 8.6 pounds)

5. **BULLS** weighing 500 pounds or more .... [353]

6. **HEIFERS** weighing 500 pounds or more ....
   - a. For **BEEF COW** replacement *(Exclude heifers that have calved)* .... [354]
   - b. For **MILK COW** replacement *(Exclude heifers that have calved)* .... [355]
   - c. **OTHER HEIFERS** weighing 500 pounds or more *(Exclude heifers that have calved)* .... [356]

7. **STEERS** weighing 500 pounds or more .... [357]

8. **HEIFER, STEER** and **BULL CALVES** weighing less than 500 pounds .... [358]

9. **TOTAL** number of **CATTLE** and **CALVES** *(Add Item 3 through 8)* .... [350]

   9a. Are all of these **CATTLE** and **CALVES** located in this State?

   ☐ YES – *Continue*   ☐ NO – *Correct answers 3–9 to include only cattle and calves in this State.*

CATTLE AND CALVES ON FEED FOR SLAUGHTER

10. **CATTLE** and **CALVES** on this place being fattened on full feed for slaughter market? .... [370]

    (If Item 10 is greater than 200 head, go to Item 12)

    11a. Total pounds of grain and concentrate feed fed *(Item 10)* cattle and calves yesterday: .... [372]
    (________ pounds/head x ________ number of head)

    11b. Total pounds of silage fed *(Item 10)* cattle and calves yesterday: .... [373]
    (________ pounds/head x ________ number of head)

STOCKERS AND FEEDERS

12. **STOCKERS** and **FEEDERS** on this place being fed a maintenance or warm-up ration? .... [371]

CALVES BORN — 3 —

14. **CALVES BORN** since July 1, 1978. *(Include those still on this farm, sold, slaughtered or died. Exclude calves purchased.)* .... [362]

DEATHS

15. **DEATHS of CATTLE and CALVES** since July 1, 1978? *(Include deaths from disease, accidents, exposure or killed by predators.)* ....
   a. Cattle .... [367]
   b. Calves .... [368]

BUTCHERINGS

16. **CATTLE and CALVES BUTCHERED** in 1978? *(Exclude animals sold alive.)* ....
   a. On this place? .... [377]
   b. For you at a custom butcher locker or slaughter plant? .... [378]

OPERATION DESCRIPTION OF LAND

Additional information is needed on your operation to assist in detecting possible duplication in reporting. Any operation description information reported in the past year may have been entered below. Please make corrections or complete.

18. Do you (the individual or operation listed on the face page) operate AGRICULTURAL LAND in a partnership or joint operating arrangement? *(Exclude landlord-tenant, cash rent or share crop arrangements.) (Check one)* ☐ YES - *continue* ☐ NO - *turn to page 4.*

19. Who are the persons in this partnership or joint land arrangement with you?

a. Name ______________________ Telephone No. __________
   (Last) *(First)* *(Middle)*

b. Address ______________________
   (Route or Street) *(City)* *(State)* *(Zip)*

c. Partnership or Operation Name ______________________

a. Name ______________________ Telephone No. __________
   (Last) *(First)* *(Middle)*

b. Address ______________________
   (Route or Street) *(City)* *(State)* *(Zip)*

c. Partnership or Operation Name ______________________

20. How many acres of land are in this partnership or joint operating arrangement? .... Acres [ ]

   a. How many of these acres were included in Item 1, page 1? .... Acres [ ]

21. How many cattle and calves are now on the Item 20 acres? .... Number [ ]

   a. How many of these cattle and calves were included in Item 9, page 2? .... Number [ ]

(Please turn to page 4)

List -4-

22. The results of this survey will be published January 30. Would you like to receive a copy? . . . . ☐ YES = 1 ☐ NO = 2 | 001

COMMENTS

Please comment on any livestock death loss, average gains, or production problems affecting your answers.

Any comments on problems or factors affecting cattle production in your area will be appreciated.

That completes the survey. Another cattle survey will be conducted in about six months and we may need to contact you again. Thank you for your help.

Reported by ______________________ Date ______________

Telephone Number ______________________ | 912
(Area Code) *(Number)* OFFICE USE

GPO 933 905

APPENDIX B. ESTIMATION AND VARIANCE FORMULAS FOR LIVESTOCK AND INVENTORY SURVEYS

B.1. Estimation Formula

$$\hat{X}_k = {}_{\mathrm{a}}\hat{X}_k + {}_l\hat{X}_k = \text{state-level expansion for } k\text{th item,}$$

$${}_{\mathrm{a}}\hat{X}_k = \sum_h \sum_i (1/P_{hi}) \sum_j {}_{\mathrm{a}}X_{hijk} = \text{state-level area-frame expansion for list incompleteness (nonoverlap) for } k\text{th item,}$$

$${}_l\hat{X}_k = \sum_h (N_h/n_h) \sum_i {}_lX_{hik} = \text{state-level list-frame expansion for } k\text{th item,}$$

where

P_{hi} = the probability of selection for ith area segment (PSU) in hth stratum,

${}_{\mathrm{a}}X_{hijk}$ = the observed value of kth item for jth tract (subunit of segment) in ith segment of hth stratum (if operator of jth tract is on list, ${}_{\mathrm{a}}X_{hijk} = 0$),

N_h = the population count for hth list stratum (or hth area stratum);

n_h = the count of usable questionnaires from hth list stratum (or sample size for hth area stratum), and

${}_lX_{hik}$ = the observed value of kth item for ith list unit from hth stratum.

B.2. Variance Formula

$$\operatorname{var} \hat{X}_k = \operatorname{var} {}_{\mathrm{a}}\hat{X}_k + \operatorname{var} {}_l\hat{X}_k,$$

$$\operatorname{var} {}_{\mathrm{a}}\hat{X}_k = \sum_h \left(\frac{N_h - n_h}{N_h}\right)\left(\frac{n_h}{n_{h-1}}\right)\left[\sum_i (P_{hi}^{-1}\, {}_{\mathrm{a}}X_{hi\cdot k})^2 - \frac{(\sum_i P_{hi}^{-1}\, {}_{\mathrm{a}}X_{hi\cdot k})^2}{n_h}\right],$$

where

$${}_{\mathrm{a}}X_{hi\cdot k} = \sum_j {}_{\mathrm{a}}X_{hijk}$$

$$\operatorname{var} {}_l\hat{X}_k = \sum_h \frac{(N_h - n_h)N_h}{n_h(n_{h-1})}\left[\sum_i {}_lX_{hik}^2 - \frac{(\sum_i {}_lX_{hik})^2}{n_h}\right].$$

APPENDIX C. SURVEY PUBLICATION

CATTLE

Released: January 30, 1979
3:00 P. M. ET

Crop
Reporting
Board
Economics, Statistics, &
Cooperatives Service
U.S. Department
of Agriculture
Washington, D.C.
20250

JANUARY 1 ALL CATTLE AND CALVES INVENTORY DOWN 5 PERCENT

All cattle and calves in the United States on January 1, 1979 totaled 111 million head, a 5 percent decrease from last year and down 10 percent from January 1, 1977, according to the Crop Reporting Board. This is the fourth consecutive year of herd reduction in the current cattle cycle.

All cows and heifers that have calved numbered 47.8 million head, down 4 percent from January 1, 1978. The number on July 1, 1978 was 48.5 million, down 7 percent from July 1, 1977.

-- Beef cows, at 37.0 million, are down 5 percent from January 1, 1978 and 11 percent below January 1, 1977.

-- Milk cows, at 10.9 million, are down 1 and 2 percent from January 1, 1978 and 1977, respectively.

Other classes on January 1 and the changes from last year and two years ago, respectively, are as follows:

-- All heifers 500 pounds and over, 16.9 million, down 5 percent and 9 percent.

-- Beef replacement heifers, 5.52 million, down 6 percent and 15 percent.

-- Milk replacement heifers, 3.94 million, up 1 percent each year.

-- Other heifers, 7.43 million, down 7 percent and 8 percent.

-- Steers weighing 500 pounds and over, 16.3 million, down 3 percent each year.

-- Bulls weighing 500 pounds and over, 2.40 million, down 6 percent and 10 percent.

-- Heifers, steers, and bulls under 500 pounds, 27.4 million, down 7 percent and 15 percent.

The estimates of cattle and calves in this report include those on feed as of January 1 shown on page 9. Operations with Cattle and all Cattle Inventory are shown by size groups on page 17. Operations with Milk Cows and Number of Milk Cows by size groups are shown on page 19.

CALF CROP DOWN 5 PERCENT

The 1978 calf crop is estimated at 43.8 million, down 5 percent from 1977 and 8 percent below 1976. Calf crop was slightly below expected levels for the year.

RELIABILITY AND ESTIMATING PROCEDURES

Primary data used in making cattle estimates in this report were obtained from probability surveys. Nationally, these surveys included information from about 60,000 farmers and ranchers sampled from livestock lists plus farm and ranch operations in 16,000 land area segments. Information was collected by mail, telephone and personal interview. Since all operations with cattle were not included in the sample, survey estimates are subject to sampling variability. This variability, as measured by the relative standard error, is less than one percent of the total cattle and calves at the national level. This means that chances are approximately 95 out of 100 that the survey estimate will be within two percent of the complete coverage value if the same procedures were used to survey all producers. The sampling variability of sampling estimates for all cows and for the calf crop is slightly larger than that for total inventory.

Survey estimates are also subject to non-sampling errors such as omissions, duplications, and mistakes in reporting and recording. These errors cannot be measured directly, but they are minimized through rigid quality controls in the data collection process and a careful review of all reported data for consistency and reasonableness.

In setting the inventory estimates, the Crop Reporting Board used survey estimates of inventory numbers, births and deaths. These survey estimates were combined with reliable check data from other sources on slaughter, imports and exports to construct a national balance sheet (on page 3), which provides an additional check on survey inventory estimates.

A P P R O V E D:

Alex P Mercure

ACTING SECRETARY OF AGRICULTURE

CROP REPORTING BOARD:
J. W. Kirkbride, Acting Chairman,
M. L. Koehn, Secretary,
D. E. Murfield, R. W. Cole,
J. P. Kreber, R. D. Latham,
R. R. Radenz, E. J. Thiessen,
S. D. Wiyatt.

CATTLE AND CALVES: TOTAL AND NUMBER BY CLASS
JANUARY 1, 1978-79

STATE	ALL CATTLE AND CALVES			ALL COWS THAT HAVE CALVED		
	1978	1979	1979 AS % OF 78	1978	1979	1979 AS % OF 78
	1000 HEAD		PERCENT	1000 HEAD		PERCENT
ALA	2130	1820	85	1100	985	90
ALAS	8.3	8.5	102	3.9	3.5	90
ARIZ	1135	1200	106	365	357	98
ARK	2120	2000	94	1170	1130	97
CALIF	4430	4700	106	1812	1790	99
COLO	3180	3090	97	929	915	98
CONN	108	101	94	60	58	97
DEL	31	30	97	17	15	88
FLA	2350	2180	93	1410	1338	95
GA	1975	1650	84	968	860	89
HAW	234	215	92	93	91	98
IDAHO	1870	1900	102	700	725	104
ILL	2950	2850	97	973	983	101
IND	2025	1750	86	780	684	88
IOWA	7800	7300	94	2180	2058	94
KANS	6000	6200	103	1800	1890	105
KY	3120	2600	83	1530	1335	87
LA	1425	1350	95	856	830	97
MAINE	132	124	94	68	65	96
MD	390	370	95	193	193	100
MASS	99	95	96	59	55	93
MICH	1470	1250	85	600	540	90
MINN	3700	3650	99	1400	1380	99
MISS	2130	1790	84	1210	1053	87
MO	6000	5550	93	2700	2560	95
MONT	2680	2607	97	1436	1480	103
NEBR	6500	6450	99	2160	2100	97
NEV	570	560	98	300	295	98
N H	74	68	92	37	34	92
N J	114	108	95	61	58	95
N MEX	1550	1500	97	651	635	98
N Y	1760	1711	97	1000	979	98
N C	1100	1080	98	561	545	97
N DAK	2050	1967	96	1080	1058	98
OHIO	2025	1850	91	853	770	90
OKLA	5900	5300	90	2300	2250	98
OREG	1490	1475	99	715	690	97
PA	1900	1840	97	902	874	97
R I	10.0	9.0	90	6.0	6.0	100
S C	690	575	83	362	305	84
S DAK	3925	3750	96	1638	1560	95
TENN	2700	2350	87	1360	1260	93
TEX	14500	13900	96	6550	6200	95
UTAH	864	810	94	397	389	98
VT	336	320	95	207	199	96
VA	1620	1550	96	750	760	101
WASH	1275	1375	108	540	570	106
W VA	550	535	97	280	270	96
WIS	4100	4100	100	2030	2038	100
WYO	1280	1300	102	595	624	105
U.S.	116375	110864	95	49748	47843	96

APPENDIX D. SELECTED USDA RESEARCH ON MISSING DATA PROCEDURES

D.1. "Missing Data Procedures: A Comparative Study" by Barry L. Ford (August 1976)

D.1.1. Conclusions

This investigation offers a comparison of six missing data procedures as applied to a specific data set taken from USDA list-frame survey. This data set is from a simple stratified sample with a large sample size in each stratum. The following procedures are applied to this data set:

(1) the double sampling ratio procedure,
(2) the double sampling regression procedure,
(3) a hot-deck procedure in which a randomly selected reported item was substituted for each missing item,
(4) the "closest" procedure in which the closest reported item was substituted for each missing item,
(5) the "two closest" procedure in which the average of the two closest report items was substituted for each missing item, and
(6) the "class" mean procedures in which the class mean was substituted for each missing item.

The most important aspect in comparing these missing data procedures is to protect against biases in the estimated means (or totals). An analysis of variance shows no significant differences among the estimated means that result in using these procedures. All the procedures reduce the relative bias that results from accepting the mean of the reported data as an estimate of the population mean. This relative bias is a result of the nonresponse rate and a difference in the variable of interest (e.g., total hogs, total cattle) between the respondents and nonrespondents. The reduction in relative bias averages 15% and varies from 8 to 26%. Considering the low correlations between the auxiliary and primary variables, this reduction is reasonable. Much larger reductions in bias could be obtained if a variable could be found with a high correlation with the variable of interest and if that variable could be easily obtained for the entire sample. Of course, there is a great deal of room for improvement over the 15% reduction, and later research should examine other procedures (including multivariate ones) to evaluate their efficiency.

An important, though secondary, concern is the estimated variances of the estimated means. All of these estimated variances, except those from the ratio and regression procedures, are underestimates of the true variances because they are generally less than the estimated variance that results with *no* missing data in the sample. Furthermore, the degree of underestimation increases as the relative bias increases. This part of the investigation clearly reveals why all of the hot-deck procedures (random, "closest," "two closest," and "class" mean substitution) may be undesirable. It does not seem reasonable to use one of the hot-deck procedures when it does not reduce the bias of the estimated mean any more than the ratio or regression procedure, yet the variance of the mean is greatly underestimated. Probably, there is an underestimation of variance in the ratio and regression procedure, but these results show that it is not nearly as large as in the hot-deck procedures.

The final result of this investigation is a recommendation of the ratio or regression procedures (the effects of these two procedures being indistinquishable). These two procedures have been more theoretically explored than the other procedures. The estimated variances of the estimated means from the ratio or regression procedure reflect better than the other procedures the true quality of the data. Finally, the ratio or regression procedure can be easily implemented into the USDA estimates by simply adding the sample control data to the computer data tapes.

D.2. "Missing Data Procedures: A Comparative Study (Part 2)" by Barry L. Ford (June 1978)

D.2.1. Summary and Conclusions

The missing data procedure now used by USDA for list-frame surveys is (1) to delete the refusals and inaccessibles and reduce the sample size accordingly and (2) to have the commodity statistician edit in values for single missing items. This procedure assumes that the missing data follow the same distribution as the reported data. To improve on this assumption and to provide consistency in editing the single missing items, this study examines six missing data procedures. All six procedures rely upon control data of a high quality. Although this high quality is not important when the missing data are single missing items, it is extremely important to the missing data procedures when there are many refusals and inaccessibles. This study shows that better control data are needed before the USDA can replace the operational procedure.

The use of better control data implies larger correlations between the control variable and the variables reported on the questionnaire. The correlations within the data set used in this study are approximately 0.30, but evidence in this report indicates that correlations should be approximately 0.60 before there is a notable improvement over the operational missing data procedure. Artificial variables having larger correlations with the control variable are incorporated into this report to compare procedures under the hypothesis that better control data can be obtained.

Of the six procedures studied in this experiment, three are slightly superior if the control data are adequate. These three candidates are the ratio procedure using balance repeated replications (BRR), the hot-deck procedure, and the hot-deck procedure using BRR. (When the BRR technique is integrated into a missing data procedure, the product gives maximum insurance against potential biases owing to replicate size, while simultaneously giving unbiased estimates of variance.) These procedures are recommended because of their simplicity or their statistical efficiency.

The statistical results in this report reveal no significant differences in the direct expansions from the missing data procedures except for the expected farrowing questions. The farrowing questions, however, present contradictory evidence. The hot-deck procedure yields the most accurate estimates of the number of expected farrowings in the first quarter and the second least accurate estimates of the number of expected farrowings in the second quarter. Thus, the farrowing questions need further investigation, using current data from several states to resolve the contradiction.

Because the statistical effects of the missing data procedures are only slightly different, the following nonstatistical considerations should also be included:

(1) Initialization: the hot-deck procedure requires initialization, whereas the hot-deck procedure using BRR and the ratio procedure (BRR) do not;

(2) Structure of the data set: the hot-deck procedure requires a randomly ordered data set whereas the other two procedures require a specific, fixed order to the data set.

Both statistical and nonstatistical comparisons will be crucial in a final decision. However, the first priority is the improvement of control data. The quality of the control data in most multiple-frame states is unknown. The correlations between control data and reported data should be monitored in these states. Control data that is adequate for stratification may not be adequate for use with a missing data procedure. Research on obtaining and constructing better control numbers should also be planned. For example, several control variables may be more efficient than just one. Good control information is necessary for a good missing data procedure.

D.3. The Effect of Refusals and Inaccessibles on List Frame Estimates by Chapman P. Gleason and Raymond R. Bosecker (August 1978)

D.3.1. Conclusions and Recommendations

Several observations concerning enumeration techniques and the impact of nonrespondents on list-frame livestock indications arose from this study.

(1) Approximately 40–50% of previous refusals can be converted through personal enumeration in the survey. However, special efforts will also have to be made to avoid losing former respondents.

(2) More work is needed to evaluate the role of individual enumerators (phone or personal) in either causing or preventing refusals and inaccessibles.

(3) Control data were of little value in estimating for nonrespondents because of poor correlations between reported and control data. However, between-survey correlations are high enough to be of assistance in estimations about previous respondents who did not respond in the current survey.

(4) Multiple-frame livestock estimates are generally biased downward because nonrespondent means on the list frame tend to be larger than respondent means.

(5) Nonrespondent means are larger for two reasons.

(a) Larger operations tend to refuse more often than smaller operations (evidenced by the increasing refusal rates for the larger size group strata and larger means for sometime refusals within strata).

(b) The proportion of operations actually having the cattle or hogs of interest is higher among refusals than it is for respondents (one reason operators refuse is because they have livestock and do not want to reveal the number).

(6) Emphasis is currently on increasing the response among large operations. This is necessary because high nonresponse rates among these strata accentuate even small differences between respondent and nonrespondent means and increase bias. However, it is no less important to further improve the response among the operations in smaller size strata because large differences between the means of respondents and nonrespondents in these strata also have a sizable impact on the bias, even with low nonresponse rates.

(7) Large changes in the means of the smaller size group strata can occur because a few positive livestock reports among the large number of zeros in these strata, coupled with large expansion factors, can have considerable effect.

(8) Because nonrespondent means differ from respondent means, charges in survey procedures regarding nonrespondents will affect comparability of survey indications between surveys.

(9) Occasional refusals have a consistently larger proportion of positive reports for the livestock of interest than do those who responded to both surveys. Generally, inaccessibles had, though not consistently, a greater proportion of zero reports. Nonrespondents in the second survey period had a larger proportion of positive reports than did first-survey nonrespondents even though those who reported in both surveys remained constant. This change from one survey period to another among nonrespondents is significant because the between-survey comparability of the estimates is affected. Consistent estimators of change plus estimators that compensate for response should be developed and implemented to control nonsampling error in livestock surveys. One possible approach is the use of a successive sampling regression or ratio estimator for that portion of the sample matched between survey periods coupled with an indication of the effect of nonresponse, such as the estimator based on supplementary evidence of livestock.

D.4. "The Use of Current Partial Information to Adjust for Nonrespondents" by Keith N. Crank (April 1979)

The study by Keith Crank (1979) is based on data from surveys in three states. The data for each state are from a stratified random sample of a list frame, and the estimate of an item total for the state list frame is the sum of the estimates for each of the strata. Suppose that for a given stratum the list frame consists of N operations of which N_1 would be respondents and N_2 would be nonrespondents if canvassed. In a survey of a sample of n operations from the N, n_1 are respondents and n_2 are nonrespondents. Suppose r of these n_1 cases are positive respondents, i.e., they report one or more hogs (or cattle) on the operation; and the average number of animals reported per positive respondent is $\bar{x}_r$. By one method or another (interviewer observation, local knowledge, etc.) it is determined whether or not there are any hogs (cattle) on the operation for n_2 of the $n - n_1$ nonrespondents. Suppose k of these n_2 nonrespondents are classified as positive; then, with the number of hogs (cattle) on the operations not being known, the status of the remaining n_3 $(= n - n_1 - n_2)$ nonrespondents cannot be determined. They are referred to as unknown nonrespondents.

The study makes the following assumption:

ASSUMPTION b: the average number of hogs (cattle) per positive respondent is the same as the average per positive nonrespondent for both known and unknown nonrespondents.

With this assumption, each of the four estimators investigated in the study has the structure

$$Y' = Np\bar{x}_r,$$

where Y' is the estimated total number of hogs (cattle) and p the estimated proportion of positive operations in the list frame. In the current procedure, p is based solely on the respondent cases, i.e., $p = r/n_1$. This estimator would be unbiased only if the proportion of positives among the nonrespondents were the same as among the respondents. For the other three estimators investigated, the study assumes that this is not the case for the nonrespondents as a group.

ASSUMPTION a2: The percentage of positive nonrespondents (i.e., percentage of all nonrespondents who are positive nonrespondents) is different from the percentage of positive respondents (i.e., percentage of respondents who are positive respondents).

To describe the three estimators, let p_1 $(= r/n_1)$, p_2 $(= k/n_2)$, and p_3 denote the proportion of positive operations among the respondents, the known nonrespondents, and the unknown nonrespondents, respectively. Because the sampling within stratum is random, the weighted average of these three proportions

$$(n_1 p_1 + n_2 p_2 + n_3 p_3)/n = (r + k + n_3 p_3)/n$$

would be an unbiased estimator if p_3 could be estimated unbiasedly. The study looks at three alternative assumptions (given in Table 1) as a basis for imputing a value for p_3. Under Assumption C1, it uses for p_3 a weighted average of p_1 and p_2:

$$p_3 = (n_1 p_1 + n_2 p_2)/(n_1 + n_2) = (r + k)/(n_1 + n_2)$$

Under Assumption C2, it takes $p_3 = p_2$. Under Assumption C3, it takes $p_3 = p_1$.

From the C.V.s in Table 1, it was concluded that each of the estimators has approximately the same precision and that the estimator to use is the one believed to have the smallest bias. The validity of each of the assumptions probably depends on the percentage of the sample that is unknown. In the Iowa study, with a nonresponse rate of 19%, the percentage of unknown nonrespondents was reduced by successive waves of effort from 7 to 2.5% of the sample. Assumption C2 appeared to be the best of the three at the 7% level and C3 at the 2.5% level. This suggests that, with relatively high percentages of unknown nonrespondents, the proportion of unknowns that is positive is likely to be closer to that of known nonrespondents, whereas with low percentages, it is likely to be closer to that of respondents.

REFERENCES

Crank, Keith N. (1979). The Use of Current Partial Information to Adjust for Nonrespondents. Economics, Statistics and Cooperatives Service, USDA, Washington, D.C.

Ford, Barry L. (1976). Missing Data Procedures: A Comparative Study. Statistical Reporting Service, USDA, Wsshington, D.C.

Ford, Barry L. (1978). Missing Data Procedures: A Comparative Study (Part 2). Economics, Statistics and Cooperatives Service, USDA, Washington, D.C.

Gleason, Chapman P., and Bosecker, Raymond R. (1978). The Effect of Refusals and Inaccessibles on List Frame Estimates. Economics, Statistics and Cooperatives Service, USDA, Washington, D.C.

CHAPTER **4**

Report on the Survey Research Center's Surveys of Consumer Attitudes

Charlotte G. Steeh
Robert M. Groves
Robert Comment
Evelyn Hansmire

1. DESCRIPTION OF SURVEY PROCEDURES

1.1. Purposes and Survey Measures

The Surveys of Consumer Attitudes are designed to monitor changes in the national economy by studying changes in consumer attitudes and expectations. The proposition that changes in consumer attitudes can indicate changes in the economy as a whole derives from the fact that the U.S. post-World War II economy has become a consumer economy. Traditional explanations of fluctuations in economic activity and in the rate of economic growth have centered on changes in the rates of business investment and government spending. However, the relatively high rate of economic growth from 1945 through the late 1960s has created affluence for the many. The income received by the majority of families is now large enough not only to satisfy minimum needs but also to permit discretionary purchases. As a result, consumers are free to program the gratification of their wants and to speed up or postpone their purchases according to the information they receive about the economy. Consumer demand is thus a function of both ability to buy and willingness to buy. Since good or bad years now depend primarily on consumer demand, attitudinal data have been able to foreshadow large changes in the economy that could not have been predicted from an analysis of such traditional economic variables as income, inventories, past purchases, or liquid assets. Thus motives,

INCOMPLETE DATA
IN SAMPLE SURVEYS
Volume 1, Part II

ISBN 0-12-363901-8

attitudes, and expectations bring about more general economic changes, and measures of willingness to buy prove especially useful in defining the points at which the economy turns toward recession or recovery.

In order for psychological variables to influence the direction and extent of changes in economic trends at a given time, attitudes must change among very many consumers in a similar manner at approximately the same time. Thus a major concern of the Surveys of Consumer Attitudes is the exploration of the process by which economic news is spread throughout the country. The rapid diffusion of news, which especially television makes possible, can cause waves of confidence and surges in spending or waves of uncertainty and widespread postponement of expenditures.

The Surveys of Consumer Attitudes began in the late 1940s as interim surveys between the annual Surveys of Consumer Finances. By 1952, the number of surveys per year had been increased to two or three and a summary measure of change in optimism or pessimism had been constructed in the form of the Index of Consumer Sentiment. From 1961 to 1977, the surveys were conducted on a quarterly basis. Since economic attitudes and expectations in the 1970s have become more volatile than they have ever been in the past, eight monthly surveys (as of 1979) supplement the more extensive quarterly surveys to provide flexible and timely data.

1.2. Population of Inference

Since the Surveys of Consumer Attitudes utilize samples of randomly generated telephone numbers, the survey population is restricted to persons who can be accessed by telephone. The survey population includes all persons 18 years or older in households with telephones in the coterminous United States. Once the sample telephone number has been answered, a question is asked, "Is this a home phone or a business phone?" For those who answer that it is a business phone, further questions are asked to determine whether or not anyone lives on the premises, even though it is a business. These questions are an attempt to include persons living in mixed residential–business structures. If six or more adults are reported living in the unit, questions are asked that attempt to implement the housing unit definition used by the Survey Research Center in personal interview surveys.[1]

The use of housing units as sampling elements in a telephone survey is subject to greater coverage problems than those encountered in personal interview surveys because of the inability to make visual observations of the sample unit. There has been no study of coverage of the survey population by the telephone survey procedures as of this date.

[1] See Appendix A for the procedures used by interviewers and for the definition of "housing unit."

1.3. Sampling Procedures

Telephone numbers in the United States have three parts: a three-digit area code, a three-digit central office code, and a four-digit number as

— — —	— — —	— — — —
Area code	Central office code	Four-digit number

The identity of all area code–central office code (AC–COC) combinations currently in service can be obtained from the Long Lines Division of A.T.&T. The data record of the A.T.&T. tape is an area code–central office code combination (e.g., 313–764); thus each record represents 10,000 distinct telephone numbers (e.g., 313-764-0000 through 313-764-9999). Stratification of the frame, therefore, cannot be accomplished below the level of the central office code, and the variables available for stratification are few. Since a systematic design is used, the stratification implicit in the design is specified by the ordering of the central office code records. The following was performed:

(1) Group together all central office codes in the same telephone exchange.

(2) Group together exchanges within the same area code.

(3) Form size categories of exchanges using number of central office codes in an exchange. These categories are proxy categories for levels of population density within the exchanges.

(4) Within each area code and each exchange size grouping, order exchanges using the geographical coordinates provided on the tapes; within codes, rotate geographical order across size groups (this results in a serpentine geographical ordering across exchange size groups—Northwest to Southeast, Southeast to Northwest, Northeast to Southwest, etc.).

(5) Order area codes geographically, keeping states and major census regions intact.

The sample design that is used is described by Waksberg[2] and is a stratified, clustered design in which primary selections are groups of 100 consecutive telephone numbers; there are 100 such clusters in every central office code (0000–0099, 0100–0199, etc.). Given this ordering of the frame, the sample is drawn by taking a random start between 1 and the selection interval specified by the desired sample size. When a central office code designated for a selection has been determined by application of the selection interval and a random start, a four-digit random number is machine-generated and appended to the AC–COC combination, yielding the ten-digit sample telephone number.[3] For

[2] Waksberg, J. (1978). Sampling methods for random-digit dialing. *Journal of the American Statistical Association* 73: 40–46.

[3] This design could be viewed as a two-stage sample: first, selecting area code–central office code combinations systematically; then, randomly sampling one secondary selection, a single telephone number in each first-stage unit.

example, if 4424 is generated, the cluster containing numbers 4400–4499 is tentatively designated for selection. At this point all clusters are given equal probabilities since all have exactly 100 numbers. These initial numbers could be termed "primary numbers." These primary numbers are then called. If they are not working household numbers, the clusters in which they fall are not selected. If they are working household numbers, their clusters are selected into the sample and a specified number of additional four-digit numbers ("secondary numbers") within the same cluster is generated. For example, if the desired cluster size is nine, eight more four-digit numbers within the same hundred series would be generated. Each of these eight would also be called, and if any yielded a nonresidential or nonworking number, it would be replaced by a newly generated four-digit number in the same hundred series, until ultimately eight more working household numbers were generated.

Let us now examine the probabilities of each number being selected into the sample. In the first stage, we sample individual clusters of 100 telephone numbers; since we discard those primary numbers that are not working household (HH) numbers, the probability of each cluster being selected is proportional to the number of working household numbers in the hundred series (to be discussed later). In the second stage, a fixed number of working household numbers is sampled from among all working household numbers in the hundred series, so that the probability of a number in a chosen cluster being selected is inversely proportional to the number of working household numbers in the hundred series. The product of the probabilities of the two stages is a constant for all numbers in working area code–central office code combinations; the sample yields an epsem design[4]:

$$(\#\text{ working HH nos.}/k) \times (b/\#\text{ working HH nos.}) = b/k,$$

where the desired cluster size b and the sampling constant k are chosen to fit the needs of the particular study.

Several properties of this design are important for our consideration.

(1) The design will always eliminate clusters that have no working numbers in them, thus increasing the proportion of working household numbers generated in each sample. In other words, the proportion of working household numbers is increased because the probability that another number in the same hundred series is a working household number is greater than the unconditional probability that a generated number is a working household number. The design takes advantage of the fact that as the central office code fills up with working numbers, numbers are generally assigned into the same hundred and thousand series (e.g., the company may begin assigning 2000 series first, filling that up before going to the next thousand series). The smaller the proportion of active hundred series in a central office code, the greater is the advantage of the cluster design.

[4] Epsem samples are those giving each population element an equal chance of selection.

(2) Departures from an epsem design can occur for two different reasons. Some clusters may not contain the desired number of working household numbers (in the preceding example, there may be a cluster of 100 numbers with only 1–8 working household numbers). In addition, a small percentage of nonworking numbers give ringing tones when dialed, and thus the investigator will not know whether to replace the number by a new sample number from the same hundred series.

(3) The sample is clustered into groups of hundred series of single central office codes, and a loss of precision attributable to the clustering will occur. There is little documentation on the causes of numbers sharing the same hundred series; they may have come into service at the same time[5] or they may share the same type of service (e.g., touch-tone telephone, party lines). In any case, the sources of homogeneity among numbers in the same hundred series seem smaller than do those for clusters in most areal probability samples.

Table 1 presents the status of all sample numbers in a fall 1976 implementation of this design. About 24% of all primary numbers generated were working household numbers. The figures in the table present the results for the numbers generated in the second stage; they show the proportion of working household numbers within clusters to be 66.5%.[6] Combining the results from the primary and secondary selections of the clustered design produces an overall 55.8% of all numbers yielding a household.

Not all unassigned numbers are connected to nonworking number recordings or operator intercepts. There are three classifications of responses that cause some problems in administering any random-digit-dialing (RDD) sample: (1) "Wrong connections" are numbers answered by people who report that their number is different from the one dialed; to verify that misdialings had not occurred,[7] at least two different calls were made to these numbers before discarding them. (2) Some numbers when dialed repeatedly yielded nothing—no ringing, no busy signal, no recording; in Table 1, these are labeled "no result from dial." (3) The last category of nonworking numbers is "fast busys." The busy signal received when an assigned number is in use is a tone interrupted 60 times per minute; a "fast busy," or "reorder," signal is the same tone interrupted 120 times per minute. Sometimes this can indicate that the circuits in the area

[5] However, it is the practice of some companies to recycle disconnected numbers so that hundred series experiencing some outmobility would receive new residents.

[6] We used the primary selections to pretest the interview schedule. In the actual field period, only the secondary selections were called. Replacement numbers were machine-generated upon nightly disposition of numbers.

[7] Series of numbers within a central office code may be "bridged" together, so that those telephones can be reached by two different numbers. For example, if the 2000 and the 9000 series are bridged, the telephone with number 787-2111 can also be reached by 787-9111. This permits the company to utilize some standard test numbers in the 9000 series without opening the entire thousand series to use. Usually the customers are completely unaware of the fact that either number can be used to call their telephones. Before an interview was taken, the interviewer verified that the sample number was actually reached.

TABLE 1

Final Disposition of Sample Numbers and Type of Nonworking Number for Stratified Cluster Sample (Fall 1976)[a]

Disposition of sample number	*All sample numbers* %	*Nonworking numbers* %
Working household numbers (answered)	65.8	—
Ring, no answer numbers (confirmed household)[a]	.7	—
Ring, no answer numbers (status unconfirmed)[a]	.0	—
Nonresidential numbers	10.5	—
Nonworking numbers	20.8	90.5
Wrong connections	1.8	7.8
No result from dial	.3	1.3
Fast busys	.1	.4
Total	100.0	100.0

[a] The total number of clusters ultimately selected was 104. Each cluster had eight working household numbers generated.

called are busy, but some nonworking numbers are also connected to these signals. The numbers so designated in Table 1 received fast busy signals at least five consecutive times throughout the month-long field period of the study.

These three categories of numbers are separated from the nonworking number group that received recordings or operator intercepts because there is a small probability that each contains some working household numbers. "Wrong connections" could be answered by people who misled interviewers; "no result from dial" numbers may have been repeatedly incompletely dialed so that the system was awaiting another digit; "fast busy" could be numbers whose local system was overloaded or malfunctioning through the whole field period. We hoped to minimize these responses, and we believe that the vast majority of these numbers were nonworking. It is important to note that among nonworking numbers a large portion were confirmed nonworking through recordings or operator intercepts (90.5%).

As noted in Table 1, some of the sample numbers yielded a ringing tone without response on each call. In the fall 1976 administration of the sample, we were able to verify the status of numbers that repeatedly yielded ringing without answer (and the final response rate was 71%). In all but a few of the primary numbers, this was done after more than 12 calls had been made to the primary sample number. Twenty primary numbers continually yielded ringing; 19 of those were nonworking numbers. Among the secondary selections, however, the results were somewhat different. Verification was made on 32 numbers, 30 of which had been called over 17 times. Seventeen, or 53.1%, of the numbers were classified as working household numbers, and four of them eventually yielded interviews. These 32 numbers represent 2.6% of all secondary sample numbers generated; they fall in 28 of the 104 clusters, four clusters having two of the numbers.

The random-digit-dialing sample just described is supplemented by rotating panels. From the summer of 1976 to the end of 1977, the specific samples reinterviewed in a particular survey were chosen in order to determine whether the responses and demographic characteristics of persons first contacted in a personal interview differed from those of people first contacted by telephone. At the beginning of monthly interviews in 1978, the panel design became systematic, and only reinterviews of random-digit-dialing samples were scheduled, at three month and twelve month intervals. Thus, any one survey included respondents being contacted for a first time, a second time (three months after the original interview), and a third time (twelve months after the original interview). Presently, respondents are recontacted only once after six months, and the total sample for one survey is usually made up of 56% new respondents, drawn independently by random-digit-dialing procedures, and 44% recontact respondents. This design permits the regular measurement of change at the aggregate and the individual level.

In eight of the twelve monthly surveys about 900 interviews are taken. In the remaining four surveys, spaced at quarterly intervals, there are approximately 1400 interviews. The larger sample size permits subsetting of data and allows detailed breakdowns by region, income, sex, and the other collected demographic variables.

1.4. Data Collection Procedures

The interviews for the Surveys of Consumer Attitudes are taken by telephone interviewers within the Ann Arbor offices of the Survey Research Center. The telephone interviewing facility consists of 18 separate, acoustically designed telephone interviewing stations, two monitoring stations, and a supervisor's station, also equipped with monitoring equipment. Twelve WATS lines are currently used in the monthly and quarterly Surveys of Consumer Attitudes, with a total of about 40 interviewers involved in data collection.

Interviewers are given about one week of training in interviewing techniques, persuasion, question delivery, and probing, along with administrative details of the job. This general training is followed by specific training in the particular survey that they will conduct. Thus, interviewers used on the Surveys of Consumer Attitudes are also used on other telephone surveys conducted by the Survey Research Center. Training is continued throughout the employment of the interviewer through the use of feedback from telephone monitors. Attempts are made to monitor at least one interview a week conducted by each interviewer. Monitors use unobtrusive listening to the interview to evaluate the interviewer's introductory comments, methods of persuasion, and use of the questionnaire. When the monitored interview has been completed, the monitor and interviewer meet to discuss the desirable and undesirable characteristics of the monitored interview. These feedback sessions are used to give attention and further training on aspects of the interviewing that pose difficulties for individual interviewers.

In order to obtain as many interviews as possible, interviewers on the Surveys of Consumer Attitudes attempt to convert initial refusals to interviews by calling the telephone household a second time. In addition, they make unlimited callbacks to sample numbers that ring but are not answered.

Refusal conversions are tried whenever a person—either the respondent or someone else in the household—refuses to be interviewed on the initial contact (Result Codes 31–33).[8] Because not all of the sample numbers that produced refusals can be contacted successfully a second time before the end of the study, most of the percentages listed in Table 2 are based on the number of completed refusal conversions, that is, on those attempts that resulted in another conversation with a person who either refused again or granted an interview. In rare cases, enough additional information was provided to permit the reclassification of the refusal to a circumstantial noninterview (Result Code 63) or a nonsample category (Result Codes 73–91). The last column of Table 2 lists the percentage-point increase in the response rate that occurred as a result of refusal conversions for each Survey of Consumer Attitudes from 1977 to 1979. It appears that, on average, refusal conversions make a minimal contribution to survey response.

The policy of unlimited callbacks produced the cumulative distributions presented in Table 3 for working household telephone numbers and in Table 4 for nonsample (both nonresidential and nonworking) numbers over the 17 surveys from November 1976 through December 1978. Surprisingly, it appears that telephone numbers in both random-digit-dialing (RDD) and recontact samples take the same amount of effort for final disposition. The median number of calls made to a number in either a RDD or a recontact sample segment is approximately three. For nonsample numbers in both types of sample segments, the median is two. Roughly 75% of all working numbers are disposed of within five calls. Nonsample numbers reach final disposition much more quickly, with only two calls required for 75% of the RDD cases and three calls for 75% of the recontact cases. Most nonsample numbers, it should be remembered, come from the RDD sample segment.

On the average, 5% of working numbers in both sample types and 3% of nonsample numbers require 16 or more calls for final disposition. The return on this effort is modest, but there does seem to be some difference based on sample type. The average percentage interviewed over the surveys from May 1977 through December 1978 for numbers requiring 16 or more calls is 23.6% for

[8] See Appendix B for definitions of result codes.

TABLE 2

Refusal Conversions by Survey (1977–1979)

Survey	Sample size (1)	*Refusal conversions* Number completed (2)	Interviews obtained (3)	Interview rate[a] (4)	Number of other interviews (5)	Increase in total response rate[a] (6)
1977						
February	2,350	151	46	30.5	1,157	2.0
May	2,753	186	51	27.4	1,317	1.9
August	2,693	91	24	26.4	1,191	.9
November	2,427	167	50	29.9	1,231	2.1
1978						
January	1,156	25	3	12.0	690	.3
February	2,376	149	46	30.9	1,231	1.9
March	1,309	66	25	37.9	768	1.9
April	1,143	64	21	32.8	723	1.8
May	2,509	257	65	25.3	1,233	2.6
June	1,141	69	18	26.1	683	1.6
July	1,302	33	5	15.2	753	.4
August	2,285	193	45	23.3	1,140	2.0
September	1,299	77	25	32.5	735	1.9
October	1,494	40	8	20.0	748	.5
November	2,756	182	58	31.9	1,402	2.1
December	1,219	125	33	26.4	737	2.7
1979						
January	1,456	105	22	21.0	862	1.5
February	2,505	284	64	22.5	1,297	2.6
Total	34,173	2264	609	26.9	17,898	1.8
Mean	1,899	126	34	26.2	994	1.7

[a] Interview rate $= \frac{(3)}{(2)} \times 100$ where the numbers in parentheses refer to the table columns.

[b] The increase is the difference $\left[\frac{(3)+(5)}{(1)} \times 100\right] - \left[\frac{(5)}{(1)} \times 100\right]$.

RDD sample segments but 40.4% for recontact segments. Thus, the policy of unlimited callbacks yields a greater return for recontact samples. Over these 15 surveys, calling RDD and recontact numbers 16 or more times increased the total survey response rate by only 1.2 percentage points on the average.

Since attrition from recontact samples is often a serious problem, special efforts have been made to minimize this effect. Table 5 describes the latest experiment carried out by the Surveys of Consumer Attitudes. Usually each successive wave of a panel is composed of only those respondents who actually granted an interview on the previous contact. From January through April 1978,

TABLE 3

Cumulative Distribution of Working Household Telephone Numbers by Calls Needed for Final Disposition, Surveys of Consumer Attitudes (1976–1978)

Number of Calls — RDD

Survey	*1*	*2*	*3*	*4*	*5*	*6*	*7*	*8*	*9*	*10*	*11*	*12*	*13*	*14*	*15*	*16+*	*Max-imum*	*Me-dian*
1976																		
November	19	39	56	65	72	77	81	85	88	90	92	92	93	94	94	100	23	3
1977																		
February	24	43	59	68	74	80	84	86	88	89	90	91	92	93	94	100	25	3
May	17	34	48	58	65	72	76	80	83	86	89	90	91	92	93	100	62	4
August	19	37	50	58	65	72	77	80	84	86	88	89	90	91	92	100	47	4
November	22	45	57	68	74	80	83	86	88	90	91	93	93	95	95	100	50	3
1978																		
January	32	58	73	80	85	88	91	92	93	94	95	97	98	98	99	100	18	2
February	23	42	57	67	75	81	84	87	90	91	92	93	94	95	96	100	44	3
March	28	46	61	71	79	82	86	89	91	93	95	95	97	98	98	100	22	3
April	27	48	61	70	77	81	85	88	91	92	93	94	95	96	96	100	42	3
May	20	39	53	63	71	75	79	82	85	87	89	90	91	92	93	100	61	3
June	26	43	58	67	74	79	83	86	87	90	90	92	93	94	95	100	34	3
July	23	40	52	60	67	73	77	79	82	85	86	87	88	90	91	100	99	3
August	14	29	40	51	58	65	71	76	80	82	84	85	87	88	90	100	39	4
September	25	43	58	67	74	79	85	88	90	92	94	95	95	95	96	100	50	3
October	24	48	61	72	80	84	89	91	93	94	96	96	97	98	98	100	40	3
November	23	39	53	64	71	78	82	86	88	90	91	93	94	94	95	100	48	3
December	22	44	60	70	79	83	87	90	91	92	94	94	95	96	97	100	32	3
Mean	23	42	56	66	73	78	82	85	88	90	91	92	93	94	95	100	—	3

Number of Calls — *Recontact*[a]

Survey	*1*	*2*	*3*	*4*	*5*	*6*	*7*	*8*	*9*	*10*	*11*	*12*	*13*	*14*	*15*	*16+*	*Max-imum*	*Me-dian*
1976																		
November	25	47	62	71	78	82	86	88	90	91	93	94	94	95	95	100	24	3
1977																		
February	24	45	60	69	75	81	85	87	90	91	93	93	94	96	96	100	25	3
May	16	32	46	57	65	71	74	78	82	84	86	88	90	91	92	100	62	4
August	23	42	55	65	72	78	82	85	87	88	90	92	93	94	95	100	39	3
November	19	36	51	61	69	76	80	83	86	88	90	92	93	94	94	100	42	3
1978																		
January	32	60	75	81	85	88	91	94	96	97	98	99	99	99	99	100	26	2
February	20	38	53	63	70	76	81	85	88	91	92	93	94	95	96	100	28	3
March	31	52	62	72	79	84	88	90	94	95	96	97	98	98	99	100	21	2
April	30	55	67	72	80	82	85	88	91	93	94	96	96	97	97	100	26	2
May	15	33	46	56	64	69	74	77	80	82	84	86	88	90	91	100	52	4
June	30	50	61	70	76	80	83	85	86	87	88	90	91	91	92	100	28	3
July	25	45	56	66	71	75	80	83	85	87	89	91	92	94	94	100	99	3
August	19	37	50	60	68	74	78	82	84	87	89	90	91	93	94	100	39	4
September	25	43	57	66	73	78	82	85	87	90	92	94	94	95	96	100	34	3
October	23	42	57	67	73	77	83	86	89	91	93	94	96	96	98	100	21	3
November	23	45	61	70	77	82	86	88	90	91	93	94	95	96	97	100	37	3
December	27	49	60	73	78	83	86	88	90	92	93	94	94	94	95	100	33	3
Mean	24	44	57	67	74	79	82	85	88	90	91	93	94	95	95	100	—	3

[a] The recontact segments for November 1976, August 1978, and November 1978 are entirely composed of respondents from personal interviews. The segments for February, May, and August of 1977 are partially composed of personal interview respondents.

TABLE 4

Cumulative Distribution of Nonsample Telephone Numbers by Calls Needed for Final Disposition, Surveys of Consumer Attitudes (1976–1978)

	Number of Calls																																			
	RDD																		*Recontact*[a]																	
Survey	*1*	*2*	*3*	*4*	*5*	*6*	*7*	*8*	*9*	*10*	*11*	*12*	*13*	*14*	*15*	*16+*	*Maximum*	*Median*	*1*	*2*	*3*	*4*	*5*	*6*	*7*	*8*	*9*	*10*	*11*	*12*	*13*	*14*	*15*	*16+*	*Maximum*	*Median*
1976																																				
November	53	79	86	89	91	92	94	94	94	95	95	95	95	96	96	100	25	1	29	46	64	75	79	79	82	82	82	86	86	86	89	89	89	100	23	3
1977																																				
February	46	77	85	87	89	92	92	94	94	94	94	94	95	95	95	100	25	2	49	69	84	88	90	92	94	94	94	95	96	97	98	99	99	100	25	2
May	34	70	75	81	84	86	87	87	89	89	90	91	92	94	95	100	33	2	36	58	70	74	75	76	81	85	85	86	88	90	90	91	91	100	42	2
August	43	69	76	82	84	86	89	90	91	92	94	95	95	97	97	100	26	2	53	67	78	84	86	89	93	93	93	96	98	98	98	98	98	100	19	1
November	41	67	77	80	84	88	90	92	94	96	97	98	99	99	99	100	25	2	43	58	70	83	87	89	89	92	92	92	95	95	95	95	95	100	20	2
1978																																				
January	50	82	87	90	91	93	94	96	97	97	98	99	99	100	—	—	14	2	35	76	87	91	91	93	98	100	—	—	—	—	—	—	—	—	8	2
February	46	74	81	84	86	89	90	91	92	93	93	94	94	95	97	100	24	2	38	70	85	87	88	88	97	97	97	98	98	100	—	—	—	—	12	2
March	51	79	84	87	88	92	93	94	94	95	97	98	99	99	99	100	18	1	57	68	82	86	89	93	96	96	96	100	—	—	—	—	—	—	10	1
April	56	76	83	86	88	92	92	93	94	94	95	95	95	96	96	100	23	1	50	73	73	80	83	87	90	93	93	93	93	93	98	98	98	100	17	2
May	46	72	79	82	83	86	88	89	90	92	92	93	94	94	94	100	45	2	30	72	78	82	82	84	88	88	92	92	92	92	94	94	94	100	23	2
June	39	70	76	79	82	86	88	89	91	93	94	95	95	96	97	100	22	2	48	62	76	81	91	95	95	95	95	100	—	—	—	—	—	—	10	2
July	46	82	89	91	93	94	94	96	97	98	98	98	98	98	98	100	34	2	19	48	63	70	74	89	93	96	96	96	96	100	—	—	—	—	12	3
August	51	71	82	84	88	89	90	91	92	93	93	94	95	95	95	100	36	1	40	71	80	88	89	91	92	93	96	96	96	96	96	96	96	100	23	2
September	50	74	80	82	84	87	88	90	92	93	93	94	95	95	95	100	32	2	20	51	61	66	76	88	90	90	90	90	95	95	95	95	95	100	35	2
October	45	72	78	81	84	86	89	90	90	91	91	92	92	93	94	100	33	2	43	60	71	74	79	83	86	86	88	95	95	95	100	—	—	—	13	2
November	43	71	78	81	84	86	89	90	92	92	93	95	96	96	97	100	28	2	48	66	80	85	88	91	94	96	96	98	98	99	99	100	—	—	14	2
December	76	86	90	92	93	95	96	96	96	97	97	97	97	97	98	100	20	1	39	48	61	65	65	65	74	78	83	91	91	96	100	—	—	—	13	3
Mean	48	75	81	85	87	89	91	92	93	94	94	95	96	96	97	100	—	2	39	62	74	79	83	86	90	91	92	94	95	96	97	97	97	100	—	2

[a] The recontact segments for November 1976, August 1978, and November 1978 are entirely composed of respondents from personal interviews. The segments for February, May, and August of 1977 are partially composed of personal interview respondents.

TABLE 5

Effect of the NIO Experiment on Panel Response and Refusal Rates

	Survey						
Parameter	*January 1978*	*February 1978*		*March 1978*	*April 1978*	*Total*	*Mean*
Date of panel origin	February 1977	August 1977	May 1977	November 1977	January 1977		
Wave (recontact)	1st	1st	2nd	1st	1st		
Number of previous NIOs	72	129	33	66	88	388	78
Number of interviews	14	30	11	10	22	87	17
Percentage of interviews	19.4	23.3	33.3	15.2	25.0	22.4	23.2
Number of refusals	23	46	3	9	4	85	17
Percentage of refusals	31.9	35.7	9.1	13.6	4.5	21.9	19.0
Total interviews	316	369	279	403	330	1697	339
Total refusals	155	299	173	241	121	989	198
Sample size	586	773	673	738	571	3341	668
Increase in panel response rate[a]	2.4	3.9	1.7	1.3	3.9	2.6	2.6
Increase in panel refusal rate[a]	3.9	6.0	.5	1.2	.7	2.5	2.5

[a] The increase is the difference between the response and refusal rates calculated including the interviews and refusals obtained from the NIO experiment and the response and refusal rates calculated excluding them.

however, the recontact sample segments also included those telephone numbers that had fallen into the "noninterview other" (NIO) category (Result Codes 62–69) when they were last called. The NIO category is used when respondents cannot be interviewed for circumstantial reasons such as illness, vacation, or language problems. In addition, numbers that rang but were never answered fall into this classification. Row 5 of Table 5 lists the percentage of these numbers that produced an interview for each survey. The average percentage interviewed over the four surveys (22.4%) compares favorably with the average on refusal conversions (26.9%) given in Table 2. Obtaining interviews from numbers previously classified as NIO increased the response rate for these recontact samples by an average of 2.6 percentage points. This exceeds the 1.8 average increase in survey response obtained through refusal conversions. Unlike refusal conversions, however, the NIO experiment also increased the refusal rate for these recontact samples by a similar average (2.5 percentage points).

This effort to increase response rates for recontact samples ended in May 1978. It was felt that the gain in response did not justify the cost. However, comparisons with the principal method for increasing response—refusal conversions—suggest that including NIO numbers in recontact samples may be more productive.

1.5. Stages of Data Processing

Because data are analyzed and presented within days of the finish of interviewing, data processing required prior to analysis must be fully automated. The flowchart in Appendix C describes the stages in detail. Some parts of this data processing (that represented by the first stages A–F) are devoted to error checking and editing the recently completed data, others (G–I) merge those data with data collected on the same individuals, and still others (J–S) adjust for missing data and introduce selection weights.

1.6. Postsurvey Adjustments

1.6.1. Household Weights

From the mid-1960s through 1971, the sample for these surveys was limited to heads of households, defined always as the husband in the case of households with married couples, and the data were thought to be self-weighting. In special cases—e.g., the husband's going out of town for the study period—a wife was allowed to become the respondent. Beginning in 1972, the sample was redefined to include all adults 18 years old and over. The household weights were designed to maintain consistency between these two types of respondent selection.

If a respondent is not a head of household or a wife, then a household weight of zero is assigned for purposes of analysis on the grounds that this person cannot be depended on to represent the attitudes of the household. The remaining individuals are then weighted by the number of adults 18 and over living in the household. If the respondent is a single head of household, then the household weight is doubled. Presumably this double weight serves as a rough correction for a differentially low response rate among this relatively young segment of the population and prevents the sample from being biased toward households containing married couples.

1.6.2. Telephone Weights

The change from personal to telephone interviewing and the introduction of a rotating panel design, both of which occurred in the summer of 1976, led to the development of telephone weights. The differential mortality associated with recontact samples and the possible bias connected with the coverage of a telephone survey indicated the need for an age- and income-based weighting scheme. Here the age-by-income bivariate distribution[9] for the current sample

[9] Seven categories of age of respondent, including missing data as a separate category, are crossed with six categories of family income, including a separate category for missing data.

(unweighted) is compared with a corresponding distribution from a recent, self-weighting personal interview sample. Individuals who fall into a cell that is underrepresented relative to the size of the "target" sample in the current sample's distribution are given a proportionately greater weight than are individuals who fall into a cell that is overrepresented. Specifically, the individual's telephone weight is obtained by dividing the proportion of the target sample in the individual's cell by the proportion of the current sample that falls in the same cell. As a result of weighting with these ratios, the current sample will show the same bivariate age-by-income distribution as the self-weighting target sample. This criterion weighting is conducted separately for the RDD and the recontact portions of the current sample since the cell-by-cell differences in response rates are not likely to be the same.

Finally, all individuals being interviewed are weighted by a multiplicative constant to adjust for differences in the probabilities of selection between the RDD and recontact portions of the sample. This last factor is calculated as a ratio of the different sampling fractions for the two parts of the sample.

1.6.3. Average Correlation Weights

Since 1977, an additional weight has been devised in which an average correlation is obtained between the answers of recontact respondents to the five index questions at the time of the previous interview and their responses in the current survey. This average correlation is subtracted from 1, and the difference is used as a weight for RDD respondents only. If attitudes are not changing, the correlation will be high and little weight will be given to the "fresh" portion of the sample. When attitudes are changing rapidly, the correlation will be low, and the responses of RDD respondents will have nearly equal weight. In practice, the average correlation in attitudes for recontact respondents over time does not differ substantially, ranging from .2 to .4, and the effect of this factor is minimal.

1.6.4. Combined Weights

All of the factors just noted are multiplied together for a final weight. Each individual's weight is then divided by the smallest weight value for any individual, so that the minimum nonzero weight is 1. The weights are then rounded and truncated to integers. Individuals with a weight in excess of 25 are reassigned a weight of 25. This generally applies to only two or three cases per survey. The average combined weight ranges between 4 and 5 (for nonzero cases). Approximately 10% of all individuals are given a weight of 0, and less than 5% of all individuals receive a weight of 10 or more.

1.7. Index Calculations

The Index of Consumer Sentiment is derived from the following five questions whose inclusion is well-grounded in theory:

1. We are interested in how people are getting along financially these days. Would you say that you and your family are better off or worse off financially than you were a year ago?
2. Now, looking ahead, do you think that a year from now you will be better off financially, or worse off, or just about the same as now?
3. Now, turning to business conditions in the country as a whole, do you think that during the next twelve months we'll have good times financially or bad times or what?
4. Looking ahead, which would you say is more likely, that in the country as a whole we'll have continuous good times during the next five years or so, or that we will have periods of widespread unemployment or depression or what?
5. About the big things people buy for their homes, such as furniture, refrigerators, stoves, television, and things like that: For people in general, do you think now is a good time or a bad time to buy major household items?

The first question measures attitudes toward personal financial progress over the past year, and the second question taps expectations of progress during the coming year. These questions are included because the meaning of income level is subjective. For example, an income increase may appear unsatisfactory if it is less than others are receiving or if it is believed to be temporary. Since consumer sentiment encompasses more than attitudes toward personal finances, the next two questions ask about the prospects for the economy over the next year and the next five years. These two questions probe general levels of confidence among consumers. The last question on current market conditions for large household durables defines yet a third dimension that helps to pinpoint changes in consumer sentiment.

The Index of Consumer Sentiment is calculated using

$$2.7 + \frac{1}{6.7558}\left[\sum_{j=1}^{5} 100\left(\frac{\sum w_i y_{ij}}{\sum w_i x_{ij}} - \frac{\sum w_i y'_{ij}}{\sum w_i x_{ij}}\right) + 100\right],$$

where

$$y_{ij} = \begin{cases} 1 & \text{if the } i\text{th respondent chooses "better" or "good" as the answer to the } j\text{th question,} \\ 0 & \text{otherwise;} \end{cases}$$

$$y'_{ij} = \begin{cases} 1 & \text{if the } i\text{th respondent chooses "worse" or "bad" as the answer to the } j\text{th question,} \\ 0 & \text{otherwise;} \end{cases}$$

$$x_{ij} = \begin{cases} 1 & \text{if the } i\text{th respondent has nonmissing data on the } j\text{th question,} \\ 0 & \text{otherwise;} \end{cases}$$

w_i is the value of the weight assigned to the ith respondent;

2.7 the adjustment used to shift from a head-only sample to a sample of heads and wives; and

6.7558 the adjustment to make spring 1966 the base year.

The w_is for individuals are currently the combined weights just described (the household weight × the telephone weight × the average correlation weight). Table 6 compares weighted and unweighted values of the index. The 2.7 addition to the aggregate index value (not the w_is for individuals) adjusts for the acceptance of wives as respondents after 1971. From the mid-1960s through

TABLE 6

The Index of Consumer Sentiment

Date by year and quarter or month	*Unweighted index* (*1*)	*Household weighted index* (*2*)	*Telephone weighted index* (*3*)	*Combined household and telephone weighted index* (*4*)	*Difference weighted–unweighted* (*4*) – (*1*)
1976					
1	84.7	NA	NA	84.5	– .2
2	NA	NA	NA	82.2	NA
3	91.9	91.3	89.6	88.8	–3.1
4	88.6	88.3	87.4	86.0	–2.6
1977					
1	89.8	91.2	NA	87.5	–2.3
2	89.8	90.2	NA	89.1	– .7
3	91.4	89.8	87.6	87.6	–3.8
4	86.4	86.5	83.2	83.1	–3.3
1978					
1	86.9	86.6	83.8	83.7	–3.2
2	89.9	88.8	86.3	84.3	–5.6
3	82.5	81.3	80.2	78.8	–3.7
4	84.6	84.9	82.3	81.6	–3.0
5	85.5	84.0	84.0	82.9	–2.6
6	79.1	80.1	79.4	80.0	.9
7	84.4	84.3	82.0	82.4	–2.0
8	82.3	82.1	78.7	78.4	–3.9
9	82.8	81.5	81.1	80.4	–2.4
10	81.8	80.8	81.1	79.3	–2.5
11	77.5	77.0	74.0	75.0	–2.5
12	68.8	68.0	66.7	66.1	–2.7
1979					
1	75.1	74.1	73.3	72.1	–3.0
2	77.8	77.0	74.9	73.9	–3.9

1972, interviews were accepted with the wife only when the husband was unavailable. Analyses of the responses of husbands and wives over six surveys conducted in 1968, 1969, and 1972 revealed that the index score for wives was, on average, 2.7 less than the index score for husbands. This constant, therefore, serves to compensate for the systematic inclusion of relatively pessimistic wives in the sample.

1.8. Sampling Variance of Index

The sampling variance of the estimator can be expressed as

$$\frac{(100)^2}{(6.7558)^2}\left\{\sum\left[\operatorname{var}\left(\frac{\sum w_i y_{ij}}{\sum w_i x_{ij}}\right) + \operatorname{var}\left(\frac{\sum w_i y'_{ij}}{\sum w_i x_{ij}}\right) - 2\operatorname{cov}\left(\frac{\sum w_i y_{ij}}{\sum w_i x_{ij}}, \frac{\sum w_i y'_{ij}}{\sum w_i x_{ij}}\right)\right]\right.$$
$$- \sum_{j \neq k} \operatorname{cov}\left(\frac{\sum w_i y_{ij}}{\sum w_i x_{ij}}, \frac{\sum w_i y'_{ik}}{\sum w_i y_{ik}}\right)$$
$$+ \sum_{j \neq k} \operatorname{cov}\left(\frac{\sum w_i y_{ij}}{\sum w_i x_{ij}}, \frac{\sum w_i y_{ik}}{\sum w_i x_{ik}}\right)$$
$$\left. + \sum_{j \neq k} \operatorname{cov}\left(\frac{\sum w_i y'_{ij}}{\sum w_i x_{ij}}, \frac{\sum w_i y'_{ik}}{\sum w_i x_{ik}}\right)\right\},$$

using the same notation as that given in Section 1.7 for the calculation of the index.

2. MISSING AND INCOMPLETE DATA

2.1. Response and Nonresponse Rates in the Surveys of Consumer Attitudes

Figures 1–4 present empirical evidence of changes in response and nonresponse rates for the Surveys of Consumer Attitudes (SCAs) over time. The response rate is the ratio of the number of completed interviews to the number of eligible sample units, and the nonresponse rate is the ratio of the number of refusals or the number of other types of noninterviews (e.g., illness, inability to speak English, not at home) to the number of eligible sample units.

Figure 1 depicts long-term change by plotting rates for those surveys in which respondents were being interviewed for the first time—either in a personal interview or, after the second quarter of 1976, by telephone as part of a RDD sample.

In Fig. 2 response rates for the telephone surveys from 1976 to 1979 have been graphed for sample segments. The response rates and the components of non-response for RDD sample segments have been plotted separately in Fig. 3. The

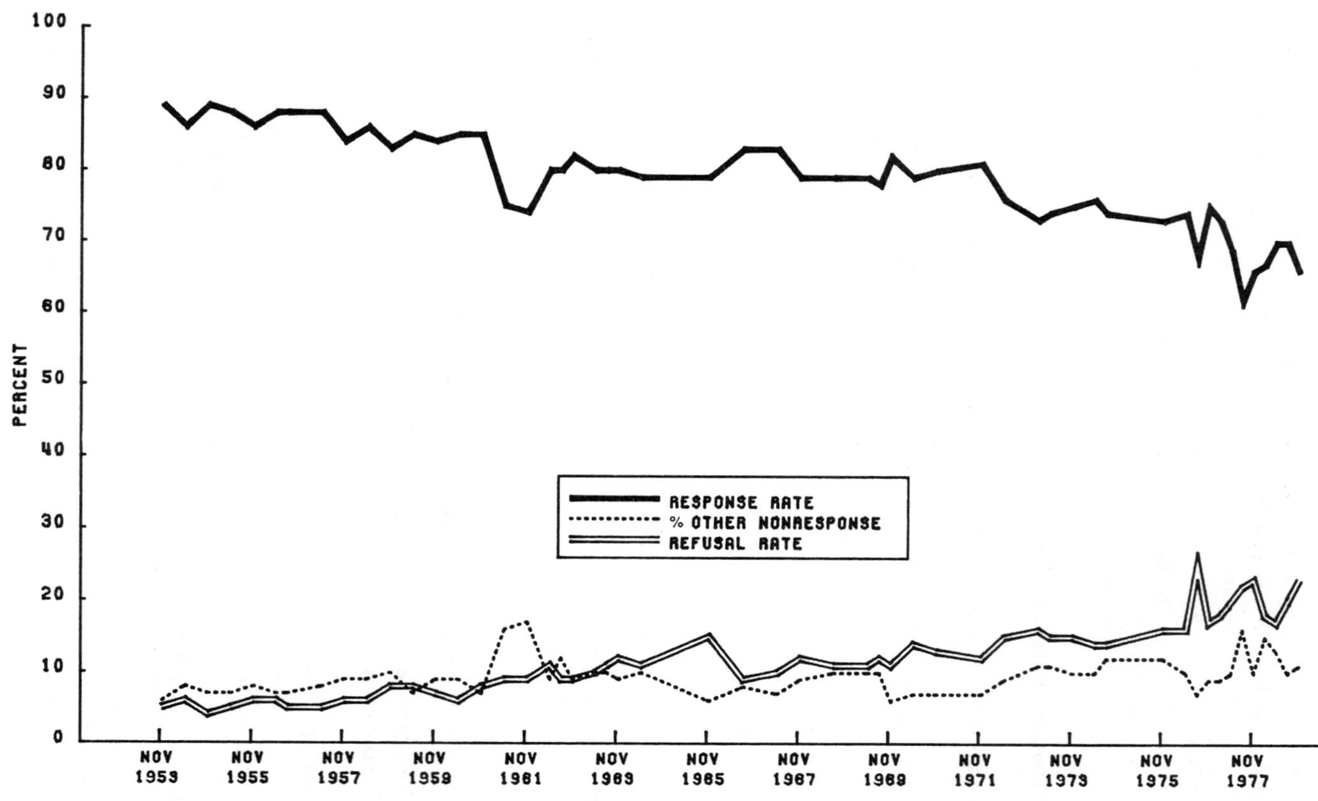

Fig. 1. *Response–nonresponse first contacts: economic surveys 1953–1978.*

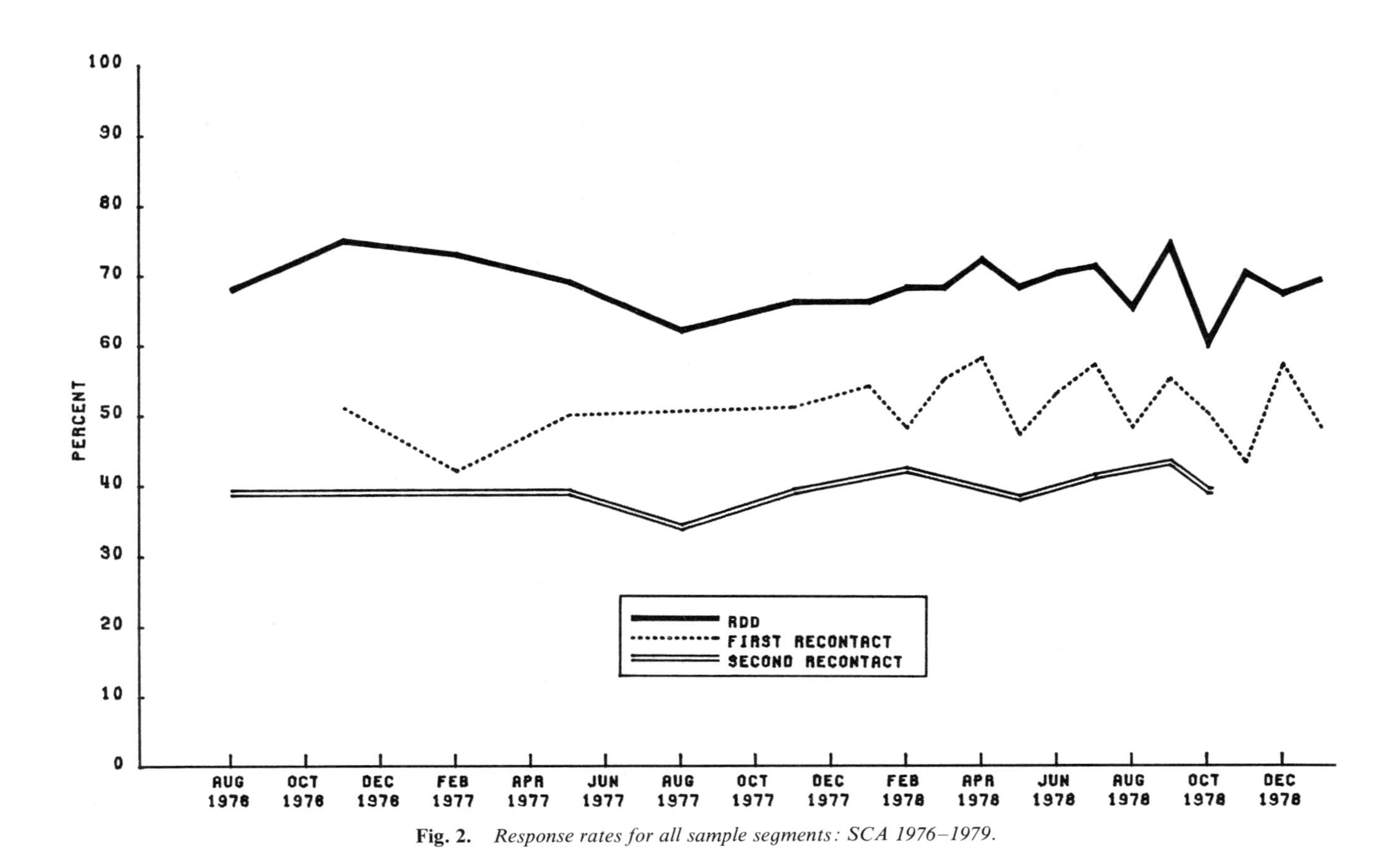

Fig. 2. *Response rates for all sample segments: SCA 1976–1979.*

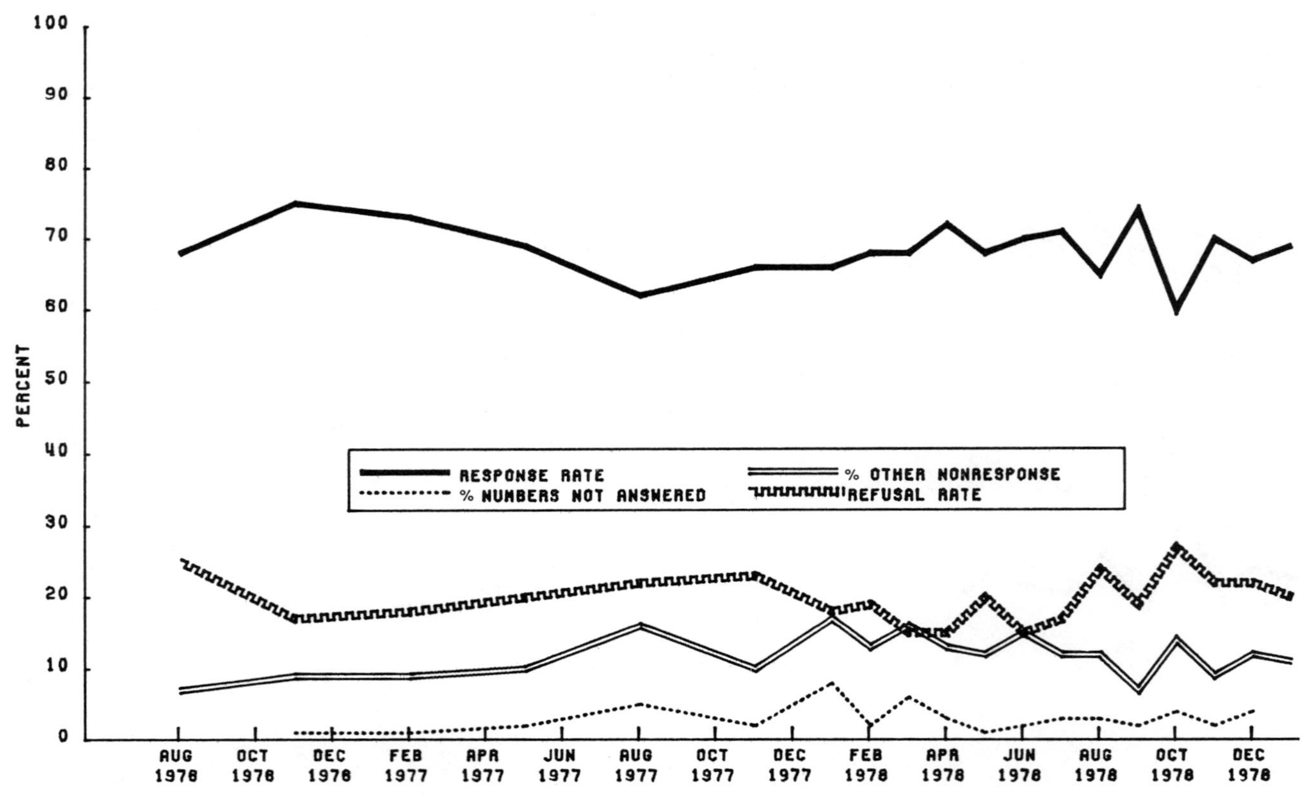

Fig. 3. *Response–nonresponse for RDD sample segments: SCA 1976–1979.*

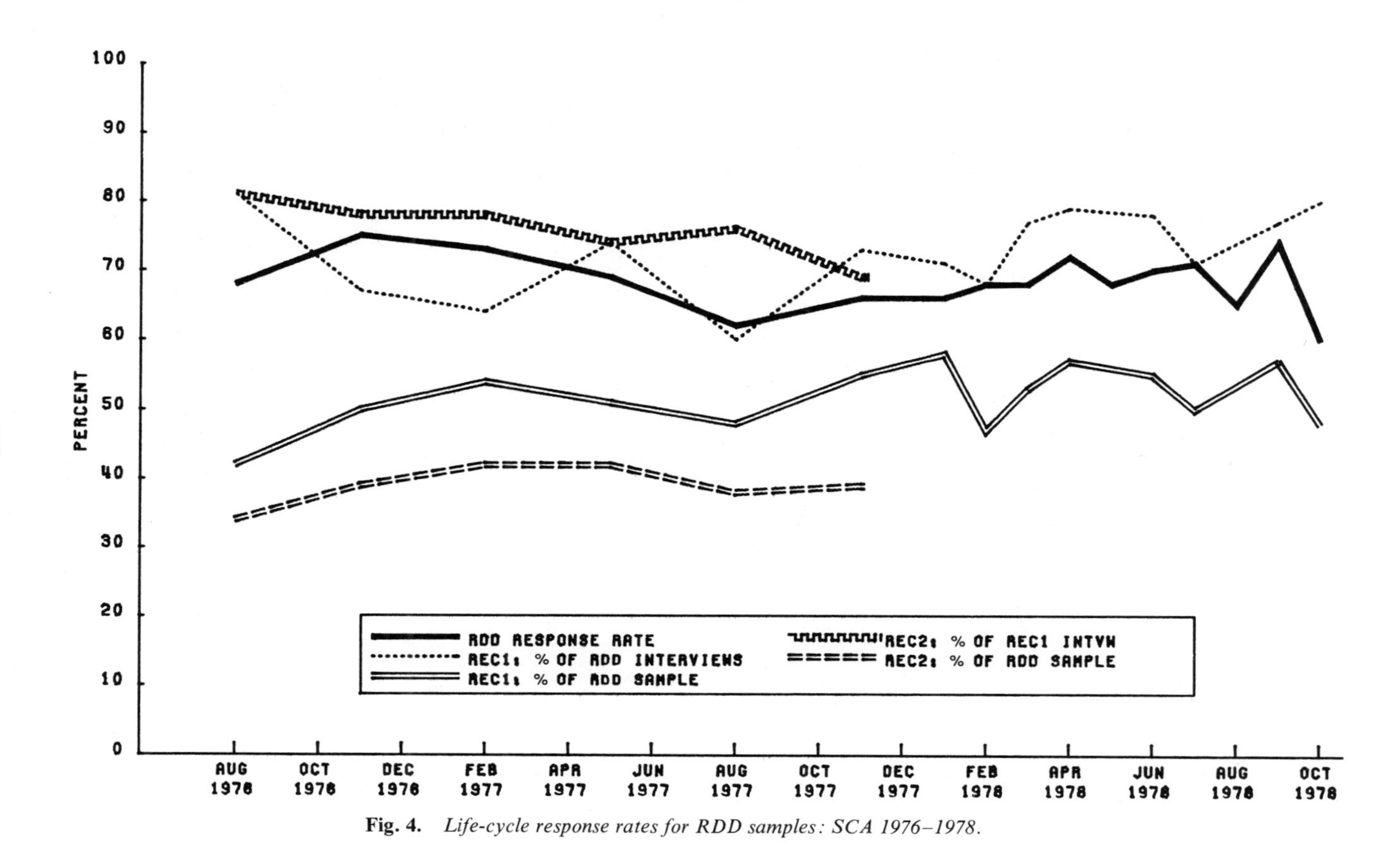

Fig. 4. *Life-cycle response rates for RDD samples: SCA 1976–1978.*

"% Other Nonresponse" category includes the precentage of numbers that ring but are never answered.

Unlike Fig. 2, which records the history of response by survey, Fig. 4 presents the patterns of response for RDD samples as panels. Thus the response rates in Fig. 4 are plotted in the month when the original RDD sample was interviewed. By beginning at the heavy black line and looking down for any point, the life cycle of an RDD sample upon reinterview can be traced. The response rates for successive waves of an RDD sample are labeled "% of RDD Sample." Figure 4 also presents recontact rates for each wave of an RDD sample. These rates are usually much higher than response rates since they are calculated from a smaller base—the number of respondents from whom another interview is desired. For most Surveys of Consumer Attitudes, only those respondents who actually granted an interview when they were previously called are designated for recontact.

Table 7 presents the results of regression analyses, using time as the independent variable. The regression coefficients measure the amount of change that occurred in the dependent variable per quarter or, in the case of RDD samples, per month. These analyses confirm the visual evidence in Figs. 1–3 that the major change in response rates has occurred over the long run. Since the

TABLE 7

Trends for Response–Nonresponse, Surveys of Consumer Attitudes (1953–1979)[a]

Dependent variable	*Regression coefficient*	*Standard error*	*N*	*Durbin–Watson*
For Fig. 1				
Response rate	−.187[b]	.014	53	1.01
Percentage of other nonresponse	.028[c]	.010	53	1.32
Refusal rate	.155[b]	.009	53	1.51
Percentage of not at home or respondent absent	−.006	.012	42	1.43
For Fig. 2				
RDD response rate	−.091	.105	19	2.43
First recontact response rate	.124	.157	17	2.64
Second recontact response rate	.120	.106	9	2.30
For Fig. 3				
RDD response rate	−.091	.105	19	2.43
Percentage of numbers not answered	.050	.065	17	—
Percentage of other nonresponse	.096	.080	19	—
Refusal rate	.016	.097	19	—

[a] The independent variable is time measured in quarters for the first analysis and in months for the other regressions.
[b] Significance level <.001.
[c] Significance level <.01.

introduction of telephone interviewing and RDD sampling, the rates of response and nonresponse have been relatively stable. The trends since 1953 for the components of nonresponse are very interesting, however. The decline in the response rate has been caused, it appears, by a rise in the refusal rate. The percentage of sample addresses where no one was home has not increased over this period. Instead, it has declined slightly. This category of nonresponse, which was not plotted in Fig. 1, could be calculated only for personal interviews.

2.2. Missing Data Rates on Index Questions

Table 8 presents the proportions of respondents who did not supply answers to each of the five questions used in the Index of Consumer Sentiment for surveys from 1953 to 1979. The proportion of respondents not answering the index questions appears to have declined since the introduction of centralized telephone interviewing in the summer of 1976. The research staff and field staff of the Survey Research Center attribute this improvement to the closer supervision of interviewers that is possible in such a facility.

TABLE 8

Percentage of Missing Data[a]

Date by year and quarter or month	*Better/worse now*		*Better/worse in year*		*12-month business*		*5-year business*		*Time to buy durables*	
	Don't know	*Not appli-cable*	*Don't know*	*Not appli-cable*	*Don't know*	*Not appli-cable*	*Don't know*	*Not appli-cable*	*Don't know*	*Not appli-cable*
1953:4	3	*	19	*	22	2	11	6	10	3
1954:2	1	1	12	1	15	2	13	10	8	5
4	1	*	14	1	15	2	14	9	8	6
1955:2	1	*	15	1	12	2	13	10	11	4
4	1	*	14	1	15	1	21	3	13	4
1956:2	2	*	12	1	13	1	19	4	?	?
3	2	*	13	1	11	1	14	6	14	2
4	1	*	14	*	15	1	21	3	17	3
1957:2	*	*	13	*	17	1	21	3	16	3
4	1	*	14	1	19	1	15	6	16	4
1958:2	2	*	14	1	22	2	10	8	17	3
4	2	*	15	1	22	1	18	5	17	3
1959:2	1	*	12	1	16	2	20	2	18	3
4	1	1	15	*	18	1	16	4	17	4
1960:1	1	*	12	1	11	2	18		13	5
2	1	*	13	1	17	2	12	6	20	4
3[b]	—	—	—	—	—	—	—	—	—	—
4	2	*	17	*	27	2	13	7	23	3

(*Continued*)

TABLE 8 (*cont.*)

Date by year and quarter or month	*Better/worse now*		*Better/worse in year*		*12-month business*		*5-year business*		*Time to buy durables*	
	Don't know	*Not applicable*	*Don't know*	*Not applicable*	*Don't know*	*Not applicable*	*Don't know*	*Not applicable*	*Don't know*	*Not applicable*
1961:1	1	*	13	1	17	2	20		12	5
2	1	*	10	1	16	1	12	8	20	3
3[b]	—	—	—	—	—	—	—	—	—	—
4	2	*	13	1	20	1	12	8	20	3
1962:1	1	*	10	1	14	1	21		15	3
2	2	*	10	1	17	1	14	9	22	2
3	2	*	12	*	20	1	11	6	22	2
4	2	*	11	*	19	1	14	8	21	3
1963:1	1	1	10	1	16	1	12	11	17	4
2	2	*	9	*	16	1	11	11	23	3
3	1	*	10	1	18	1	14	10	20	3
4	2	*	10	1	19	1	14	9	18	2
1964:1	1	*	11	*	12	1	11	9	18	2
2	1	*	9	1	17	1	11	7	19	2
3[b]	—	—	—	—	—	—	—	—	—	—
4[b]	—	—	—	—	—	—	—	—	—	—
1965:1	1	*	10	*	14	1	12	6	19	3
2[b]	—	—	—	—	—	—	—	—	—	—
3	1	*	12	*	18	2	12	6	14	2
4	1	*	9	*	16	1	11	7	19	1
1970:1	2	*	11	1	15	1	6	7	14	2
2	2	*	12	1	15	1	8	5	16	1
3	1	*	12	1	15	2	8	9	15	1
4	1	*	13	1	17	1	5	9	10	2
1971:1	1	*	13	*	15	1	8	12	13	1
2	1	*	14	*	17	2	11	16	17	1
3	1	*	15	*	16	1	7	14	12	1
4	1	*	16	*	21	2	8	14	11	2
1972:1	1	*	11	*	16	3	10	19	15	1
2	*	*	11	*	17	1	9	14	17	1
3	1	*	8	*	17	2	11	16	15	2
4	1	*	8	*	18	2	9	24	20	1
1973:1	1	*	9	*	16	2	10	16	15	1
2	1	*	10	*	14	1	9	13	14	2
3	1	*	9	*	15	*	9	12	16	1
4			13	*	15	2	9	15	15	1
1974:1	1	*	9	*	10	1	6	7	14	*
2	1	*	10	*	12	1	8	14	16	2
3	1	*	10	*	14	*	8	15	16	*
4	1	*	11	*	12	1	7	17	13	1
1975:1	*	*	9	*	6	*	6	16	14	1
2	*	*	7	*	8	3	6	20	14	1

TABLE 8 (*cont.*)

Date by year and quarter or month	*Better/worse now*		*Better/worse in year*		*12-month business*		*5-year business*		*Time to buy durables*	
	Don't know	*Not applicable*	*Don't know*	*Not applicable*	*Don't know*	*Not applicable*	*Don't know*	*Not applicable*	*Don't know*	*Not applicable*
3	*	*	7	*	9	3	7	18	15	1
4	*	*	12	*	12	2	10	19	15	1
1976:1	1	*	10	*	10	2	11	20	21	2
2	1	*	11	*	16	1	14	16	18	4
3	1	*	10	*	12	4	13	19	14	3
4	1	1	9	1	13	3	7	23	11	1
1977:1	1	*	5	1	11	7	—[c]	15	—[c]	—[c]
2	*	*	4	*	10	6	—[c]	15	—[c]	—[c]
3	*	*	4	*	9	5	—[c]	18	—[c]	—[c]
4	*	2	5	1	8	4	—[c]	10	—[c]	—[c]
1978:1	*	*	4	*	6	2	5	5	4	1
2	*	1	5	1	10	6	6	11	10	3
3	1	*	5	1	5	5	8	5	6	3
4	1	*	3	*	8	5	3	10	5	1
5	*	1	3	*	5	3	4	9	8	4
6	*	*	2	*	5	3	3	7	5	2
7	*	1	3	1	6	3	4	4	3	1
8	*	*	3	*	5	5	3	7	10	2
9	1	*	5	*	8	3	4	8	7	1
10	1	*	6	*	9	4	4	9	4	2
11	1	*	4	1	10	2	7	9	11	2
12	1	*	8	1	9	4	8	11	7	1
1979:1	1	*	3	1	6	6	4	5	4	1

[a] Asterisks denote less than 1%.
[b] No survey taken.
[c] Data not available.

APPENDIX A. PROCEDURES USED BY INTERVIEWERS TO DETERMINE HOUSING UNITS AND RESPONDENTS

Survey Research Center
The University of Michigan

SECONDARY NUMBER COVER SHEET

1. Hello, my name is ______________. I'm calling from the University of Michigan, in Ann Arbor. Here at the university, we are currently working on a study for the Survey Research Center. First of all I need to be sure I've dialed the right number.

   Is this . . . . [Label with phone number]

2. (IF NOT CLEAR) Since this telephone number has been generated by a computer, I do not know whether this number is for a business or a home. (Which is it?)

   [1. BUSINESS] [2. HOME] [3. BOTH]

   ↓ (1. BUSINESS)

   2a. Does anyone live there on the premises?

   [1. YES] [5. NO]

   ↓ (1. YES) — (5. NO) END CONTACT

   2b. Do (they/you) have another phone number in the residence or do (they/you) use this number for personal business?

   [1. HAVE OTHER] [2. USE THIS]

   (1. HAVE OTHER) END CONTACT — (2. USE THIS, 2. HOME, 3. BOTH) ↓

3. As I said, we are conducting this study from The University of Michigan. Our interview concerns many interesting topics, including the way the economy is going. I would like to interview someone in your household, and in order to determine whom I need to interview, I'll need a listing of the members of your household—not their names, just their sex and age and relationship to you. → GO TO P. 2

4. Call Record

Call Number	01	02	03	04	05	06	07	08	09	10	11	12	13	14	15	16	17
Date																	
Time																	
Result																	
Ier No.																	

** ALL REFUSALS AND NONINTERVIEWS MUST BE EXPLAINED ON PAGE 3.

5. Length of Interview ____________
(Minutes)
6. Let's start with you—how old are you? (Are you male or female?) Now I'd like the sex and age and relationship to you of each of the other members of your household who are *18 years of age or older*.

A Relation to Informant	A^2 Relationship to Respondent	B Sex	C Age	D Eligible Person Number	E Respondent

7. Now I'd like the sex and age and relationship to you of each of the members of your household who are 17 or younger.

SELECTION TABLE E_1	
If the number of eligible persons is:	Interview the person numbered:
1	1
2	2
3	3
4	3
5	3
6 or more	5

8. You've said there are (REPEAT LISTING), does that include everyone living there at the present time? (IF NO, CORRECT ABOVE.)
9. Now I will use a selection procedure—I'm going to number the people in your household to determine whom we need to interview—(it will just take a second . . .)
10. The total number of eligible persons is (__) so I am to interview person #__ who is (RELATION TO INFORMANT).
11. Before we start, I would like to assure you that the interview is completely voluntary; if we should come to any question which you don't want to answer, just let me know and we'll go on to the next question.

HOUSING UNITS

Housing units are defined by two general criteria: the fact that the living quarters are *used separately* by their occupants, and by certain *physical characteristics* of those quarters.

Operational Definition of a Housing Unit (HU)

A housing unit is a room or group of rooms *occupied or vacant and intended for occupancy* as *separate living quarters.* In practice, living quarters are considered separate and therefore a housing unit when:

the occupants live and eat apart from any other in the building,

AND THERE IS EITHER

direct access from the outside or through a common hall,

OR

complete kitchen facilities for the *exclusive use* of the occupants, regardless of whether or not they are used.

Occupants Live and Eat Apart

Occupants are considered to be *living apart* from any other group in the building when they own or rent different living quarters. They are considered to be *eating apart* when they provide and prepare their own food or have complete freedom to choose when and where they eat and they do not have to pay a fee for meals whether or not they are eaten.

Housing units may be occupied by a single family, or by two or more family units.

If a single building is occupied by persons or groups of people who live and eat apart from each other, the living quarters of each person or group are to be evaluated in accordance with the other HU criterion of direct access or complete kitchen facilities.

Direct Access

Living quarters have direct access if there is EITHER:

an entrance directly from the outside of the building;

OR

an entrance directly from a common hall, lobby, or vestibule, used by the occupants of more than one unit or by the general public.

Access is not direct if, as the only means of access to the unit, the occupants must pass through another person's living quarters.

Complete Kitchen Facilities

A unit has complete kitchen facilities when it has: an installed sink with piped water, *and* a range or cook stove, *and* a mechanical refrigerator. Portable cooking equipment does not qualify as a range or stove, nor is an ice box equivalent to a mechanical refrigerator. All kitchen facilities must be located in the building, but they need not be in the same room.

Kitchen facilities are for the exclusive use of the occupants when the living quarters are rented (or bought) with the understanding that the renter (or buyer) is not required to share kitchen facilities with other tenants.

Examples

Most types of living quarters that meet the HU definition are easily identifiable. These are single family houses, row houses, town houses, duplexes, flats, garden-type apartments, apartments over commercial structures, high-rise apartments, and mobile homes.

EXCLUDED QUARTERS

Certain types of *institutional and transient or seasonal facilities* which are never to be listed are called excluded quarters.

Transient or Seasonal

Transient or seasonal quarters are to be excluded if:

There are *five or more* units (whether beds, rooms, suites, cabins, trailers, tent or trailer sites, or boat moorings) operated *under a single management;*

AND

When the living quarters are EITHER:

—*transient;* that is, *more than 50 percent* of the units are normally occupied or intended for occupancy by persons who usually stay less than 30 days or who pay at a daily rate;

OR

—*seasonal;* that is, the facility is closed during at least one season of the year.

Transient or seasonal quarters to be excluded are: nonstaff accommodations in missions, flophouses, or Salvation Army shelters; accommodations for guests in *transient* hotels, motels, Y's, residential clubs, or resort establishments; vacation accommodations in campgrounds; transient trailer parks or marinas; and bunkhouses, cabins, trailers, and tents for migratory or seasonal workers.

Exceptions within Excluded Quarters

In facilities containing excluded institutional, transient or seasonal quarters, check carefully for any living quarters which are occupied or intended for occupancy by residential owners or employees and their families. Staff quarters which meet the HU criteria are to be classified accordingly.

Institutionally operated student housing which meets the HU definition, such as married student apartments or trailers, is not to be excluded.

UNCLASSIFIED QUARTERS

Unclassified quarters are living quarters that do not clearly meet the HU criteria, yet which are not clearly to be excluded. For these unconventional situations, a determination will be made by the Sampling Section.

HOUSING UNIT DETERMINATION SHEET

For unusual housing situations where 6 or more adults are living in a single housing and/or there is an unusual housing situation, i.e., Live-in Hotel, Rooming House, Dormitory, Military Barracks, etc.

1. IF NOT ALREADY CLEAR Could you tell me what kind of living quarters these are (that is, do you live in a dormitory, a hotel, or what)?

- ☐ Group quarters for students in dormitories, fraternities, sororities, co-op (but not married student housing)
- ☐ Quarters for the religious in cloistered, i.e., convents, monasteries
- ☐ Patient quarters in hospitals, rest homes, or nursing homes
- ☐ Nonstaff accommodations in Missions, Flophouses, or Salvation Army Shelters
- ☐ Inmates quarters in mental institutions
- ☐ Barracks on a military base

↓

TERMINATE CONTACT

☐ Hotel or Other

↓

1a. Are more than half of the hotel rooms occupied by permanent residents?

YES

NO → TERMINATE CONTACT

GO TO QUESTION 2

OTHER, EXPLAIN: ____________________

____

____

____

GO TO QUESTION 2

2. IF NOT CLEAR Do you expect your stay there to last more than 30 days?

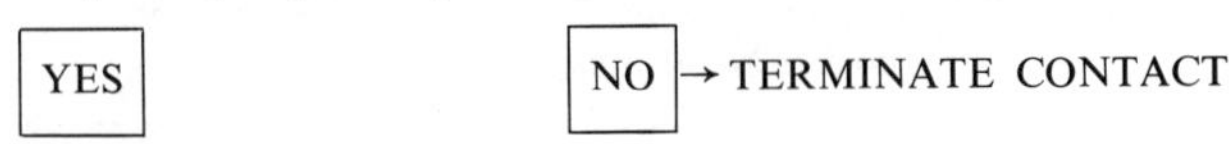

GO TO QUESTION 3

3. Do you provide and prepare your own meals?

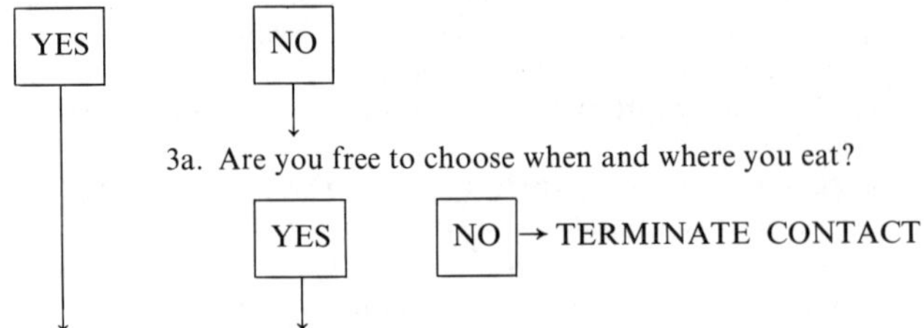

4. Do you have to pay a fee for meals whether or not you eat them?

IF MEALS PROVIDED AS PART OF SALARY, ANSWER IS YES

5. Does your unit have complete kitchen facilities that you do not have to share with other units?

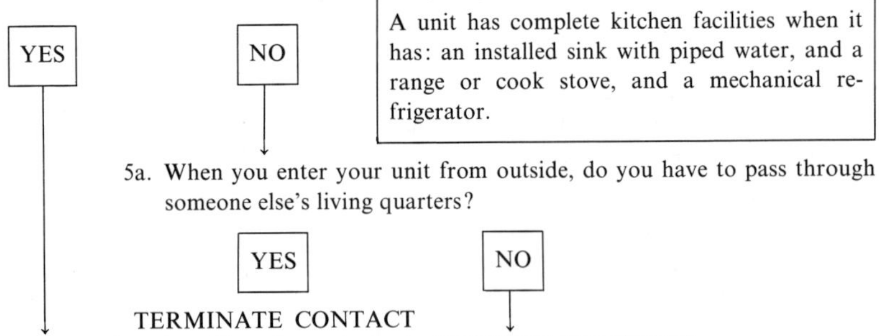

6. Is this phone number shared by other units in the (TYPE OF HOUSING)?

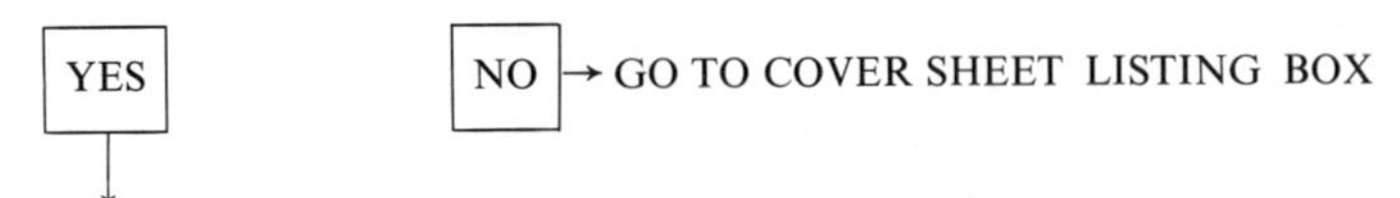

6a. How many units share this phone number? ________

Thank you. I must consult with my supervisor before going on. We will call you again later.

APPENDIX B. DEFINITIONS OF RESULT CODES

RESULT CODES

(revised 5/1/79)

No Final Disposition

ANSWERED

11. Appointment with R
12. Call back; best time for R known
13. Call back; best time for R not known
14. No qualified informant reached
22. First wrong connection

NOT ANSWERED

41. Ring, no answer
42. Regular busy signal
43. Complete silence
**44. Fast busy signal or strange noise

Final Disposition

INTERVIEW - ALL SAMPLES

01. Complete interview
05. Partial interview - study ok

NON-INTERVIEW - ALL SAMPLES

31. Initial refusal by R
32. Initial refusal, not by R
33. Initial refusal, R undetermined
34. Refusal, never reached for conversion
35. Initial refusal, breakoff no conversion
51. Final refusal by R
52. Final refusal, not by R
53. Final refusal, R undetermined
55. Final refusal, breakoff on conversion
++56. Final refusal, no conversion attempt
++62. NI - never answered
63. NI - circumstantial
65. NI - incomplete
++67. NI - lost
69. NI - grid filled

NON-INTERVIEW - RDD ONLY

++68. NI - NS, end of study replacements

NON-SAMPLE - RDD ONLY

73. Final Silence
**74. Final strange noise or fast busy
82. Second wrong connection
84. Non-working number
85. Non-residential number
86. Out of primary area
91. No eligible R

NON-SAMPLE - RECONTACT ONLY

75. Wrong # for R, no new # available
76. R's # is no longer in service

++These codes are for administrative use only, and should never be used by interviewers

**Six-attempt procedure should be used for strange noise and fast busy signal; Verify by calling the # twice. Confirm by waiting 3 days after initial 2 calls and trying twice again. If 2 more 44's are obtained, wait 3 more days and try again. If sixth call on sixth day is a strange noise or fast busy, code 74.

APPENDIX C. STAGES OF DATA PROCESSING

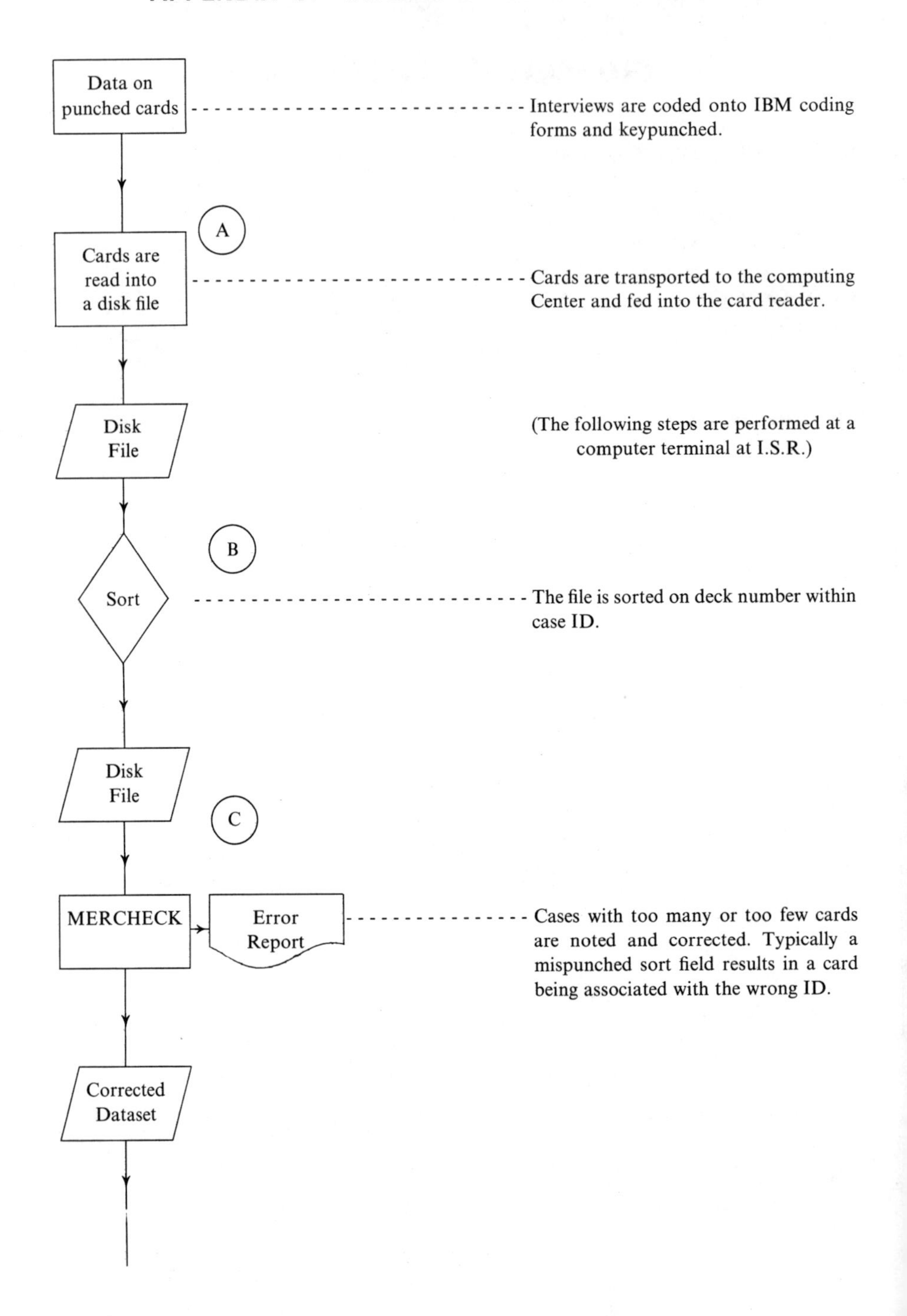

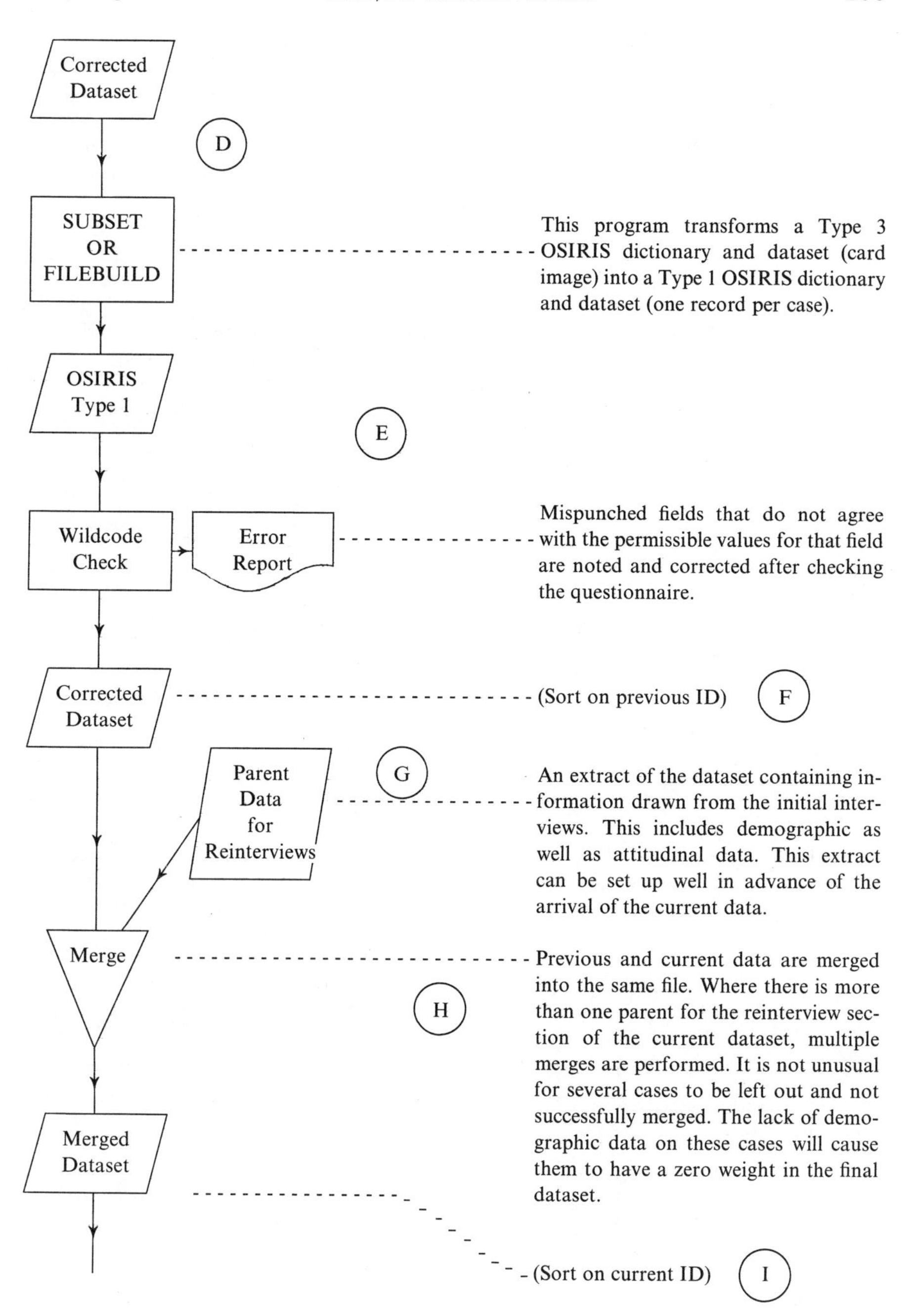

Note: The form of the dataset that is made available through the Direct Data Entry (DDE) system is equivalent to that which results from steps A thru E.

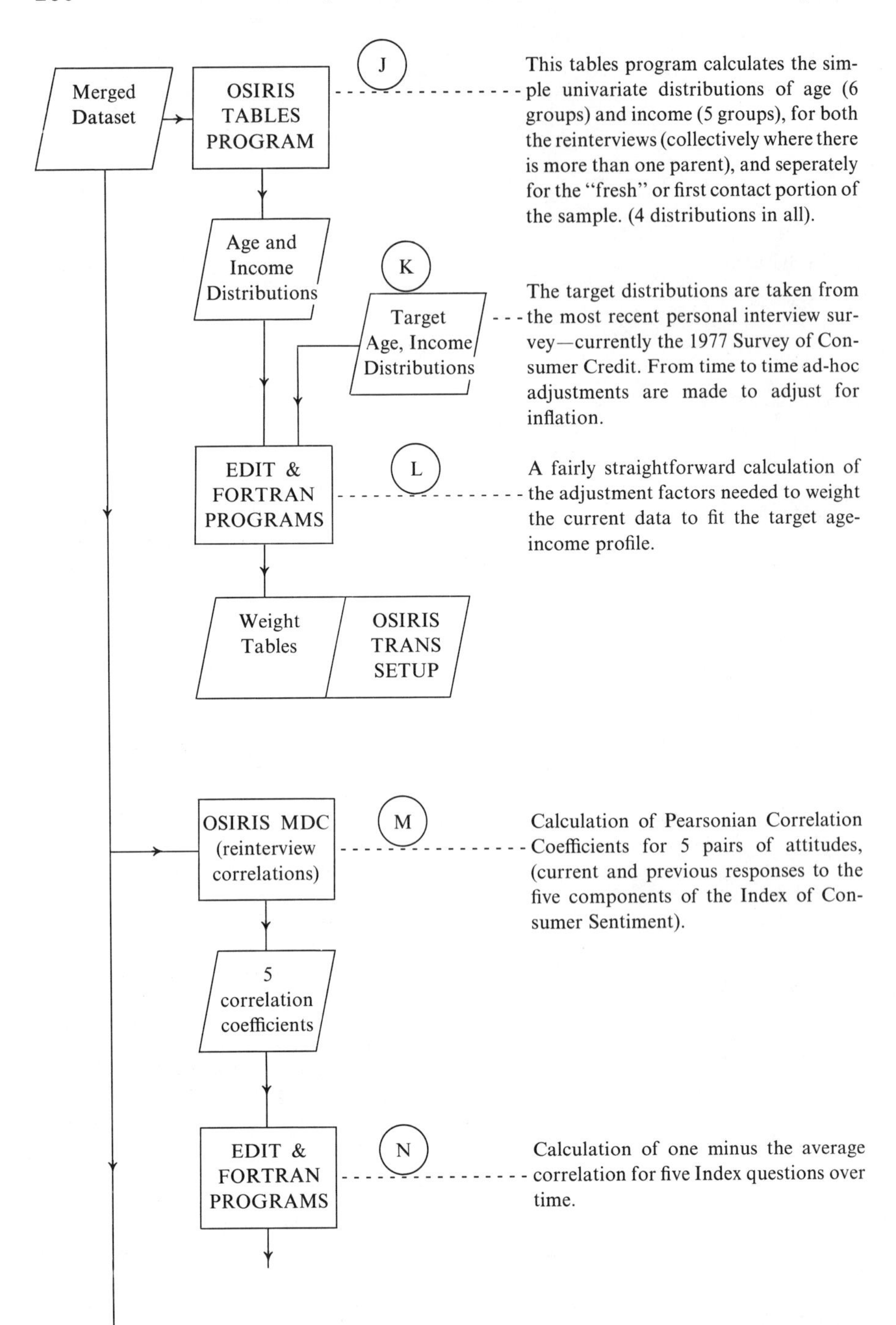
Merged Dataset
OSIRIS TABLES PROGRAM
J
This tables program calculates the simple univariate distributions of age (6 groups) and income (5 groups), for both the reinterviews (collectively where there is more than one parent), and seperately for the "fresh" or first contact portion of the sample. (4 distributions in all).
Age and Income Distributions
K
Target Age, Income Distributions
The target distributions are taken from the most recent personal interview survey—currently the 1977 Survey of Consumer Credit. From time to time ad-hoc adjustments are made to adjust for inflation.
EDIT & FORTRAN PROGRAMS
L
A fairly straightforward calculation of the adjustment factors needed to weight the current data to fit the target age-income profile.
Weight Tables
OSIRIS TRANS SETUP
OSIRIS MDC (reinterview correlations)
M
Calculation of Pearsonian Correlation Coefficients for 5 pairs of attitudes, (current and previous responses to the five components of the Index of Consumer Sentiment).
5 correlation coefficients
EDIT & FORTRAN PROGRAMS
N
Calculation of one minus the average correlation for five Index questions over time.

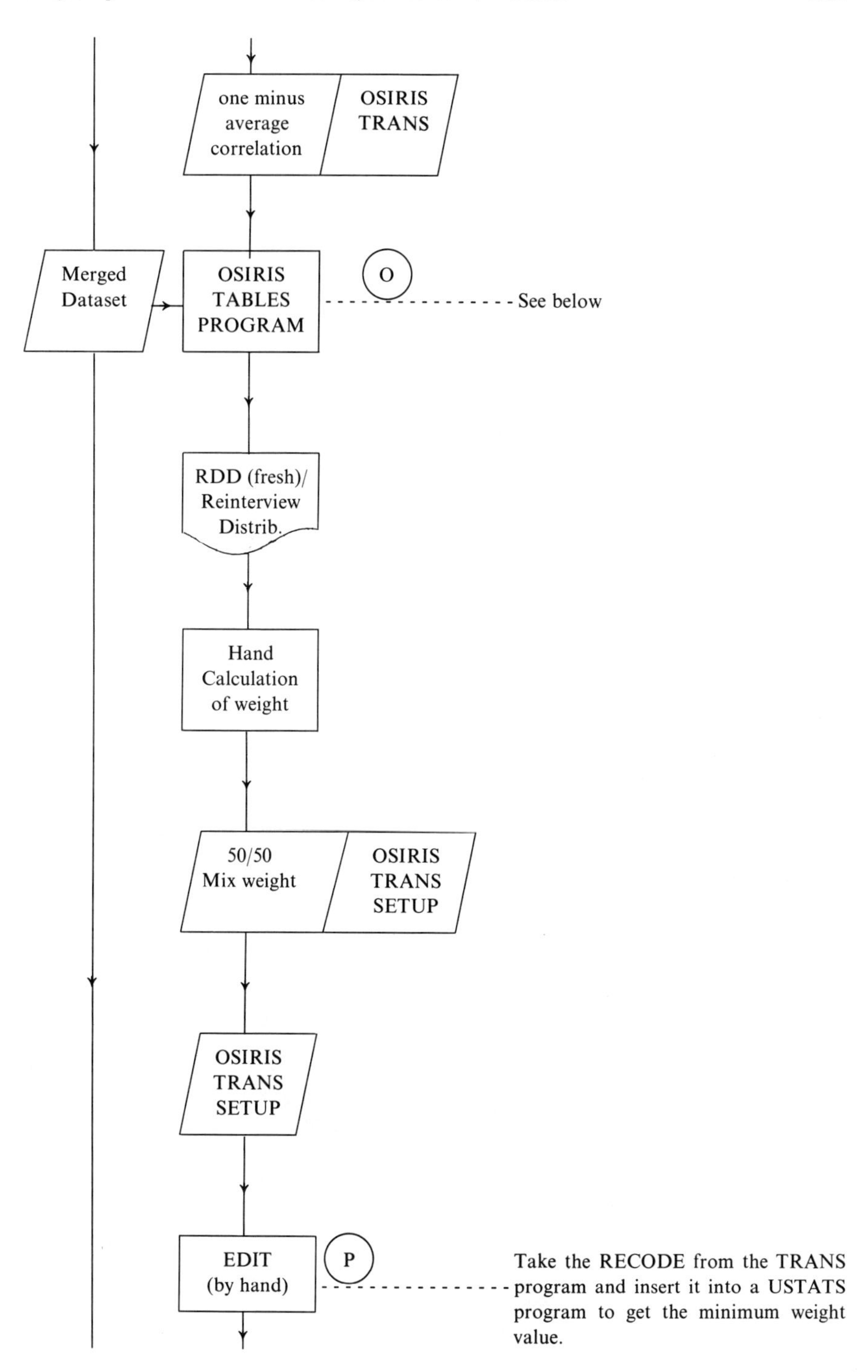

Take the RECODE from the TRANS program and insert it into a USTATS program to get the minimum weight value.

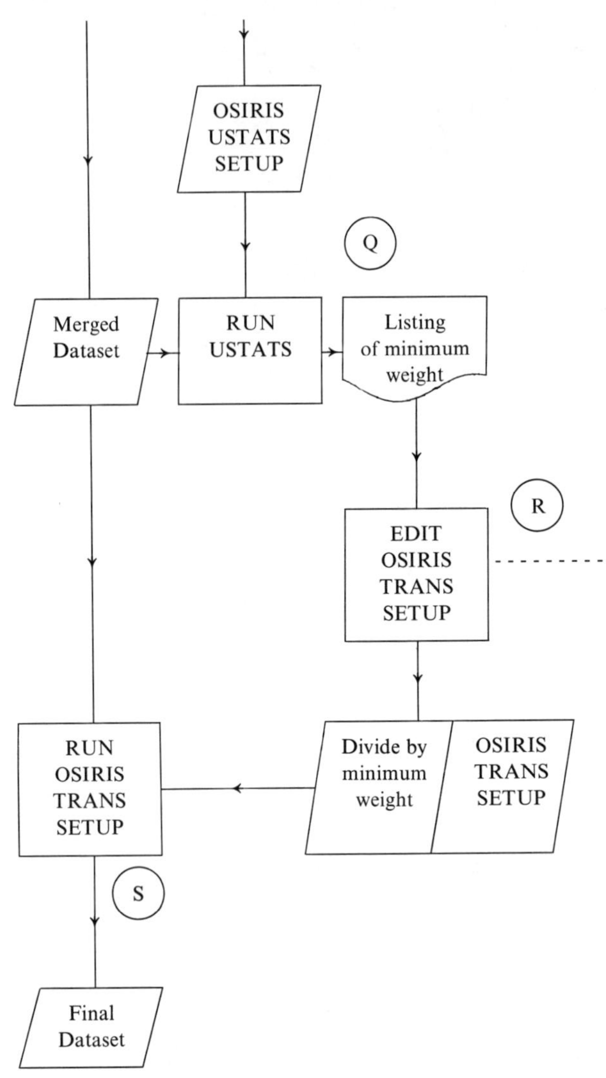

Alter the TRANS RECODE so that each weight value gets divided by the smallest weight to give a new minimum weight of one. This will give an average weight that is roughly constant from survey to survey.

CHAPTER **5**

Treatment of Missing Data in an Office Equipment Survey

Richard Valliant
Richard Tomasino
Morris H. Hansen

1. INTRODUCTION

The office equipment survey discussed in this chapter is a survey of selected types of office equipment used by all classes of nonresidential establishments. Thus in addition to comprehensive coverage of private business establishments of all types, the survey covered churches, schools, hospitals, government establishments at all levels, nonprofit institutions, and other types of nonresidential activity but excluded family-type farms. The survey utilized a multistage area probability sampling approach with variable sampling fractions for various sizes and classes of establishments. Multistage area probability sampling was used because of the unavailability of lists. Conducted by Westat, Inc., it was perhaps the first probability sampling survey ever attempted in the United States with such comprehensive establishment coverage.

This survey was selected as a case study on the treatment of missing data because it illustrates the treatment of several types of incomplete data, namely,

(1) incompleteness due to undercoverage of the survey target universe;
(2) incompleteness due to nonresponse from identified sampled establishments or from questionnaires; and
(3) incompleteness due to missing items on otherwise usable schedules.

INCOMPLETE DATA
IN SAMPLE SURVEYS
Volume 1, Part II

ISBN 0-12-363901-8

Specific steps were taken during the interviewing and processing phases of the survey in an effort to control each of these types of incompleteness to acceptably low levels.

2. DESCRIPTION OF THE SURVEY

The survey sample design and methods were developed and pilot tested in the first half of 1977, the fieldwork on mapping and sample selection was done in the fall, and the collection of the questionnaires was substantially completed by the end of the year. The entire operation was done under a very tight time schedule.

2.1. Goals and General Approach

Broadly, the goal of the survey was to provide information on office equipment for a sample from the population of all nonresidential economic and social units in the 48 coterminous states of the United States, other than households and household-type farms, to the extent that the units were identifiable establishments. Information was obtained on specified types of individual office machines, by brand and model, and on the environments in which the machines were located and used, as well as on other characteristics.

The first-stage sampling units were counties or small groups of contiguous counties (such as Standard Metropolitan Statistical Areas). Postal ZIP Codes were sampled within the sampled counties, and then area segments were sampled within the sampled ZIP Codes. Within each of the sampled area segments a complete canvass and listing of establishments was made, and establishments included in the survey were subsampled from these lists, with varying subsampling rates depending on size and type of establishment. Trained interviewers visited each establishment in an effort to obtain the desired information.

Although the first type, undercoverage of the target universe, might have been substantially or totally explainable by sampling variability, it was felt to be caused primarily by underlisting of establishments in sampled areas. Some such undercoverage had been anticipated on the basis of other widespread survey sample experience. During field listing, a check of listers' work was made to reduce the amount of underlisting. In making the survey estimates, a ratio estimation procedure was used in which appropriate adjustments were made in sample weights in an effort to reduce the effects of undercoverage and, at the same time, to reduce sampling errors. For the second type of incompleteness, i.e., unusable schedules, attempts were made to keep the incompleteness to an acceptably low level by using callbacks, both in person and by telephone. In data processing, weight adjustments were made for respondents, which, in effect, imputed for the total nonresponses. Missing-data items, the third type of incompleteness, were imputed for several key items in the survey, using a version of the hot-deck method. The specific steps taken to deal with each of the three types of incompleteness are described in some detail in Section 3.

Completed questionnaires were obtained from 7568 establishments in the survey, with a response rate from the establishments initially selected for the sample of 88%. The sampled establishments were located in 410 sample segments, selected from within 79 first-stage sampling units.

A goal of the study was to achieve estimates for a number of summary measures for equipment with expected average sampling variability (coefficients of variation) no greater than 10%. In order to obtain information needed in the design of the questionnaire and the sample to meet the specified goals, Westat conducted a pilot study covering approximately 600 establishments in three primary sampling units to test procedures. On the basis of the pilot study, rough advance estimates of expected sampling variability from the survey were made for certain summary estimates, including estimates for all machines, for principal individual types of machines, and for a measure of the total amount of machine usage. It was estimated that the average relative expected sampling variability for the principal individual types of machines would be no more than 5–7% and would be approximately 10% for total machine usage. Of course, sampling errors were expected to be larger—sometimes much larger—for more detailed characteristics or for statistics by broad kinds of business or establishment size classes. Because the information collected is proprietary, specific office equipment data items that were collected are not listed in this chapter.

2.2. The Sample Design

2.2.1. The Sample of First-Stage Units

As indicated earlier, the sample was selected in several stages. The first-stage, or primary sampling, units used in the selection of the sample are individual counties or small groups of adjacent counties. The first-stage unit was of an aggregate size that allowed an interviewer to travel from home and back daily with reasonable efficiency and without requiring any overnight travel to do work throughout the area. Such areas include Standard Metropolitan Statistical Areas (SMSAs), consisting of whole counties as defined by the Office of Management and Budget (or in New England, whole-county approximations to SMSAs), which account for about two-thirds of the population of the United States.

Some of the SMSAs are large enough to be self-representing; these include such metropolitan areas as New York and Chicago, down to areas of the size of

Washington, D. C., and Baltimore. A total of 25 such self-representing units was included in the sample and accounts for the 18 largest SMSAs (some SMSAs such as New York, Chicago, and Detroit were considered as being composed of more than one first-stage sampling unit for administrative purposes).

Outside of the large self-representing SMSAs, about half of the U.S. population resides in other SMSAs and about half in non-SMSA counties. In these non-self-representing areas, the first-stage sampling units were individual SMSAs, individual counties, or groups of adjacent counties. These units were stratified by SMSA or not, by region, by percentage of the population in urban areas, and by measures of socioeconomic status. Then samples of these were selected, with the probability inclusion in the sample being proportional to their population size. About half of the selected non-self-representing first-stage sampling units were SMSAs and half were not.

A total of 79 first-stage units was selected for inclusion in the sample; about a third of them accounted for the large self-representing SMSAs, a third represented other SMSAs, and a third represented non-SMSAs. The population residing in the sampled first-stage units actually accounts for about 40% of the total population of the United States.

This sample of first-stage sampling units had already been selected and used by Westat for other sample surveys. There would be substantial advantages, from the point of view of effective performance of fieldwork in the survey, if first-stage sampling units could be used in which trained staff already existed. Consequently, the existing sample for which such staff existed was examined to see if it would be an appropriate sample of first-stage units for the survey. This examination was done by determining what the estimated sampling variability would be if these first-stage units were covered completely, i.e., if *all* of the various types of establishments in these areas were to be included in a sample census. A number of characteristics shown in the *County and City Data Book* for 1972 was used, and variances were estimated by applying the appropriate weights and variance-estimating equations to the data for these first-stage units. On the basis of these results, it was concluded that it would be advantageous to use Westat's existing sample of 79 first-stage units because contributions to sampling errors from the first-stage units included in the sample would not be too large and there were substantial advantages in the use, to the extent feasible, of already trained and tested field personnel.

2.2.2. The Sample of Second- and Third-Stage Units

The second-stage sampling units were individual ZIP Codes or, sometimes, a group of several adjacent ZIP Codes. ZIP Codes were chosen as second-stage sampling units, even though they did not always represent well-defined areas, because approximate information was available on the client's customers by ZIP Codes and on the distribution of business establishments by ZIP Codes from the Bureau of the Census for the 1975 *County Business Patterns* (CBP). The pilot study had shown that such information was highly effective in reducing sampling errors as compared with sampling other types of areas within counties for which such information was not available. The two data sets were complementary in the sense that the customer information tended to cover sectors of the nonresidential population that are not covered by CBP.

A first step was to identify the ZIPs that were to be associated with each sampled county. Since the ZIPs do not observe county boundaries, the procedure used in sampling ZIPs was to associate each ZIP uniquely with a county, and thereby with a first-stage sampling unit. The Postal Service identifies each ZIP with a post office and each post office with the county in which it is located. Using a list of ZIPs and the county location of the post office serving each, which was obtained from the Postal Service, each ZIP was then associated with the county of the post office for sampling purposes.

Using the two sources of data for ZIP Codes, two preliminary measures of size, which were related to the magnitude of establishment activity, were attached to each ZIP in the selected first-stage units. After the two sets of data were examined by ZIPs, some ZIPs were consolidated as necessary with geographically adjacent or nearby ZIPs to obtain ZIPs or ZIP groups that contained a minimum measure of size. In rural areas, this sometimes involved consolidation of a considerable number of ZIPs, whereas in urban-type areas, it sometimes involved consolidation of a primarily residential ZIP or ZIPs with a ZIP whose size was larger than the minimum. The purpose of making such consolidations was to define as second-stage sampling units areas that had a minimum amount of business activity and thus to increase the efficiency of the second-stage sampling units. Each ZIP group was then assigned a combined measure of size based on the maximum of the two normalized preliminary measures of size. From this measure, a final measure of size was assigned that identified the desired number of segments for the ZIP, i.e., the number of segments into which the ZIP would be divided if the ZIP were selected for the sample.

Once the measures of size (in terms of segment counts) were assigned to the ZIPs, a sample of approximately five ZIPs was selected in each first-stage sampling unit. One segment was then to be selected from each ZIP. The ZIP selections were done separately within each first-stage unit, with probabilities of selection proportionate to the measures of size assigned to each ZIP. These probabilities, considered jointly with the probabilities of selecting the first-stage unit, were such that each potential segment had an equal chance of being included in the sample and were also such that a sample of about five segments would be selected within each of the 79 sampled first-stage units, except in large self-representing first-stage units in which more segments were generally selected.

Unique or special ZIPs were to be distinguished from area ZIPs. Unique ZIPs are ZIPs that are specifically assigned to an institution of some type, a building, a governmental unit of some kind, a company, a university or a part of a university, or an establishment or group of establishments with some organizational connection. Establishments covered by unique ZIPs were often, but not always, geographically contiguous. There is no common pattern. Unique ZIPs are so designated by the post office; others are referred to as area ZIPs, and they generally represent areas for which postal service is provided, as distinguished from institutions or organizations.

Unique ZIPs occurred under many circumstances, ordinarily represented important establishments or activities, and were given special treatment in the sampling. These ZIPs were assigned measures of size in the same manner as were area ZIPs, based jointly on information available from CBP and from marketing records. However, they were retained in the sample, at this stage of sampling, without subdividing them and choosing a segment of some type within. Later, at the time of interviewing, it was sometimes found necessary or desirable to do some subsampling, but this was done with special information obtained at the time of interview for sampled unique ZIPs.

The initial sampling of unique ZIPs with probabilities proportional to their measures of size, and retaining the entire unit in the sample, meant that some of these came into the sample with relatively high probabilities or, occasionally, with certainty within the selected first-stage sampling units. This was desirable from a sampling point of view because of the large amount of office equipment activity involved.

2.2.3. The Fourth Stage of Sampling (Listing of Individual Establishments and Subsampling)

The final stage of sampling in the area segments involved another major field operation—a field canvass—to identify and list each establishment located in each segment and categorize it roughly by size and type. The classifications were ordinarily assigned by the lister from observation, without inquiry, and were very rough and tentative, but were quite adequate for sampling purposes. The larger establishments were all retained in the sample. Smaller establishments were subsampled with subsampling fractions varying from one-twelfth to one-half, depending on the size and kind of activity as identified by the lister. The size and kind of business reflected in the tabulations were obtained, subsequently, by personal interview.

It would have been feasible, but not cost-effective, to canvass the sampled segments and include all of the establishments in the sample without introducing the fourth stage of sampling. In fact, in the pilot study a variation of this procedure was used for comparison with a subsampling procedure, and it was found to be substantially more efficient to carry through the field listing and subsampling. Covering all of the segments, instead of listing and subsampling, would have increased the number of establishments to be interviewed by approximately a factor of six but would not have importantly improved the sample since the types of establishments that were subsampled were those of less interest because they were relatively low users of office equipment. The added cost for covering all establishments would have been much greater than the benefit to the survey. The sampling fractions were chosen on the basis of approximate costs and variances estimated from the pilot study and by optimum sample-design considerations.

For each segment, a map and detailed written boundary description were prepared for field use during the listing. The listers were instructed on systematic ways of canvassing an area in an effort to ensure that all parts of a segment were covered. On receipt of listings from the field, checks were made against segment descriptions, field counts, and other sources in an effort to ensure that the correct areas had been adequately listed.

For each establishment, the field agent listed the name, address, kind of business, and employment size class. "Kind of business" was a simple description, such as clothing store, insurance agency, or machine shop, and required no formal classification, such as a Standard Industrial Classification (SIC) code, on the part of the lister. "Employment size class" was recorded in one of five broad categories—under 10, 10–49, 50–99, 100–499, and 500 or over—based on the lister's perception of size. The listers were not to inquire in the establishments to obtain information, but they did use directories in the lobbies of office buildings and made inquiries of office-building receptionists or management offices. In cases in which the kind of business or size or both were not readily apparent and in which an educated guess could not be made, telephone calls were made by the listers to secure the information.

At Westat, the listings were checked for acceptability, and any apparent duplicates were removed (such as separate listings of the same establishment on different floors of an office building). Each establishment was then assigned to a subsampling category, depending on the kind of business and employment size of the establishment as perceived by the lister. The perceptions of the listers were not required to be very precise. The final codes for kind of business and size were assigned on the basis of the more detailed information obtained for those in the sample. Establishments from the listings were classified as either high or low interest (depending on their likelihood of having some type of office equipment activity), using the listers' descriptions. The low-interest category consisted primarily of retail and personal-services establishments. (The lower use of equipment by such establishments was anticipated and was confirmed in the pilot study.) Thus only a rough description, such as "retail store," was needed to make the high/low classification. Three employment size classes—less than 10, 10–49, and 50 or over—were used that, when crossed with the two interest classes, formed the subsampling categories. In addition, all of a specified type of establishment that were of special interest in the survey were treated as one category regardless of size.

After assignment to the subsampling categories, systematic samples of the establishments within each category were taken at rates that were determined to be approximately optimal based on data from the pilot test and other sources. The rates themselves were chosen to provide an anticipated sample of about 7000 or more completed interviews.

2.3. Data Collection Procedures

For purposes of field organization, the sample design involved 79 first-stage sampling units that were divided into five regions. Each region was supervised by a full-time regional supervisor, who was supported by an assistant. Interviewers reported directly to their supervisor, who, in turn, reported directly to the home office. This overall organization provided close monitoring and control of all aspects of the data-collection effort.

In order to implement the field organizational plan, the supervisors and assistant supervisors were trained in a separate five-day training session, which was followed by a series of three intensive three-day training sessions for the interviewers. As part of the training, an audiovisual presentation was made to familiarize the interviewers with office equipment and environments. Each session was designed to maximize the trainees' active involvement and participation in the learning process, and each was conducted by the survey sponsor's and Westat's professional staff.

The materials used for the data-collection efforts were of two types: reference and assignment. The reference materials included a listing and interviewing manual, which provided a thorough explanation of all aspects of the questionnaire. As their assignment materials, interviewers received a segment folder for each assigned segment containing maps, copies of the completed listing sheets, as well as the necessary reporting forms. This list was a computer-printed sheet indicating the name and address of each establishment at which an interview was to be conducted. In addition, each questionnaire was prelabeled with the name and address of the sampled establishment. An interviewer was assigned one segment to work at a time, with the release of assignments based on the quality of work done by each interviewer and the interviewer's speed of work.

2.4. Data Processing

Data processing on the survey included all activities performed on questionnaires from the time they were received by regional field supervisors until the production of clean data tapes. These activities included regional and home-office receipt, coding, and manual editing of the questionnaires, coding verification, keypunching and keypunch verification, computer editing and correction or adjustment of the data, and the preparation and formatting of the data tapes for analysis.

Receipt of closed-out cases occurred first on the regional level, as interviewers mailed to their regional supervisors their completed work for the week. The supervisors maintained a check-in log, which was a computer-generated listing by segment within primary sampling units (PSUs), of each sampled establishment in the region. Each case was then edited for completeness and accuracy, and for 10% of the cases, a part of the interview was verified through a telephone call to the establishment.

Cases were then sent to the home office for another series of receipt and editing procedures. The home-office receipt operation was set up at Westat's office to receive incoming questionnaires from the field. This first step in the home-office data-processing chain was to

(1) ensure the return of all cases in the sample;
(2) provide a preliminary edit of key items in the questionnaire;
(3) determine the status of each case in order to calculate response rates; and
(4) control the dissemination of questionnaires through the remaining in-office activities (e.g., telephone nonresponse conversion or computer data processing).

Manual editing and coding were combined into one operation. This edit/code operation was carried out simultaneously for each questionnaire. As coders went through the questionnaires, coding them for keypunch, they also thoroughly edited the instruments for consistency of the responses. Editing and coding were standardized by development of a coding manual, the format of which included general coding instructions, question, column, and computer card deck number references for every item, a description of the content of each item (usually the question itself), all code values and their definitions, and all manual editing instructions.

Each coder used on the project was put through a two-day in-person training program during which both general coding conventions and specific instructions for coding the questionnaire were taught. This training was accompanied by close monitoring and review of several practical exercises on coding the questionnaire.

Coding and editing of the questionnaires were separated into two major divisions—the coding of items that related to categorizing an establishment according to its Standard Industrial Classification (SIC) code and the coding of the remainder of the questionnaire. Since SIC coding was an important and complex part of the study, a consultant experienced in industry classifications was employed on the project solely to work with SIC coding and its related matters. Establishments for which information was too limited to allow assignment of an SIC were referred to Dun & Bradstreet.

Each coder's work—both SIC and remainder—was independently verified on a 100% basis. All coding situations that were identified but did not follow the rules specified in the coding manual were referred to the supervisor. This was considered an important procedure to ensure the reliability of the data by seeing all decisions were made consistently, recorded, and incorporated into the coding manual when necessary.

In formatting the questionnaire, the data-processing staff gave special attention to its use by keypunch operators. The locations of items to be keypunched were made consistent throughout the questionnaire, items to be recoded had coder boxes added to eliminate confusion, and areas of the questionnaire that could not be formatted appropriately were coded onto separate transcription sheets. All keypunch work was 100% verified to ensure keypunching accuracy.

A complete machine-edit program was developed, incorporating edit specifications provided by the survey sponsor. Three types of edit programs were developed:

(1) value or range checks, which reviewed each data item for allowable code values;

(2) consistency or logic checks, which examined the logic of the relationships in the data; and

(3) special checks, which were designed to further review the data collected on equipment usage and brand/model codes against other criteria.

Questionnaires were divided into several batches to facilitate processing, and all three programs were run for each batch of questionnaires. The data were run through as many cycles of the edit programs as were needed to yield clean data.

3. TREATMENT OF INCOMPLETENESS

3.1. Efforts to Reduce Incomplete Data in Data Collection

The survey questionnaire was divided into the three parts described as follows:

Part I: questions relating to the establishment, such as employment size, kind of business, and number of pieces of office equipment;

Part II: questions relating to the locations of the equipment; and

Part III: questions relating to individual pieces of equipment.

To be considered usable or complete, a schedule must have had Part I completed at the minimum. As part of the field interviewing up to four attempts were made to complete each interview in person. Recontacts resulted if an establishment was closed, a knowledgeable respondent was not available, or a respondent did not have time to complete an interview in one session. If a sampled establishment refused to cooperate or if a complete interview was not obtained for some other reason during the field-interviewing phase, the establishment was later recontacted by telephone and an attempt made to administer a shortened version of the questionnaire.

Of the 9449 cases initially in the sample, 883 were determined after field inquiry to be ineligible. The ineligibles included vacancies, seasonal establishments, certain establishments that were not within the boundaries of a sampled segment, duplicate listings, and similar cases. "Duplicate listing" refers to an establishment that may have been listed twice by a field lister because it appeared to be two establishments but that at the time of enumeration turned out to be one establishment. (Probabilities of selection were adjusted for establishments for which such listings were identified, whether or not they were duplicated in the final sample.) Of the remaining 8566 eligible establishments, complete interviews were obtained in the field from 6854, or 80%. The remaining 1712 were referred to the central office for follow-up by telephone. A shortened form of the questionnaire, containing items felt to be of primary importance, was administered by telephone to establishments that were nonrespondents. Because of time restrictions, 1201 establishments were contacted by telephone rather than the full set of 1712 field nonrespondents. Establishments that were omitted were ones of lower priority, which typically had fewer than 10 employees and were expected to have low usage of office equipment. Through telephone recontact, 714 field nonrespondents, or 59% of those recontacted, were converted to complete interviews, giving a total of 7568 interviews completed either in person or by telephone for an overall response rate of 88% of eligible establishments.

A less intensive follow-up of small establishments would not be a desirable procedure if the initial sampling fractions for large and small establishments had in fact been approximately optimum and if the initial response rates had been about the same. In fact, given the total sample size, the sampling rate for segments was somewhat lower than the optimum rate for the large establishments but was larger than the optimum for the small establishments. Consequently, the smaller establishments were somewhat oversampled in the initial selections as compared with the larger establishments, so that it was appropriate to give priority to the larger ones in the telephone follow-up effort.

3.2. Measurement of Incomplete Data

Tables 1–6 give various unweighted and weighted tabulations of complete interviews[1] and nonresponses. The weights used are the basic weights described

[1] Throughout this chapter, "completed interviews" are defined as cases for which specific sets of key items were completed sufficiently to accept the return as usable, possibly with a limited amount of item imputation.

TABLE 1

Disposition of Sample Cases before and after Telephone Follow-up in an Office Equipment Survey

Disposition of case	*Unweighted number of establishments*				*Weighted[b] number of establishments (000s) after telephone follow-up*	
	Before telephone follow-up		*After telephone follow-up*			
	Number	*Eligible (%)*	*Number*	*Eligible (%)*	*Number*	*Eligible (%)*
Total sample cases	9449	—	9449	—		
Total eligible cases	8566	100	8566	100	4747	
Complete interviews	6854	80	7568	88	4149	88
Field	6854	80	6854	80	3881	82
Telephone	—	—	714	8	268	6
Nonrespondents	1712	20	998	12	598	13
Refusal	892	10	489	6	271	6
Unable to contact	392	5	257	3	203	4
Breakoff	106	1	59	1	13	—
Other noninterview	322	4	193	2	111	2
Ineligibles	883		883			
Vacant	462		462			
Seasonal	47		47			
Other	374		374			

[a] Note that the dash represents values of less than 1.
[b] Basic weights only are used; effect of ratio estimates is not included.

in Section 3.4.1 and are the inverses of the selection probabilities of the establishments. Table 1 gives the breakdown of the sampled establishments among different categories of respondents and nonrespondents. Before the telephone follow-up, 80% (unweighted) of the interviews were completed in the field. The telephone follow-up converted an additional 8% (unweighted) to "completes.)" The corresponding weighted numbers were 82% complete in the field and 6% complete by phone. By telephone, about 45% of the field refusals, 34% of the field noncontacts, 44% of the field breakoffs, and 40% of the other field noninterviews were converted to "completes" (unweighted). Of the number that were ultimately nonrespondents, about one-half were refusals (both weighted and unweighted) and the remainder were other types of nonresponse.

Within employment size classes, as shown in Table 2, the breakdown of respondents and nonrespondents is similar for both weighted and unweighted counts because the selection probabilities within a size class were designed to be constants within three broad kind-of-business classes and to differ only among the broad classes for the <50 employee establishments. Establishments in the 50+ class received the same basic weight with only a few exceptions. For the

TABLE 2

Numbers of Establishments by Disposition and Employment Size Class

Final disposition	Size class									
	10		*10–49*		*50–99*		*100+*		*Total*	
	Number	*Percent*	*Number*	*Percent*	*Number*	*Percent*	*Number*	*Percent*	*Number*	*Percent*
UNWEIGHTED										
Completes										
Field	4,046	81.0	1,786	81.6	477	67.1	500	74.5	6,809	79.5
Telephone	298	6.0	250	11.4	103	14.5	108	16.1	759	8.9
Nonresponse										
Refusal	309	6.2	83	3.8	68	9.6	29	4.3	489	5.7
Breakoff	21	.4	17	.8	14	2.0	7	1.0	59	.7
Unable to contact	211	4.2	23	1.1	16	2.3	7	1.0	257	3.0
Other	109	2.2	31	1.4	33	4.6	20	3.0	193	2.3
Total	4,944	100.0	2,190	100.0	711	100.0	671	100.0	8566	100.0
Percent of total	58.3		25.6		8.3		7.8		100.0	
WEIGHTED										
Completes										
Field	3,075,703	82.0	642,810	84.4	93,373	67.8	69,148	73.3	3,881,034	81.8
Telephone	164,287	4.4	69,173	9.1	21,417	15.6	13,567	14.4	268,443	5.7
Nonresponse										
Refusal	224,500	6.0	29,100	3.8	12,933	9.4	4,500	4.8	271,033	5.7
Breakoff	7,000	.2	3,300	.4	1,600	1.2	800	.8	12,700	.3
Unable to contact	189,619	5.1	8,000	1.0	3,600	2.6	1,700	1,8	202,919	4.3
Other	91,500	2.4	9,600	1.3	4,785	3.5	4,600	4.9	110,485	2.3
Total	3,752,609	100.0	761,983		137,708	100.0	94,315	100.0	4,746,614	100.0
Percent of total	79.1		16.1		2.9		2.0		100.0	

TABLE 3

Number of Establishments by Disposition and Industrial Group (Unweighted)

Industrial group	*Completes*		*Nonresponse*					
	Field	*Telephone*	*Refusals*	*Breakoff*	*Unable to contact*	*Other*	*Total*	*Grand total (%)*
1 Agriculture, forestry, fishing, mining								
Number	93	10	3	0	9	2	117	1.4
Percent	79.5	8.5	2.6	0	7.7	1.7	100.0	
2 Construction								
Number	219	26	15	0	16	4	280	3.3
Percent	78.2	9.3	5.4	0	5.7	1.4	100.0	
3 Manufacturing								
Number	1019	219	60	23	29	29	1370	16.0
Percent	74.4	16.0	4.4	1.7	1.5	2.1	100.0	
4 Transportation, communication, utilities								
Number	330	40	19	4	10	9	412	4.8
Percent	80.1	9.7	4.6	1.0	2.4	2.2	100.0	
5 Wholesale trade								
Number	537	46	46	2	14	14	659	7.7
Percent	81.5	7.0	7.0	.3	2.1	2.1	100.0	
6 Retail trade								
Number	1500	95	105	2	22	32	1756	20.5
Percent	85.4	5.4	6.0	.1	1.3	1.8	100.0	
7 Finance, insurance, real estate								
Number	702	82	40	5	27	18	874	10.2
Percent	80.3	9.4	4.6	.6	3.1	2.1	100.0	
8 Services								
Number	2113	217	176	21	133	76	2736	31.9
Percent	77.2	7.9	6.4	.8	4.9	2.8	100.0	
9 Government								
Number	296	24	25	2	6	9	362	4.2
Percent	81.8	6.6	6.9	.6	1.7	2.5	100.0	
Total								
Number	6809	759	489	59	257	186	8566	100.0
Percent	79.5	8.9	5.7	.7	3.0	2.2	100.0	

TABLE 4

Numbers of Nonresponding Establishments by Industrial Group and Employment Size

	Size					
SIC group[a]	*<10*	*10–49*	*50–99*	*100+*	*Total*	*Percentage of total*
Unweighted						
1	11	1	1	1	14	1.4
2	20	10	2	3	35	3.5
3	55	27	26	24	132	13.2
4	15	12	9	6	42	4.2
5	44	21	7	4	76	7.6
6	115	23	13	10	161	16.1
7	73	9	6	2	90	9.0
8	294	49	52	11	406	40.7
9	23	2	15	2	42	4.2
Total	650	154	131	63	998	100.0
Percentage of total	65.1	15.4	13.1	6.3	100.0	
Weighted						
1	6,000	400	100	400	6,900	1.2
2	12,000	5,000	700	800	18,500	3.1
3	13,700	6,300	3,300	3,000	26,300	4.4
4	15,800	3,300	1,800	800	21,700	3.6
5	29,600	7,100	1,500	500	38,700	6.5
6	119,500	11,600	1,900	2,400	135,400	22.7
7	44,600	2,500	800	1,300	49,200	8.2
8	255,519	13,400	11,600	1,500	282,019	47.2
9	15,900	400	1,218	900	18,418	3.1
Total	512,619	50,000	22,918	11,600	597,137	
Percentage of total	85.8	8.4	3.8	1.9		

[a] See Table 3 for SIC group identification.

<10, 10–49, 50–99, 100+, the weighted response rates were 86, 93, 82, and 87%, respectively. For the 50+ employee establishments, a larger proportion was completed by telephone than for the <50 employee cases. The telephone follow-up effort was directed primarily at larger establishments. For the <10 and 10–49 size classes, 4.4 and 9.1%, respectively, were telephone completes, but for the 50–99 and 100+ classes, 15.6 and 14.4%, respectively, were telephone completes. In each size class, about half of the nonresponses were refusals.

By industrial group, as in Table 3, the response rates are very similar for the different groups, although slightly lower for establishments in the service group. The division of unweighted completed interviews between field and telephone is also similar for the different groups at about an 80% field/8% telephone split except for manufacturing, for which the split was 74/16 and for retail trade at 85/5.

TABLE 5

Numbers of Contacts and Contacts per Responding Establishment by Employee Size Class

	Size				
Contacts	*<10*	*10–49*	*50–99*	*100+*	*Total*
Unweighted					
Total contacts	6486	3166	977	1227	11,876
Contacts/estab.	1.5	1.6	1.7	2.0	1.6
Weighted					
Total contacts (000s)	4706	1073	191	160	6,131
Contacts/estab.	1.5	1.5	1.7	1.9	1.5
Total respondents	4344	2036	580	608	7,568

Tables 4 gives counts of nonrespondents by industrial group and size, the distributions of which may be compared with the counts of all establishments by industry and size in Tables 2 and 3. The proportion of nonrespondents in the 100+ employee class is about the same as the proportion of all establishments in that class. For the <10 and 50–99 employee classes, the proportions of nonrespondents are greater than the proportion for all establishments, whereas for the 10–49 employee class, the proportion of nonrespondents is less than for all establishments.

TABLE 6

Numbers of Contacts and Contacts per Responding Establishment by Industry Group

	Unweighted		*Weighted*	
SIC group[a]	*Number of contacts*	*Contacts per establishment*	*Number of contacts (000s)*	*Contacts per establishment*
1	140	1.4	86	1.4
2	380	1.6	182	1.5
3	2,107	1.7	371	1.6
4	608	1.6	233	1.5
5	913	1.6	436	1.5
6	2,180	1.4	2050	1.3
7	1,225	1.6	608	1.5
8	3,777	1.6	1983	1.6
9	538	1.7	174	1.6
Total	11,876	1.6	6131	1.5

[a] See Table 3 for SIC group identification.

The proportions of nonrespondents and all establishments in Tables 3 and 4 in the different industry groups are very similar except for retail trade and services, in which the proportions of nonrespondents are, respectively, 16.1 and 40.7%, unweighted (22.7 and 47.2%, weighted) versus the proportions in retail and service for all establishments of 20.5 and 31.9%, unweighted.

Tables 5 and 6 give the number of contacts made per responding establishment by size and industry group. By size, the average number of contacts (unweighted) ranges from a low of 1.5 for the <10 class to 2.0 for the 100+ class. By industry, the unweighted averages range from 1.4 for agriculture, forestry, fishing, and mining to 1.7 for manufacturing and government and for weighted averages from 1.4 to 1.6.

3.3. Data Processing of Incomplete Data

Two types of incomplete data for which imputations were made were unusable schedules (total nonresponses) and missing data items on schedules that were otherwise usable.

The procedure used for adjusting for total nonresponse was a weight-adjustment procedure that will be described in detail in Section 3,4. During initial data processing, total nonresponses were included and a major effort was made to obtain an employment-size category and SIC code in sufficient detail for each establishment. Weight adjustments were then made to basic weights for responding establishments within size/SIC groups to impute for total nonresponses.

A second type of imputation that was made was for missing data items. Although a questionnaire may have been accepted as a whole as a complete response, some of the completes did have missing data items for which imputations were made. Rather than imputing for all missing items on a questionnaire, six key items were chosen by the survey sponsor as requiring imputation when the data were missing:

(1) environment type,
(2) environment purpose (general or specialized),
(3) environment usage,
(4) machine brand,
(5) machine process,
(6) machine tenure (owned or leased), and
(7) a measure of the amount of machine usage.

Other items for which missing data were not imputed included categories, such as "don't know," "no answer," or similar choices, that could be tabulated separately when making estimates.

The general method of item imputation used is known as a hot-deck procedure in which missing items are given values based on other similar complete questionnaires that had values for the items. To obtain an imputed value, the establishment with the missing item was clerically matched with another establishment whose major characteristics matched as closely as possible those of the establishment with the missing item. Specific imputation rules and priorities were specified. To facilitate following these rules and priorities, several successive batches of ordered lists of establishments were prepared, containing both clean questionnaires with responses for all of the specificed items intermixed with questionnaires for which item imputation was required. Ordinarily, but not always, the item to be imputed would be in an adjacent questionnaire in an ordered listing. The characteristics used in forming the matches were, from lowest to highest level of importance, industry (defined by groups of SIC codes), employee size class, number of office machines in the establishment, environment type, and machine type. With rare exceptions, a questionnaire was used only once for imputation of a particular item.

Because the closeness of the match was governed by the number and types of complete questionnaires available, relaxations of the matching criteria were sometimes necessary. If an establishment in the desired industry group was unavailable, for example, an attempt was made to find a match in a more broadly defined industry group. If that relaxation failed to produce a match, the size class for which a match was desired was broadened. A specific order for the relaxation of the matching criteria was established in order to ensure uniformity in the implementation of the procedure.

Imputation was done in such a way that no edit specifications, as previously described in Section 1, were violated. To ensure that no violations actually occurred, cases containing imputed values were reedited by computer after imputation. All imputed values were identified by special flags on the data tape, although no indication was given of the specific data record from which the imputed value was taken. As shown in Table 7, the amount of item imputation required varied from 1.6% of all responses for machine process to 7.1% for a measure of total environment usage. The method of variance estimation used, which will be described in the next section, did not reflect the effect on variances

TABLE 7

Numbers of Imputations and Imputations as Percentages of Total Responses for Different Items

Item	*Total number of responses*	*Number of responses imputed*	*Percentage imputed*
Environment type	8,589	325	3.8
Environment purpose	8,589	300	3.5
Environment usage	8,589	612	7.1
Machine brand	11,602	297	2.6
Machine process	11,602	189	1.6
Machine tenure	11,602	537	4.6
Machine usage	11,602	671	5.8

of the item imputation, although, as noted, the procedure could have been adapted to do so. In variance estimation, all imputed values were treated as equivalent to actual responses. The impact on variables of item imputation, using the hot-deck procedure, was to increase the variance aggregated or average estimates for the item approximately in proportion to the proportion of responses imputed, and thus the effect on variances was small.

3.4. Weighting, Estimation, and Variances

Associated with each establishment for which a completed questionnaire was obtained was a weight used in all estimates. Computation of each weight was done in several stages and reflects the selection probability of an establishment, the adjustment for complete nonresponse, and two stages of ratio estimation made to reduce both the sampling errors of estimates and the net effects of underlisting establishments.

3.4.1. Basic Weights

The basic weight attached to each establishment depended on the selection probability of the establishment, which was determined in most cases by the overall probability of selecting area segments (.01), and the subsampling size/interest category into which the establishment was classified subsequent to listing. Cases that were certainty selections within first-stage units (PSUs) or within ZIP groups or cases in unique ZIPs that were not subsampled have weights that depend on the probability of selection of the first-stage unit and the ZIP group containing them, not on the subsampling category. The sample of area segments was designed to be self-weighting, with each segment weight being 100. Within a segment, establishments were subsampled at the rate $1/w_k$, where k indicates a particular size/interest subsampling category. The basic weight of a particular establishment j in PSU h, segment i, and category k was $w''_{hijk} = 100w_k$. For the cases that were PSU or ZIP-group certainties or unique ZIPs that were not subsampled, the weights were calculated as the inverse of the overall establishment selection probabilities and thus they reflect the stages involved in their selection.

3.4.2. Imputation for Total Nonresponse

Upward adjustments were made to the basic weights for responding establishments to account for nonresponding establishments. A nonresponse could occur for the following reasons:

(1) The interviewer was unable to enter because of security problems.
(2) The establishment refused to cooperate.

(3) The establishment broke off interview before completion of the questionnaire.

(4) The interviewer was unable to contact establishment (after existence of establishment was confirmed).

(5) Other types of noninterview were encountered (after existence of establishment was confirmed).

Other cases were classified as either nonestablishments or ineligibles because they were vacant, seasonal and not active at time of survey, or out of segment. No nonresponse adjustment was or should have been made for nonestablishments or ineligibles.

The adjustments were made within classes defined by SIC code and employment size such that each class had a minimum of about 75 respondents, although this constraint was relaxed for some categories. Special efforts were made to obtain at least industrial classification (SIC) and employment size class information even for nonrespondent establishments by using three sources of information. First, the interviewer was asked to make efforts to fill in supplemental information based on observation or by inquiry at the establishment or in the neighborhood concerning the kind of activity and employment size of the establishment. Also, the lister was to have entered such information based on observation and occasionally based on limited inquiry. Finally, for nonrespondent establishments for which such information did not seem reasonably adequate for approximate classification, arrangements were made with Dun & Bradstreet (D & B) to provide any information they had in their records. The names, addresses, and telephone numbers, when available, were provided to D & B for about 800 establishments (including some with acceptable questionnaires but with inadequate employment or SIC information). They reported whether or not these establishments were included in their records and, if so, the employment size class and SIC they had assigned to the establishments. They found and reported such information for about 60% of the establishments sent to them. Library and other sources of information were used for any remaining large establishments.

Information from the sources just mentioned provided the basis for classifying all nonrespondent establishments by employment size and SIC for use in adjustment for nonresponse, although sometimes the classification was assigned on the basis of limited information. The initially adjusted weight for respondent j in PSU h and segment i in particular SIC/size class l was

$$w'_{hijkl} = (w''_l / w''_{l\mathrm{R}}) w''_{hijkl},$$

where the numerator w''_l is the sum of the basic weights for all eligible establishments in the sample that fell in the SIC/size class l, whether respondent or not, the denominator $w''_{l\mathrm{R}}$ is the sum of the basic weights for respondents in the class, and w'_{hijkl} is the basic weight for the particular respondent. Such weight adjustments had the effect of bringing the estimated number of establishments within each of the weight-adjustment classes up to what it would have been had responses been obtained from all establishments in the sample and, in effect, imputed responses for the eligible establishments that were nonrespondent in the sample.

An alternative procedure for adjustment for nonresponse might have been considered in which the nonrespondents were adjusted following similar procedures, but only within those returns that were not initial respondents or only within the returns from which responses were obtained by telephone. The philosophy for such an adjustment would be that the nonrespondents are more like the later respondents than they are like all respondents. Studies of this alternative in other surveys have shown that the assumption of greater similarity to late respondents than to all respondents is often not supportable. Moreover, by adjusting within all respondents, more adjustment groups and therefore, more homogeneous adjustment groups (with respect to the measures used for defining the adjustment groups) can be obtained. Also, adjustment within the smaller number of cases that would have been involved by using only the late respondents would have substantially increased variances, with no assurance of reduced biases and, in fact, some presumption of increased biases. Consequently, the decision was made to make the adjustments by the procedure described.

3.4.3. Ratio Estimation

Two stages of ratio estimation were used. The first-stage ratio estimates had no effect on the average level of aggregated estimates, other than as a result of random-sampling variability, but they increased the level of estimate for some SIC/size classes and decreased it for others. This estimate is based on figures derived from the preliminary 1975 *County Business Patterns* summary tape for five-digit ZIP Codes made available by the Bureau of the Census. It was made separately for each of a number of groups identified, again, on the basis of SIC and employment size class (these were different groups from those used for adjusting for nonresponse). Two figures were obtained from the 1975 *County Business Patterns* summary tape for each group m: X_m, the total establishments in the United States in group m, and X'_m, the estimated total in the same group m based on data from the same preliminary CBP tape for those ZIPs sampled in the office equipment survey. The counts in each sampled ZIP were weighted by the reciprocal of the probability of selection of the ZIP, aggregated over all sampled ZIPs for each group, and then adjusted so that over all groups combined the total of the estimated counts would equal the total count from the tape, i.e., so that

$$\sum_m X'_m = \sum_m X_m.$$

Note that if one of the groups m identifies manufacturing establishments with 10 or more employees and if the sampled ZIPs for the survey tend to have too many such establishments in them as a result of sampling variability, then the X'_m will be larger than the X_m, and vice versa. The purpose of this first-stage ratio estimate is to reduce the effects of such sampling variability, moderately, by adjusting for such differences. The estimate was made by adjusting the weights for establishments classified in group m from the survey by factors X_m/X'_m.

This adjustment was made for selected groups that account for most of the establishments in the survey. It was not made for all groups because CBP excludes, for example, the federal government and general activities of other levels of government. The size/SIC classes for which the first-stage ratio estimates were made and the adjustment factors are shown in Table 8. These factors were computed in advance of obtaining any sample results because they depend only on information in the preliminary 1975 CBP summary tape for all ZIPs and for the ZIPs selected for the sample, not on observed results from the field survey.

These groups, which were chosen because they were large enough to be reasonably represented in the survey, account for the bulk of the CBP coverage in industries in which CBP has substantially complete coverage. Note that the first-stage ratio estimate factors vary around 1 and do not tend to change the level of the estimates aggregated over all groups, but they will moderately alter the distribution among the size classes and kinds of business identified by the groups. Note also that the variability in the factors shown in Table 8 provides an indication of the sampling variability involved in the sample of ZIPs. The use of first-stage estimates will reduce some of the sampling variability moderately, but because they reflect only the two initial stages of sampling, they will by no means substantially eliminate such sampling variability. Also, since there was

TABLE 8

Codes and Employment Size Class for SIC Groups

Approximate description of SIC group	*SIC codes*	*Employment size class*	
		<10	10+
Construction	15–17	.949	.937
Manufacturing	20–39	1.001	.941
Transportation, communication, utilities	41–42, 44–49	1.079	.972
Wholesale trade	50–51	.983	.948
Retail trade	52–59	1.031	.994
Finance	60–67	.948	.938
Personal and selected other services	70, 72, 75, 76, 78–80	1.009	1.008
Business and legal services	73, 81	.966	.890
Educational and social services	82–84, 86	1.220	1.189

extensive subsampling within the sampled ZIPs, the combined sampling variability, including the effects of sampling segments from the ZIPs and subsampling of establishments within segments, was larger than illustrated by Table 8.

The second-stage ratio estimate adjustment is an adjustment made partly to reduce sampling variability, but primarily to reduce biases by adjusting the level of the estimates from the survey. The net effects of underlisting and the omission of a relatively small number of ZIPs from the lists from which the ZIPs were sampled call for some adjustment in the level of estimates to partially or substantially offset any biases from such causes. This second-stage ratio estimation procedure adjusted results in the survey by SIC and employment size class using establishment counts from the published 1975 CBP report. Four adjustment factors were computed, with each having the form Y_n/y'_n, where Y_n is the number of establishments with 10+ employees in SIC class n (classes different from those used for the first ratio-estimation procedure or the total nonresponse adjustment) from the published 1975 CBP report and y'_n is the survey estimate of the number of privately owned establishments with 10+ employees in SIC class n.

Three of the adjustments were made for classes of establishments for which the survey and CBP were felt to have approximate comparability of coverage, as shown in Table 9. The fourth factor was an overall adjustment derived from the three classes combined and was applied to all of the remaining establishments. The establishments with under 10 employees were omitted in computing

TABLE 9

CBP Counts, Survey Estimates, and Adjustment Factors for SIC Groups

Approximate description of SIC group	*SIC codes included in group*	*Total CBP establishment counts*[a]	*Estimated total establishments from survey*[b]	*Adjustment factor*
Manufacturing	20–39	319,380	260,800	1.224
Transportation, communication, and utilities	41–42, 44–49			
Finance, insurance, and real estate	60–67	169,502	163,000	1.040
Business and legal services	73, 81, 89			
Educational and social services	82–84, 86			
Retail trade	52–59	369,947	324,300	1.141
Personal services	70, 72, 75, 76, 78–80			
All industries combined		858,829	747,900	1.148

[a] From the 1975 CBP (as published), with unclassifiable establishments distributed among industries.
[b] After first-stage ratio estimate and adjustment for total nonresponse.

the adjustment factors because of lack of comparability in employment reporting between the survey and the CBP for smaller establishments. In particular, the survey defined sole proprietors and family workers as employees and CBP does not. The fourth adjustment factor was applied to all establishments not covered by the first three factors, including establishments with under 10 employees, government and government-owned establishments, and all other classes of establishments, since it was intended primarily to offset any effects of incompleteness of listing and of the possible omission of some ZIPs from the universe list and since omissions would affect all classes of establishments. The estimates from the survey, the corresponding figures from published CBPs for 1975, and the ratios of these, which served as adjustment factors, are shown in Table 9.

The second-stage ratio estimation adjustment factor for the three groups combined was 1.148, and its estimated coefficient of variation was 10.2%. This level of sampling error is consistent with expectations, and sampling error could explain the total adjustment; however, the adjustment is in the expected direction, and we assume that undercoverage was a contributing factor to the adjustment. The adjustments have the effect of reducing the sampling errors of the final estimates, as well as of adjusting for such undercoverage.[2]

The first and second stages of ratio estimation and the adjustment for total nonresponse had the effect of adjusting the basic weights as follows:

$$w_{hijklmn} = \frac{w_l''}{w_{l\mathrm{R}}''} \frac{X_m}{X_m'} \frac{Y_n}{y_n'} w_{hijklmn}''$$

where $w_{hijklmn}$ represents the final weight for the jth establishment in PSU h, segment i, subsampling category k, and adjustment groups l, m, and n.

Using the final weights, the estimated totals have the form

$$x' = \sum_{h,i,j,k} w_{hijk} x_{hijk},$$

where the subscripts l, m, and n have been dropped for convenience and where x_{hijk} is the observed value of some characteristic for a particular responding

[2] After this chapter was prepared, results became available from a repeat of the office equipment survey a year later. In this repeat survey, the same segments were used and were recanvassed to update the listings. The recanvassing made use of the initial listings and identified establishments existing at the time of the original survey that were missed and still in existence at the time of the repeat survey. The dependent relisting procedure did indeed identify a considerable number of additional establishments that should have been listed initially. Certain data for 1977 as well as for 1978 were obtained for those establishments, and the 1977 nonresponse adjustments and ratio estimates were recomputed. The effect of adding these establishments to the 1977 survey would have increased the responses by approximately 11% (without changing the number of nonrespondents). Inclusion of these establishments in the 1977 survey would have reduced the average coverage adjustment to approximately 5% instead of approximately 15% as then reported. Thus although sampling errors could plausibly have accounted for the full nonresponse adjustments made in 1977, the coverage adjustments were, in fact, substantially the result of underlisting as had been presumed.

establishment. Similarly, ratio estimates, such as means and percentages, have the form $r = x'/y'$, where x' and y' are estimated totals.

3.4.4. Variance Estimation

The variances of estimates from the survey were computed by a replication method known as balanced half-sample replication (McCarthy, 1969). This involves identifying the establishments included in each of a specified number of balanced half-samples (24 in this case), repeating the nonresponse and the two ratio-estimation adjustments separately and independently for each of the 24 half-samples, as well as for the total sample, preparing estimates (tabulations or computations of items for which variances were estimated) for each of the half-sample replicates, and then computing the variances among the half-sample estimates. This particular method of variance estimation was chosen primarily because, for the complex estimation procedures used, derivation and computation based on explicit variance formulas would have been time-consuming and would have been required separately for each of the types of various derived statistics that might be computed from the data, such as percentages or medians. The method of balanced half-sample variance estimation obviates the need to derive explicit variance formulas for each type of estimate or desired statistic and approximately reflects the type of PSU design and the various weight adjustments employed. The simplicity and generality of the half-sample method was an important advantage. As adapted for this study, this method did not reflect any contribution to variance due to item imputation, although the method could have been used to do so. Separate item imputations would have been necessary for each half-sample, but because of a shortage of time and a feeling that the item imputations did not contribute importantly to the variances of most statistics, this was not undertaken.

In forming pairs of sample units to use in defining half-samples, units were collapsed together in much the same way as would have been used in forming pairs for computing collapsed strata estimates of variance. For non-self-representing PSUs, the units represented in the half-samples were groups of PSUs themselves; for self-representing PSUs, the units were groups of sample segments. The groups were identified in pairs, which were determined by the stratification used in drawing the sample, and one group from each pair was included in each balanced half-sample replicate. In accordance with the general balanced half-sampling procedure, the particular set of groups included in a half-sample was determined by a square orthogonal matrix of dimension equal to the number of pairs that were formed.

The weight for each establishment in each of the 24 half-sample replicates and in the full sample was computed and included on the data tape, so that variances could be computed readily for any desired estimates, whether simple tabulations or any kind of more or less complex, derived statistics.

For a particular half-sample, the adjusted weight for an establishment that is in a particular nonresponse adjustment class l and in SIC/size groups m and n is

$$w_{\text{HS}hijk} = w''_{hijklmn} \frac{2w''_{l\text{HS}} X_m Y_n}{w''_{l\text{RHS}} X'_{\text{HS}m}\, y'_{\text{HS}n}},$$

where

$w''_{hijklmn}$ is the basic weight for establishment j in size/interest class k in PSU h, segment i, and adjustment groups l, m, and n;
$w''_{l\text{HS}}$ the sum of the basic weights for all sampled establishments in nonresponse adjustment class l in half-sample HS;
$w''_{l\text{RHS}}$ the sum of the basic weights for responding establishments that are in nonresponse adjustment class l and in the half-sample;
$X'_{\text{HS}m}$ the estimated total establishments in SIC/group m using the 1975 CBP tape counts for the ZIP groups in the half-sample (after adjustment so that $\sum X_m = \sum X'_{\text{HS}m}$);
X_m the total establishments in group m from the entire 1975 CBP ZIP tape;
Y_n the total establishments in SIC/size group n from the 1975 CBP; and
$Y'_{\text{HS}n}$ the estimated total establishments in SIC size group n from the survey data for establishments in half-sample HS after the half-sample adjustment for nonresponse and the first ratio adjustment.

An estimated total from half-sample HS was of the form

$$x'_{\text{HS}} = \sum\sum\sum\sum w_{\text{HS}hijk} x_{hijk},$$

where as estimated ratio had the form

$$r_{\text{HS}} = x'_{\text{HS}}/y'_{\text{HS}}.$$

Based on the half-sample estimates, the estimated variance of a particular estimated total x' was

$$\hat{\sigma}^2_{x'} = \frac{1}{24} \sum_{\text{HS}=1}^{24} (x'_{\text{HS}} - x')^2,$$

where x' is the estimated total derived from the full sample. Similarly, the estimated variance of a ratio r was

$$\hat{\sigma}^2_r = \frac{1}{24} \sum_{\text{HS}} (r_{\text{HS}} - r)^2.$$

REFERENCES

Bureau of the Census (1973). *Country and City Data Book 1972* (*A Statistical Abstract Summary*), U.S. Department of Commerce, Washington, D.C.

Bureau of the Census (1978). *Country Business Patterns 1975, U.S. Summary CBP-75-1*, U.S. Department of Commerce, Washington, D.C.

McCarthy, P. J. (1969). Pseudo replication: Half samples. *Review of the International Statistical Institute* 37(3): 239–264.

CHAPTER **6**

Annual Survey of Manufactures*

William G. Madow

1. OBJECTIVES OF THE ANNUAL SURVEY

The total industrial statistics program of the U.S. Bureau of the Census consists of the quinquennial Census of Manufactures; the Annual Survey of Manufactures (ASM); the current survey program of more than 100 monthly, quarterly, and annual commodity surveys; the monthly survey of Manufacturers' Shipments, Inventories, and Orders; and other related surveys conducted either for other agencies or on an occasional basis. Complementary to this program is the *County Business Patterns* (CBP) series (published annually and jointly by the Bureau of the Census and the Social Security Administration), which presents data by industry group and county for numbers of establishments, payrolls, and employment.

The content of the ASM is broadly comparable to that of the quinquennial Census of Manufactures, for it provides, annually, estimates of all the principal measures yielded every 5 years by the censuses. Data are published for the following items from both the census and ASM:

(1) number of employees (total, production workers, and other employees);

(2) payroll (total, production worker wages, and other employees' salaries and wages);

* This case study is based on Ogus and Clark (1971), Cole and Altman (1977), and Clark and Kusch (1979). The revisions made in the sample design beginning with the 1979 Annual Survey of Manufactures are summarized in Waite and Cole (1980). A revised editing procedure is being developed (Greenberg, 1982). A draft of this case study was reviewed by Donald F. Clark of the Bureau of the Census.

INCOMPLETE DATA
IN SAMPLE SURVEYS
Volume 1, Part II

ISBN 0-12-363901-8

(3) total plant-hours of production workers;
(4) total cost of materials;
(5) total value of shipments;
(6) capital expenditures (total, for new plant, for new equipment, and for used plant and equipment separately);
(7) quantity of purchased electricity;
(8) value of inventories (total, finished products, work in progress, and materials);
(9) value of shipments and receipts (total and by individual product class, including separate figures on receipts for contract work and sales of products bought and sold as such); and
(10) value added by manufacture. (This key output measure is calculated as the difference between total receipts and the total cost of materials and related inputs plus the algebraic difference between the end and beginning of year finished and work-in-process inventory values.)

In the censuses, product data are compiled for each of nearly 10,000 individual products and for approximately 1600 product classes, which are homogeneous combinations of the individual products. For example, creamery butter shipped in bulk and creamery butter shipped in consumer packages are two different products within the product class of creamery butter; upholstered sofas, chairs, and rockers are each products included in the product class of upholstered wood furniture. In the ASM, product data are obtained only at the product class level. Physical quantity data are also compiled in the census for approximately 5000 of the individual products but are not requested in the ASM because of variations among the units of measure for different products included within the same class.

The data from the ASM on general statistics (employment, payrolls, value added, total capital expenditures, etc.) are presented in considerable industrial and geographical detail.

The Standard Industrial Classification (SIC) Manual, issued by the U.S. Office of Federal Statistical Policies and Standards, defines three levels for classifying manufacturing industries:

(1) major industry groups made up of 20 broad (two-digit code) categories such as food and kindred products, chemicals and allied products, and electrical machinery, equipment and supplies;
(2) industry groups made up of 150 intermediate (three-digit code) product categories, such as meat products, dairy products, industrial inorganic and organic chemicals, drugs, electrical industrial apparatus, household appliances, and electric lighting and wiring equipment;
(3) industries made up of 451 (four-digit code) product categories such as creamery butter, natural and processed cheese, alkalies and chlorine, industrial gases, motors and generators, and industrial controls.

The ASM presents a comprehensive set of data in full industrial detail at the national level and a similarly comprehensive set of related statistics geographically in somewhat less industrial detail.

The estimates from the ASM of individual product class shipments are currently limited to national totals. Information for all significant aspects of manufacturing, as compiled every 5 years from the complete Census of Manufactures, is also available for the intercensal years, partly from the ASM, partly from CBP. The estimates provided by the ASM permit extrapolation of the census totals for a large number of measures, whereas the CBP data permit extrapolation in terms of key employment figures for fine geographic detail. The ASM implicitly provides, also, measures of change of trends between the census years. These are considered of primary importance; accordingly, some comparable data for both the current and previous year or years are given in most of the tables. As a further recognition of the importance of change, year-to-year relatives, as well as the supporting totals, are shown in the product class tables, where space is available for presenting such measure explicitly.

Although the principal objectives of the ASM are to provide the basic estimates just described, the survey also serves the following supplementary purposes:

(1) It provides a file of current information on all large manufacturing establishments in the country.

(2) It reduces the Census of Manufactures workload for both the ASM panel and the Bureau of the Census. The ASM respondents find they are better prepared to complete census reports because they are already familiar with many of the concepts and definitions, many of the inquiries are indentical, and they have reported the comparable data for the identical inquiries in the past several annual surveys. Additional gains are realized from the application of editing tests that take advantage of the companies' prior year ASM data in checking their census reports and from more complete and more easily controlled coverage for the large companies that dominate both the ASM and the census.

(3) The ASM panel is a convenient vehicle for collecting supplementary census information that is not needed on a complete basis.

(4) The ASM estimates are used as benchmarks for the important monthly survey: Manufacturers' Shipments, Inventories, and Orders. This key series, which reveals current trends in the manufacturing sector, is closely tied to the ASM to improve the reliability of the monthly estimates and to afford a degree of refinement in industrial detail unobtainable without such a benchmark.

(5) It provides annual time series of many related economic variables, both for individual establishments and for groups of like establishments, that can be used for a variety of special economic studies.

(6) It provides otherwise unobtainable lists of establishments entering into new product lines that should be covered in one or another of the current commodity surveys.

2. THE DESIGN AND SELECTION OF THE ASM SAMPLE

A sample design attempts to maximize the amount of useful information obtained for a fixed cost or, conversely, to minimize the cost, having specified the minimum amount of information that a survey must provide. In applying this principle to ASM sample, the characteristics of the universe, the operating conditions under which the survey is conducted, and a number of special restraints importantly affect the design.

The following are the principal considerations that determined the choice of the sample design for the ASM:

(1) The manufacturing universe has an extremely skewed distribution, with a relatively small proportion of establishments dominating the aggregates. Therefore, all large units were included in the panel with certainty, and the sample selection probabilities for the smaller units were intended to correlate well with their size.

(2) A general rule was adopted that aimed at making the relative error of each estimate inversely proportional to the square root of its importance.

(3) Emphasis was placed on controlling the reliability of product class estimates, which are inherently more difficult to estimate than general statistics totals.

(4) Two restraints were imposed on the sample design because of operating convenience, coverage considerations, and related data needs: (1) to define the manufacturing company[1] rather than the establishment as the sampling unit and (2) to define an arbitrary certainty stratum based on employment size.

(5) The sampling system should permit the exclusion of smaller companies in the panel from subsequent samples (without biasing those samples) in order to distribute the reporting burden equitably. Such rotation should be made periodically (that is, after each complete census) rather than yearly because of the greater costs and control problems associated with frequent changes.

(6) No restraint had to be observed on the number or fractional character of the sampling probabilities used.

(7) The total sample size was set at approximately 60,000 establishments, taking account of the quality of the estimates and the cost for a sample of that size.

To begin, the manufacturing universe is geographically scattered and comparatively small. Its 350,000 establishments contrast sharply with the 65 million households, 3 million farms, and 1.5 million retail establishments in the United States. Fortunately, satisfactory sampling frames are available from the quinquennial Census of Manufactures and are supplemented by Social Security Administration (SSA) lists of new manufacturers. In addition, the collection of

[1] This has been changed in the 1979 revision. See following discussion.

reports by mail is feasible since the data items to be reported are generally available from the records normally kept by manufacturing companies. The latter two features permit considerable flexibility in selecting the sample. In particular, area sampling in order to reduce collection costs is not required. Instead, each sampling unit can be considered individually.

That each unit can be considered individually is of major importance because the manufacturing size distribution is "L" shaped. It has an extremely long tail stretching to the right, so that relatively few large units dominate the aggregates. This is illustrated by the abbreviated table (Table 1), showing the contribution of the smallest and largest manufacturing establishments to national aggregates.

For both the Census of Manufactures and the ASM, establishments are the basic reporting units. Heretofore, in the ASM, the company has been the sampling unit, with selected companies required to report for all of their manufacturing establishments. This rule was adopted for several reasons.

(1) It greatly simplified operating controls on the sample.

(2) It provided a basis for representing in the sample newly built establishments of multiunit companies.

(3) It provided data needed for supplementary company studies dealing with mergers, consolidation, etc.

With the advent of the Bureau of the Census' Company Organization Survey (COS), these reasons are no longer compelling. The COS annually updates the multiunit company universe, identifying changes in company organization brought about by mergers, acquisitions, new plants, and plants that have ceased operations.

Beginning with the 1979 ASM, therefore, an establishment (rather than a company) sample has been utilized. Gains in sample efficiency occur because establishments are given a chance of selection proportional to their individual

TABLE 1

Distribution of Establishments, Employees Payroll, and Value Added in Manufacturing by Establishment Size Class[a]

	Percentage of all manufacturing total 1972			
Employment size class	*Establishments*	*Employees*	*Payroll*	*Value added*
1–4	35.9	1.1	.9	1.1
5–9	14.9	1.7	1.5	1.5
⋮	⋮	⋮	⋮	⋮
1000–2499	.5	12.5	14.0	14.3
2500 and over	.2	16.2	21.7	20.7

[a] *Source:* U.S. Bureau of the Census (1972).

measures of size. In particular, small establishments of large companies are no longer included in the sample with certainty. Design and selection with the company as the sampling unit are discussed in the remainder of this section. Since 1979, the same types of sample design and selection have been used, but with the establishment as the sampling unit.

Large companies must be included in every sample with certainty if the survey is to be successful, but the sample of smaller companies can be changed periodically, and this has been done. New samples have been selected after each complete Census of Manufactures, or approximately every 5 years. For the earlier surveys, each new sample was selected independently of the previous one. As a result, several hundred small companies were selected by chance to report in two or more successive panels. In order to eliminate this kind of reporting burden, the requirement has been imposed that the small companies selected for a given ASM panel should be systematically excluded from the next panel without biasing the estimates.

This requirement of spreading the reporting burden among the smaller companies might be met by frequently rotating the smaller companies, that is, by dropping some and adding others every year. No yearly rotation procedure has been adopted, largely because of the additional expense and additional control problems entailed.

The problem was attacked subject to the additional restraints that each company should have a single probability of being selected (which would apply to all items it reports), that each company in the frame should be sampled independently, and that sampling should be with probability proportional to size (PPS), where size is defined as a variance measure determined from the company's base-period product-class values.

Use of a single probability for all items reported by a given sampling unit is the usual practice. The alternative, to use different probabilities for different items and perhaps to request different information from different establishments, was not seriously considered. This PPS sampling with a large certainty stratum was adopted on the theory that it would more nearly approximate the objective of minimizing the variances of the estimates than would other methods when sampling from the extremely skewed distributions that characterize the manufacturing universe. The decision to sample independently, however, departed from the more standard practice of earlier surveys in which systematic PPS sampling had been used.

Since the 1959 ASM, for the selection of new sample panels, each unit was sampled independently of the selection or nonselection of every other unit or combination of units, with the probability of selection varying from unit to unit. For this reason, the ASM sampling procedure is termed *Poisson sampling*, and that term will be used throughout this chapter.

With Poisson sampling, a linear unbiased estimate of the total Y is given by

$$Y' = \sum_{s} Y_h W_{h'}, \tag{1}$$

where s is the particular set of units selected, Y_h the current value of unit h, and W_h the sampling weight, which is the reciprocal of the probability of selection p_h of unit h.

The variance of the estimated total Y' for Poisson sampling is

$$\sigma^2(Y') = \sum_{h=1}^{N} Y_h^2 W_h^2 p_h q_h = \sum_{h=1}^{N} Y_h^2 q_h / p_h. \tag{2}$$

In terms of the sampling weight W_h, the variance can be written as

$$\sigma^2(Y') = \sum_{h=1}^{N} (W_h - 1) Y_h^2. \tag{3}$$

With Poisson sampling, the total number n' of units selected is a variable. The expected sample size n is

$$n = E(n') = \sum_{h=1}^{N} p_h. \tag{4}$$

The variance of the sample size is

$$\sigma^2(n') = \sum_{h=1}^{N} p_h q_h. \tag{5}$$

Also, unbiased estimates of $\sigma^2(Y')$ are given by

$$\sigma_s^2(Y') = \sum_{s} Y_h^2 q_h / p_h^2 \tag{6}$$

Equations (1)–(6) may be obtained by writing

$$Y' = \sum_{h=1}^{N} W_h a_h Y_h, \tag{7}$$

where $a_1, \ldots, a_N$ are independent random variables with probability that $a_h = 1$ equal to p_h, where p_h is the probability of selecting h, and with probability that $a_h = 0$ equal to $q_h = 1 - p_h$, where q_h is the probability that h is not selected $(h = 1, \ldots N)$. Then $W_h = 1/p_h$ follows from the requirement that Y' be unbiased, and equations (2)–(6) follow from the properties of the random variables $a_1, \ldots, a_N$.

Poisson sampling provides flexibility in drawing nonduplicating samples that can yield unbiased estimates. With systematic PPS sampling, nonduplicating samples can be selected only if the original order of the sampling frame is maintained and the probabilities are not changed or are all changed proportionally. These restraints are necessary with systematic sampling because the probabilities of selecting particular different units in successive samplings are not independent of their order. Both restraints are seriously objectionable in the ASM program for both theoretical and practical reasons. Because the topic will not be discussed further here, the reader is referred to Ogus and Clark (1971).

2.1. General Approach for Assigning Measures of Size to Individual Sampling Units

If the expected sample size n is fixed, then the variance $\sigma^2(Y')$ of a sample estimate is minimized when $p_h = tY_h$, that is, when the sampling probability assigned to each unit h is proportional to its Y_h value. This condition also minimizes n when $\sigma^2(Y')$ is fixed. This principle was applied, in an approximate way, in assigning probabilities for the ASM.

Most of the estimates in the ASM are developed by the "difference estimate" formula

$$Y'' = D' + X, \tag{8}$$

where $D' = (Y' - X')$ is the sample estimate of the change from the last census and X the total from that census.

The variance of the difference estimate has the same form as that of the unbiased estimate Y', given in (1) and can be written as

$$\sigma^2(Y'') = \sigma^2(D') = \sum_{h=1}^{N} (W_h - 1)D_h^2, \tag{9}$$

where $D_h = (Y_h - X_h)$ is the value of the change for company h.

Also, $\sigma^2(Y'')$ is minimized when

$$p_h = t|D_h|, \tag{10}$$

that is, when the probability of selecting unit h is proportional to the absolute value of the unit's change D_h. (For simplicity of notation, the absolute value symbol $||$ will be omitted in the remainder of this chapter.)

To use PPS sampling, a measure of size must be assigned to each sampling unit, and according to Eq. (10), the optimum measure is one that is proportional to the change in the value of the unit. For the ASM, because of the great stress placed on the product-class estimates, the most appropriate measure of size was taken to be one that is related to the changes D_{ch} in product-class shipments. This was done in two stages. First, measures of size D_{ch}, were derived for each product class reported by each company. These product-class measures were then combined into total company measures of size D_h in order to observe the restraint that a single probability of being selected should apply to all items that a company reports (Ogus and Clark, 1971).

2.2. Designation of Arbitrary Certainty Companies

All companies with one or more establishments having 250 or more employees were to be included with certainty in the survey panel. In applying this criterion, establishments were included whose "average" employment (average of the four quarterly production-worker figures plus the March "all other" employee figures) or whose November production-worker plus March "all other" employee figures equaled or exceeded 250. The reason for adding the second version (November plus March) of the rule was to anticipate the growth of companies that would not qualify as certainty companies based on average 1972 total employment but that were likely to qualify in the following year. The effect of this type of rule was to make the sample panel conform more closely to specifications over the intercensal period that lay ahead.

Of the 312,000 establishments in the manufacturing universe in 1972, about 38,000 were in the certainty stratum as a result of these certainty cutoff rules. Of the 38,000 certainty establishments, approximately 2000 were single-unit companies, and the remaining 36,000 establishments were accounted for by some 4500 multiunit companies.

After the elimination of the arbitrary certainty companies from the sampling frame, further analysis was used to determine a certainty cutoff for the remaining companies. Companies with measures of size that placed them above this certainty cutoff were also included in the sample with certainty, and these companies were termed "analytical certainties." The remaining companies made up the noncertainty stratum, and their selection for the survey will be described in the next subsection.

In the equation $p_h = tD_h$, the proportionality constant t depends on the desired noncertainty sample size n and on the cumulative measure of size of the noncertainty universe $\sum^N D_h$, that is, $t = n/\sum^N D_h$. Since $p_h = tD_h = (n/\sum^N D_h)D_h$, the measure of size D_h and the corresponding $\sum^N D_h$ can be found for which $p_h = 1$. All p_hs of larger size define analytical certainty companies, and for all smaller D_hs, $p_h = tD_h$ is the appropriate probability of selection.

2.3. Selection of Noncertainty Companies

With every company in the noncertainty sampling universe assigned a final p_h five-decimal-place value, a Poisson sampling operation, giving each company an independent chance of selection, was then applied as follows:

(1) Using a random-number-generator program, a random number R_h between .00000 and .99999, inclusive, was developed for each noncertainty company.

(2) Each such random number R_h was added to its corresponding p_h. Whenever $p_h + R_h \geq 1.00000$, the company was designated as selected for the sample; when $p_h + R_h < 1.00000$, the company was designated as not selected.

2.4. Nonresponse Rates

The rate of nonresponse has been increasing among the single-unit companies over the past several years, with the highest nonresponse rate occurring among the smaller single units. The Table 2 illustrates this and shows the multiunit response rates as well.

This nonresponse occurs despite the fact that the ASM is a mandatory survey, and despite a rigorous follow-up program. After the initial mailing of the report form, early in the year, delinquent establishments receive up to four follow-up letters with increasingly severe language. The fourth mailing is certified, with another copy of the report form included. The largest delinquent establishments, with employment of 1000 or more, are also contacted by telephone.

The consequence of this increasing rate of nonresponse among single-unit companies is the necessity for a greater amount of imputation in order to produce universe level estimates. Starting with payroll data obtained from the Internal Revenue Service and the Social Security Administration, all other data for a delinquent establishment are imputed using average relationships among data items that depend on the establishment's four-digit industry classification. These imputed data have associated with them an unknown amount of bias, which is not directly measurable.

Whether the increasing complexity of the ASM report form has been the cause, or a major cause, of increased nonresponse is not known. An opinion questionnaire was briefly considered, to try to determine the causes. This approach was discarded in favor of a direct attempt to reduce nonresponse by the experimental use of a short form for the smaller single units.

TABLE 2

Receipts by Total Number of Employees (TE) and Survey Year

	Response rate (%)						
Description	*1969*	*1970*	*1971*	*1973*	*1974*	*1975*	*1976*
Single units							
Total	82.1	77.1	81.7	77.9	72.6	73.7	73.4
Fewer than 50 TE (includes employment unknown)	78.1	75.2	79.2	72.4	69.8	65.8	66.7
1–4 TE	77.7	71.7	78.7	72.4	64.9	55.4	69.3
5–9	83.0	74.6	83.0	78.7	69.9	69.7	68.6
10–19 TE	85.6	73.5	84.0	82.9	73.5	77.0	75.7
20–49 TE	86.6	79.5	85.5	83.4	76.4	76.5	78.5
≥50 TE	89.7	86.2	86.6	87.2	78.8	81.9	82.3
Multiunits							
Total	89.5	84.5	83.0	89.2	90.4	92.3	89.7

3. AN EXPERIMENT ON THE EFFECT OF A SHORT FORM IN INCREASING RESPONSE OF SMALLER COMPANIES

The primary objective of the ASM short form experiment (Cole and Altman, 1977) was to determine whether the present rate of nonresponse among the smaller single units could be reduced by using a short form.

A secondary objective of this experiment was to determine the extent to which the errors of the survey estimates would be affected by using a short form.

Two kinds of nonresponse were considered: receipt (or unit) nonresponse and item nonresponse. The receipt nonresponse rate is defined as the ratio of the number of report forms not returned, or returned completely blank, to the number of report forms that should have been returned. The item nonresponse rate is the ratio of the number of times an item entry is not made to the number of times that an item entry should have been made. Item nonresponse is defined relative to the group of establishments that provide receipts and not to the entire universe of establishments. An establishment that is out of business for the survey year has no data entries to make. This out-of-business establishment does have an obligation to return the report form and indicate the status. Such establishments contribute to the receipt response rate but do not affect the item response rates.

Item nonresponse is of greater interest in the ASM than receipt (unit) nonresponse. Although a higher receipt response would usually imply a higher item response, this is not necessarily the case. Two groups of establishments with similar receipt response rates may have considerably differing response rates for individual items.

Because the short form experiment began with the 1975 survey year, the second year of reporting for the ASM panel, interest centered on those establishments that did not respond to the standard form in 1974, and the experiment in its first year was confined to companies that had been mailed the standard form in 1974.

The experiment was made more inclusive in its second year (the 1976 survey year) by adding samples of manufacturing births that received an ASM report form for the first time. This group will not have whatever preconditioning effect the first year experimental group may have had that was due to the standard form they received in 1974. The effect of short form usage may not be apparent in a single year. Over the five-year life of the ASM panel, nonresponse rates for the short form and the standard form might initially be similar but then start to diverge.

The secondary objective of this study is expected to be more difficult to attain. This is to make an assessment of the effect of short form usage on the response errors of the estimates. If higher rates of item response were associated with the short form, then the estimates for those items should be improved, assuming that the quality of the reported data, after survey processing, is comparable for short forms and standard forms. The rate of response for items not included on the short form is, of course, zero. Use of the short form deliberately raises the total amount of imputation for these items. The accuracy of such imputation depends partially on how well the related "control" items are reported on the short form. If these basic items were better reported on short forms and the additional items (those not included on the short form) were not very well reported on the standard form, there might be no loss, or an insignificant loss, in the quality of the estimates for these items.

To make comparisons of the errors of the estimates resulting from short-form and standard-form usage requires consideration of the relative amounts of imputation involved for different items and of the relative biases of these imputations.

3.1. Experimental Design

The short form was designed with the ASM standard form serving as a base. This was done to minimize the processing problems associated with an additional form. The clerical processing, data keying, computer editing, and analyst correction operations were very similar for both the short form and the standard form. Several of the items on the standard form were not included on the short form, either because they were highly correlated with other items on the short form, or because they were "detail"-type inquiries that did not seem appropriate for a short-form inquiry. These items are supplemental labor costs, plant hours, quantity of electricity, detailed fuel consumption, and method of valuation for inventories. In addition, for the items that were included on the short form, with the exception of value of products shipped and employment, only the totals were requested.

The basic experiment consisted of comparing the response rates for the short form and for the standard form for two groups of establishments: those that did not respond for the 1974 survey year (1974 delinquents) and those that did respond (1974 respondents). Only the smaller (total employment less than 50, or unknown) single-unit establishments from the 1974 ASM were included in the experiment.

Four experiment panels were selected by systematic sampling:

(1) Panel 1 consisted of 326 1974 ASM delinquents to be mailed the short form.

(2) Panel 2 consisted of 348 1974 ASM delinquents to be mailed the standard form.

(3) Panel 3 consisted of 343 1974 ASM respondents to be mailed the short form.

(4) Panel 4 consisted of 363 1974 ASM respondents to be mailed the standard form.

Following the mailing of the forms, each of the four panels was subject to the same follow-up procedures as was the remainder of the small single-unit establishments in the ASM panel.

3.2. Experimental Results: 1975

In order to analyze the effect of the short form, receipt response tabulations were kept for each of the four panels. An item response tabulation was also maintained. A receipt was defined as being a returned report form with at least one item completed or a form indicating that the establishment had gone out of business. To be considered a receipt, out-of-business establishments were not required to supply any data but they were required to indicate their status. Out-of-business establishments were not included in the base of the item response tabulation. Forms that were returned but were found to be completely blank were not counted as receipts for tabulation purposes.

A few establishments considered not to be in scope of the ASM (non-manufacturers and manufacturers with no paid employees) were dropped entirely from the experiment. These establishments were excluded from the mailout count, receipt count, and the item response count.

Table 3 shows the receipt rates for each of the four panels. In addition, the item response rates for the panels are also shown. The item response rate is defined here as the percentage of receipts that contained a response for the particular item. In analyzing the tabulation, two comparisons were made:

(1) Panel 1 versus panel 2 (1974 delinquents: 1975 short-form response versus 1975 standard-form response) and

(2) Panel 3 versus panel 4 (1974 respondents: 1975 short form response versus 1975 standard form response).

The null hypothesis we wish to test is that the short form will produce no better response rates than does the standard form. This will be tested against the alternative hypothesis that the short form will produce better response rates. In the first comparison dealing with 1974 delinquents, the basic receipt response rate was slightly higher for the short form (27.0 versus 22.9%). Under the assumption of the null hypothesis, the Student's t test was used for testing the statistical significance of the difference between the response rates at the 5% level.

$$t = (\hat{p}_1 - \hat{p}_2)/\{\hat{p}\hat{q}[(1/n_1) + (1/n_2)]\}^{1/2}, \tag{11}$$

where

$\hat{p}_1$ is the proportion of 1975 short form receipts from 1974 delinquents,
$\hat{p}_2$ the proportion of 1975 standard form receipts from 1974 delinquents,
$n_1 = 326$ (number of in-scope establishments in panel 1),
$n_2 = 348$ (number of in-scope establishments in panel 2),

TABLE 3

Total Receipt and Item Response Percentages for the Four Panels

		Item response for receipts							
Panel	*Total receipts*	*Total employment* (TE)	*Annual payrolls* (AP)	*Cost of materials consumed* (CM)	*Inventories* (INV)	*Gross value of assets* (GVA)	*Rental payments* (RP)	*Capital expenditures* (CE)	*Value of products shipped* (VP)
1. 1974 delinquents: 1975 short form	27.0	90.9	89.8	86.4	81.8	82.9	72.7	73.9	78.4
2. 1974 delinquents: 1975 standard form	22.9	60.0	58.8	55.0	50.0	50.0	47.5	51.2	47.5
3. 1974 respondents: 1975 short form	86.0	98.0	98.3	94.2	90.1	92.5	85.8	85.4	93.0
4. 1974 respondents: 1975 standard form	82.4	95.7	95.3	93.3	85.0	88.3	73.7	85.3	85.7

$\hat{p} = (x_1 + x_2)/(n_1 + n_2)$ is the weighted response rate for panels 1 and 2,
$\hat{q} = 1 - \hat{p}$ the weighted nonresponse rate for panels 1 and 2.
x_1 the number of responses from panel 1, and
x_2 the number of responses from panel 2,

The procedure that was followed was first to compute a value of t using Eq. (11). This calculated t is then compared with a tabular value of t with $n_1 + n_2 - 2$ degrees of freedom. The tabular value of t for all tests in this paper is 1.65. The null hypothesis of $p_1 \leq p_2$ will be rejected only if the calculated t is greater than the tabular value for t (1.65). The calculated t for this comparison is equal to 1.23; therefore, we fail to reject the null hypothesis and we are not able to conclude that there is a significant difference between the receipt response rates for panels 1 and 2.

There was a noticeable improvement, however, in the item response rates for the short form versus the standard form for establishments that had been delinquent in 1974. If the t test is used to test for significant differences in item response rates between panels 1 and 2 (at the 5% level), the hypothesis that $p_1 \leq p_2$ is rejected for each of the eight data items (see Table 4). This indicates that the differences between the item response rates of panels 1 and 2 are significant. From these tests, the conclusion is that the short form is not more likely to generate a receipt from a prior-year delinquent, but if there is a receipt, the short form is more likely to generate a more complete response for those items common to the short form and the standard form.

The comparison involving the 1974 respondents (panel 3 versus panel 4) produced similar results. Although the receipt response rates were not significantly different, there were significant differences in the item response rates for six of the eight data items (see Table 5). For these panels also, the conclusion is that the short form is not more likely to generate a receipt from a prior-year respondent, but the short form is likely to generate a more complete response for items common to the short and standard forms.

3.3. Conclusions

The results of the short-form experiment can be used to estimate the effect on overall item response of mailing the short form to all single units in the ASM

TABLE 4

t Values: Panel 1 versus Panel 2

Panel	*Receipt*	*TE*	*AP*	*CM*	*INV*	*GVA*	*RP*	*CE*	*VP*
Panel 1 versus panel 2	1.23	4.69	4.63	4.49	4.37	4.53	3.37	3.04	4.15

TABLE 5

t Values: Panel 3 versus Panel 4

Panel	*Receipt*	*TE*	*AP*	*CM*	*INV*	*GVA*	*RP*	*CE*	*VP*
Panel 3 versus panel 4	1.31	2.78	2.27	.49	2.04	1.89	3.98	.37	3.37

with fewer than 50 employees. For each item, the estimated overall response rate is derived as follows:

$$I = (1974 \text{ receipt response rate})(R_1)(\text{item}_1) + (1974 \text{ delinquency rate})(R_2)(\text{item}_2),$$

where

R_1 is the receipt rate for 1974 respondents (1975),
R_2 the receipt rate for 1974 delinquents (1975),
item_1 the rate at which the item was reported (1975) for 1974 respondents who responded in 1975, and
item_2 the rate at which the item was reported (1975) for 1974 delinquents who responded in 1975.

The estimated item response rate for the short form is then

$$I_1 = (.70)(.860)(\text{item}_1) + (.30)(.270)(\text{item}_2)$$

and for the standard form is

$$I_2 = (.70)(.824)(\text{item}_1) + (.30)(.229)(\text{item}_2).$$

Table 6 shows the estimated item response percentages for the short form items if (1) the short form is used for all small single units (TE < 50) or (2) the standard form is used. From the table, it can be seen that the expected item response rate on the standard form for the small single units in the ASM panel is 54.2%. This requires that data for an item to be imputed for 45.8% of these establishments. If the short form were to be used, the expected item response rate would be 62.2%. This would result in item data being imputed for only

TABLE 6

Expected Item Response Percentages for Small Single Units

Form type	*Average item response*	*TE*	*AP*	*CM*	*INV*	*GVA*	*RP*	*CE*	*VP*
Short	62.2	66.4	66.5	63.7	60.8	62.4	57.6	57.4	62.7
Standard	54.2	59.3	59.0	57.6	52.4	54.4	45.8	52.7	52.7

37.8% of these establishments. As a percentage of the ASM totals, the expected approximate amount of imputation for small single units would be reduced from 6.4 (.458 × .14) to 5.3% (.378 × .14), with the expanded use of the short form, since, on a weighted basis, these establishments account for 14% of the estimated total employment.

Since no detail data are collected on the short form, there would be an increase in the amount of imputation for detail. With the standard form, detail data is imputed for an average of 45.8% of the establishments; with the short form, this detail would always be imputed. For 62.2% of the establishments, this detail would be imputed from reported totals. As a percentage of the ASM estimates, the average detail item imputations would increase from 6.4 (.458 × .14) to 14% (1.00 × .14) with the short form. Of this 14%, 8.7% would be imputed from reported totals, with the remaining 5.3% being fully imputed.

4. PREPARATION OF DATA RECORDS—THE GENERAL STATISTICS EDIT PROGRAM

In the ASM, data for both general statistics items, such as payroll, employment, cost of materials, and value of shipments, that are common to all manufacturing firms and shipments by individual class of product are collected. For about 60,000 establishments in the ASM sample panel, some 60 general statistics data items are now collected each year. For about 10,000 small, single-unit companies in the panel, a short form is mailed on which 10 general statistics data items are collected.

Most of these general statistics items are edited in the General Statistics (GS) Edit program. The remainder, which include detailed fuels consumed, depreciation, retirements, and values of products exported, have been added to the ASM report form since the GS edit program was developed and are edited in separate programs. In total, 43 data items are tested in the GS edit program (see Clark and Kusch, 1979).

4.1. Some Results of the General Statistics Edit

Before discussing the edit program, let us consider the proportions of items selected by the computer edit for further action and the disposition of these items.

A computer edit "diary" tabulation that is produced each year summarizes actions taken by the computer edit and by the analysts as a result of reviewing the edit referrals. Table 7 summarizes this information for the 1975 ASM at the all-industry level for selected items. The same kinds of information are tabulated for each of the 450 four-digit manufacturing industries. The table

TABLE 7

Program and Analyst Actions by Item as Percentages of the Total Number of Records Tabulated for the 1975 Annual Survey of Manufactures General Statistics Edit

Computer/analyst action	*Accept*		*Blank impute*		*Change*	
	No change	*Change*	*No change*	*Change*	*No change*	*Change*
Total employment (TE)	83	—	11	2	3	1
Production workers (PW)	82	—	12	2	3	1
Total salaries and wages (SW)	86	—	10	2	1	1
Total cost of materials (TCM)	81	—	12	2	3	2
Total value of inventories (TI)	81	—	12	2	3	1
Total capital expenditures (TCE)	82	—	14	1	1	1
Total value of shipments (TVS)	81	—	13	2	1	2
Gross value of assets (GVA)	79	—	15	2	2	1

[a] All values in this column are less than .5%.

shows, for example, that the total value of shipments (TVS) reported is accepted by the edit in 87% of the cases, is changed by the edit to a value the analyst did not later change in 7% of the cases, is changed by the edit to a value the analyst did not accept and further changed in 2% of the cases, is imputed from blank and not changed by the analyst in 13% of the cases, and is imputed from blank and then changed by the analyst in 2% of the cases.

For the eight items shown in Table 7, the analyst change rate for data changed by the computer edit, excluding imputation from blanks, is about 35%; that is, of 20,000 GS edit changes made to these items, about 7000 further changes were made by analysts.

It cannot be inferred, however, that 65% of these edit changes are acceptable to the analysts. The analysts do not have all the edit changes available for review, since edit referrals are generated by such changes only when the specified size cutoffs are exceeded. In the 1976 ASM, approximately 6000 establishment records were referred because of GS edit actions. A total of 20,000 records were referred in all, however, because of other referrals for items not processed by the GS edit. Many of the smaller GS edit changes are thus incidentally available for analyst review. Of the 7000 analyst changes just mentioned, it appears that most are to edit-changed data that were not the cause of the edit referral. (A 1976 ASM tabulation [unpublished] shows that analysts made only 800 changes to edit-changed data for these eight items in cases in which the amount of the change was large enough to generate the edit referral. The analyst change rate for this group was 26%.)

The point of this discussion is to indicate the difficulty of obtaining an evaluation of the edit based on analyst actions in actual survey processing. Although the analyst change rates for just the large edit changes can be determined, this is not the case for the smaller edit changes, for which it appears that there may be higher rates of change.

Analyst correction and acceptance rates such as these have been taken as a general indicator of whether or not the edit is working well for particular four-digit industries. If most of the computer edit changes available for review were, in turn, changed by the analysts (in particular, if the originally reported figure was usually restored), we would conclude that the data were being "overedited" and that the test tolerances were too tight. Similarly, many analyst changes to data that were not changed by the edit program would be taken as evidence of "underediting."

4.2. Description of the Preparation of Data Records

As the current year's reports are received, one for each manufacturing establishment in the survey, they are subjected to prekeying screening and coverage checks to resolve problems tha are due to changes in ownership and grossly deficient reporting. Reports are then keyed and processed through a series of computer edits, including the GS edit, on a flow basis. In three earlier edits—"assembly and housekeeping," "coverage and match," and "magnitude"—establishment records can be rejected completely if they have certain basic inconsistencies that cause them to fail the tests in these edits. When this occurs, the establishment record must be corrected to remove the cause of its rejection and then must be put through these edits again. Once records have reached the GS edit, however, they can no longer be rejected. In the GS edit, data problems may be resolved by adjustment of data, or they may not be resolvable. In either case, if the changes made by the edit are larger than specified cutoffs or if the data values in unresolved edit problems are large, edit referrals are generated. These edit referrals, which show for the entire establishment record the item values prior to and after the GS edit, are reviewed by survey analysts. After corrections from this review of the edited records are carried back to the establishment records, preliminary survey tabulations are developed, and analyst review and corrections of these tabulated totals are required before survey publications are issued.

The purposes of the GS edit can be enumerated at this point:

(1) to supply "missing" data, that is, to supply data when a blank entry is judged to be not acceptable;

(2) to identify all items for which reported data are unacceptable and to make adjustments to these data so that they are no longer classified as unacceptable by the edit; and

(3) to provide for analyst review a set of referrals that indicates what actions were taken by the edit in identifying and in attempting to resolve the larger data problems.

Just prior to the GS edit, the magnitude edit is performed. This edit is primarily concerned with whether two key items, total salaries and wages and total value of products shipped, are generally consistent with each other and with the establishment's prior-year values for these items or, if there is no prior-year record, consistent with employment. Just one set of test limits is used for this purpose. Inconsistencies in these basic items that cannot be resolved by this edit result in rejects for analyst resolution. It is assumed that were such gross inconsistencies to remain in the record the GS edit could not operate effectively.

The GS edit functions primarily by performing a series of ratio tests, employing test limits that vary by four-digit industry. These are "simple" ratios, involving just two items of current-year data, and "ratios of change," which consist of the simple ratio for the current year divided by the corresponding simple ratio for the previous year.

For each test ratio, there is a four-digit industry average $\bar{R}$ and associated upper and lower tolerance limits UL and LL. Some test limits are "statistical" in that they have been computed as the industry average plus or minus a multiple of the standard deviation of the ratio. (Multiples of 3.5 and 5 have commonly been used to control the test failure rate at a 2 to 3% level.) Other test limits are termed "logical" in that they define a range within which the test ratio must logically lie (for example, a detail-to-total ratio cannot exceed unity) or a range within which the ratio is judged very likely to lie. Many test ratios for which statistical limits were initially developed have been changed to logical limits because of the ineffectiveness of the statistical limits. The industry average $\bar{R}$s, which are used for imputation, are updated periodically (annually in the case of ratios affected by inflation). Test limits are revised less often, usually as a consequence of particular problems with the results of the GS edit. The ratio averages and limits are constant for each four-digit industry, so that small establishments have test limits that are the same as those of large establishments. Although this is not always appropriate, a large-scale analysis of census data indicates that for the majority of ratios and industries these constant values are appropriate.

The general approach used in the GS edit is to perform the ratio tests in a predetermined sequence. The order of testing reflects an assessment of the relative reporting reliability of each item and the existence of two "chains" of data, one associated with payroll (payroll, employment, plant-hours) and one with shipments (shipments, cost of materials, inventories). Payroll and shipments are first tested against each other. This increases the likelihood of detecting joint errors in these items and decreases the possibility of missing consistent errors in a chain of data.

The edit is designed to identify and correct just one defective item at each step. If the ratio being tested does not lie within its tolerance limits, the edit assumes that just one component (numerator item or denominator item) is defective and determines which component is most likely to be in error. This is done by computing an adjusted probability measure for each component. This measure is a function of (1) the a priori reliability of the component, whether the other ratios containing the component are presently passing or failing their tests, (2) the a priori reliabilities of the other items contained in those other ratios, and (3) any edit flags previously assigned to the component.

Attempts are then made, by trial data adjustments, to improve the value of the possibly defective item. The criterion for improvement is a reduction in the failure score associated with the item, which depends on the number of its failing ratio tests and on preassigned measures of importance for each test ratio. After ratio testing, any imbalances between sums of detail items and their totals are corrected by allocation and raking procedures.

The GS edit consists of a series of subroutines whose purpose is to edit those "general" data items that are comparable in definition over the whole range of manufacturing industries. Essentially, there are two types of establishment records coming into the GS edit. The first is a data record where most, if not all, the data items have been reported. In this situation, the purpose of the edit is to detect and correct errors in the reported data and to impute data for those blank or zero items where a zero value is not acceptable. The other type of record is one in which virtually all the data items are blank, that is, unreported, and for which it is desired that the edit simply impute data for all blank items. The edit distinguishes type 2 from type 1 by the presence of a flag (full impute), which can be either manually assigned prior to the keying of the schedule or computer-assigned in one of the series of edits that precedes the GS edit. The presence or absence of this flag results in that type of record being sent to a unique subroutine.

4.3. Illustration of Ratio Testing, Determination of Defective Component, and Imputation

The following example illustrates the ratio testing done in the GSEDIT subroutine, the identification of the defective component, and the search for the best possible impute. For the purpose of this example, we assume that year-to-year testing is not required, i.e., only the simple ratios are tested, and that the data are all reported so that there is no flag value associated with any of the items. Let us assume that our establishment belongs to industry 3241 (Cement, hydraulic) and that the data that concern us are as follows:

TE = 120	CR = 0	GVB = 4000(000)
SW = 1200(000)	CF = 500(000)	GVE = 2000(000)
WW = 1150(000)	CEE = 500(000)	BR = 0
OW = 50(000)	CCW = 0	MR = 1000(000)
TCM = 13,000(000)	TE = 6000(000)	TVS = 4500(000)
CP = 12,000(000)	GVBA = 6000(000)	VR = 400(000)

where

TE	total number of employees	TEI	total inventories at end of year
SW	total salaries and wages	GVBA	gross values of book assets at end of year
WW	production workers' wages	GVB	gross value of buildings
OW	all other salaries and wages	GVE	gross value of machinery and equipment
TCM	total cost of materials and services used	BR	payments for buildings
CP	cost of materials, parts, etc.	MR	payments for machinery
CR	cost of resales	TVS	total value of products shipped
CF	cost of fuels consumed	VR	value of resales
CEE	cost of purchased electricity		
CCW	cost of contract work		

4.3.1. Ratio testing

Table 8 shows the first five ratios that would be tested by the subroutine along with the industry average $\bar{R}$, the lower limit LL, the upper limit UL, and the actual ratio value R based on the preceding data.

The edit tests each ratio sequentially. The first ratio that fails is TCM/TVS, whose value lies above the upper limit.

4.3.2. Determination of Defective Component[2]

To ascertain whether TVS or TCM is the defective component, we compute for each the reliability measure $f(x)$, the probability measure $Q(x)$, and the cumulative failure score $F(x)$. In this case, x is either TVS or TCM. We consider all the ratios that contain TVS and TCM. These are shown in Table 9, along with the appropriate values $\bar{R}$, LL, UL, R, and fc, the failure count, assigned to each ratio as an indicator of the importance of the ratio.

[2] Explanation and derivation of formulas are given in Appendixes A and B.

TABLE 8

Test Data for Specified Ratios

Ratio	$\bar{R}$	*LL*	*UL*	*R*
1. SW/TWS	.211	.010	.498	.267
2. SW/TE	12.680	1.000	35.000	10.000
3. WW/SW	.769	.000	1.000	.958
4. CW/SW	.232	.000	1.000	.042
5. TCM/TVS	.371	.007	1.000	2.889

We also require the a priori probabilities (of being defective) for each of the items in the ratios of Table 9, which are presented in Table 10.

We examine TVS first. From Table 9, we see that ratios 5, 6, and 9 fail. Thus we compute the reliability $f(\text{TVS})$ as

$$f(\text{TVS}) = \frac{q(\text{TVS})}{p(\text{TVS})}\left[\frac{q(\text{SW})}{p(\text{SW})}\cdot\frac{p(\text{TCM})}{q(\text{TCM})}\cdot\frac{p(\text{TEI})}{q(\text{TEI})}\cdot\frac{q(\text{GVBA})}{p(\text{GVBA})}\cdot\frac{q(\text{BR})}{p(\text{BR})}\cdot\frac{p(\text{MR})}{q(\text{MR})}\cdot\frac{q(\text{VR})}{p(\text{VR})}\right]$$

$$= 508,$$

so

$$Q(\text{TVS}) = 1 - \frac{1}{1 + f(\text{TVS})} = 1 - .001965 = .998035$$

TABLE 9

Test Data for Ratios Containing TVS and TCM

Component	*Ratio*	$\bar{R}$	*LL*	*UL*	*R*	*fc*
TVS	1. SW/TVS	.211	.010	.498	.267	11
	5. TCM/TVS	.371	.007	1.000	2.889	6
	6. TEI/TVS	.168	.005	.573	1.333	4
	7. GVBA/TVS	1.634	.010	6.000	1.333	3
	8. BR/TVS	.001	.000	.200	.000	1
	9. MR/TVS	.002	.000	.200	.222	1
	10. VR/TVS	.014	.000	.750	.089	1
TCM	5. TCM/TVS	.371	.007	1.000	2.889	6
	16. CP/TCM	.534	.000	1.000	.923	1
	17. CR/TCM	.035	.000	1.000	.000	1
	18. CF/TCM	.282	.000	1.000	.038	1
	19. CEE/TCM	.141	.000	1.000	.038	1
	20. CCW/TCM	.011	.000	1.000	.000	1

TABLE 10

Probabilities of Item Defect

Item	*P*	*Item*	*P*
SW	.027	CCW	.091
TVS	.028	TEI	.084
TCM	.058	GBVA	.122
CP	.091	BR	.148
CR	.091	MR	.148
CF	.091	VR	.091
CEE	.091		

and

$$F(\text{TVS}) = 6 + 4 + 1 = 11.$$

For TCM, only ratio 5 fails, so we have

$$f(\text{TCM}) = \frac{q(\text{TCM})}{p(\text{TCM})}\left[\frac{p(\text{TVS})}{q(\text{TVS})}\cdot\frac{q(\text{CP})}{p(\text{CP})}\cdot\frac{q(\text{CR})}{p(\text{CR})}\cdot\frac{q(\text{CF})}{p(\text{CF})}\cdot\frac{q(\text{CEE})}{p(\text{CEE})}\cdot\frac{q(\text{CCW})}{p(\text{CCW})}\right]$$
$$= 46543,$$

so

$$Q(\text{TCM}) = 1 - .000021 = .999979$$

and

$$F(\text{TCM}) = 6.$$

How is F calculated? The failure score $F(x)$ is the sum of the failure counts [relabeled now as fc] for ratios, containing item X, that fail (see Appendix A for discussion.)

Since $Q(\text{TVS}) < Q(\text{TCM})$, TVS is determined to be the defective component. Because TVS is a total, the program would first attempt to substitute the sum of detail for TVS if the sum were a different value. We assume that the sum is not different, so this substitution is not viable. Rounding will not be attempted since TVS is never rounded by the GS edit. So we resort to imputation in an attempt to find a more acceptable value. The program stores the values $Q(\text{TVS})$ and $F(\text{TVS})$ for later comparisons.

4.3.3. *Imputation*

Each ratio containing TVS will be used to impute a new TVS value. The sequence of imputing is the same as that for the ratio testing. At each stage of imputation, the edit decides if this latest value is better than the previous best value by comparing in order the failure scores $F(x)$, the probability measure

$Q(x)$, and the absolute deviations $\Delta(x)$ of the imputed value from the original value.

Imputation is based on the relation

$$\text{TVS(IMP)} = \text{TVS}'(\bar{R}/R)^{\pm 1},$$

where the exponent is positive if TVS is in the numerator of the ratio and negative if TVS is in the denominator and TVS′ denotes the value of TVS considered best up to this point of imputation and is the value used in the computation of R in the preceding expression. If the other component of the ratio containing TVS is zero, so that $\bar{R}/R$ is not well defined, then that ratio is not used to impute a value.

4.3.3.1. Impute 1: Ratio 1 (SW/TVS). As imputation begins, the original value of TVS is considered best. Therefore

$$\text{TVS}_1 = \text{TVS(IMP)}_1 = 4500(.211/.267)^{-1} = 5694.$$

Then TVS_1 is tested in all ratios containing TVS. Ratios 5 and 6 fail, which yield

$$F(\text{TVS}_1) = 10,$$

$$Q(\text{TVS}_1) = .999941,$$

$$|\Delta(\text{TVS}_1)| = |5694 - 4500| = 1194.$$

Since $F(\text{TVS}_1) < F(\text{TVS}')$, we set $\text{TVS}' = 5694$.

4.3.3.2. Impute 2: Ratio 5 (TCM/TVS). We have

$$\text{TVS}_2 = 5694(.371/2.283)^{-1} = 35{,}038.$$

This value satisfies each ratio, therefore,

$$F(\text{TVS}_2) = 0,$$

$$Q(\text{TVS}_2) \doteq 1.000000,$$

$$|\Delta(\text{TVS}_2)| = |35{,}038 - 4500| = 30{,}538.$$

Then $F(\text{TVS}_2) = 0 < F(\text{TVS}')$ implies that TVS_2 is preferred over TVS′, so we set $\text{TVS}' = \text{TVS}_2 = 35{,}038$. This value will be used in the next imputation.

4.3.3.3. Impute 3: Ratio 6 (TEI/TVS). We have

$$\text{TVS}_3 = 35{,}038(.168/.171)^{-1} = 35{,}663.$$

Since TVS_3 will pass all ratios,

$$F(\text{TVS}_3) = 0,$$

$$Q(\text{TVS}_3) \doteq 1.000000,$$

$$|\Delta(\text{TVS}_3)| = |35{,}663 - 4500| = 31{,}163.$$

Thus $F(\mathrm{TVS}_3)$ and $Q(\mathrm{TVS}_3)$ are exactly the same as for TVS′. However, since TVS_3 deviates more from the original value than does TVS′, we retain TVS′ = 35,038 as a better value.

4.3.3.4. Impute 4: Ratio 7 (GVBA/TVS). We have

$$\mathrm{TVS}_4 = 35{,}038(1.634/.171)^{-1} = 3667.$$

Ratios 5, 6, and 9 fail when using TVS_4, so

$$F(\mathrm{TVS}_4) = 11,$$

$$Q(\mathrm{TVS}_4) \doteq .998031,$$

$$|\Delta(\mathrm{TVS}_4)| = |3667 - 4500| = 833.$$

The failure score for TVS_4 is higher than for TVS′, therefore we do not accept it as a better value.

4.3.3.5. Impute 5: Ratio 8 (BR/TVS). Since BR = 0, we do not impute a value for TVS based on this ratio.

4.3.3.6. Impute 6: Ratio 9 (MR/TVS). We have

$$\mathrm{TVS}_6 = 35{,}038(.002/.029)^{-1} = 508{,}051.$$

Since TVS_6 fails ratio 1,

$$F(\mathrm{TVS}_6) = 11,$$

$$Q(\mathrm{TVS}_6) \doteq .999998,$$

$$|\Delta(\mathrm{TVS}_6)| = |508{,}051 - 4500| = 503{,}551.$$

The failure score is increased for TVS_6, so it is disregarded as an acceptable value.

4.3.3.7. Impute 7: Ratio 10 (VR/TVS). We have

$$\mathrm{TVS} = 35{,}038(.014/.011)^{-1} = 27{,}530.$$

Since TVS_4 satisfies all ratios,

$$F(\mathrm{TVS}_7) = 0,$$

$$Q(\mathrm{TVS}_7) \doteq 1.000000,$$

$$|\Delta(\mathrm{TVS}_7)| = |27{,}530 - 4500| = 23{,}030.$$

Thus $F(\mathrm{TVS}_7)$ and $Q(\mathrm{TVS}_7)$ are the same as for TVS′, but they differ from the original value less than does TVS′. Therefore, we replace TVS′ by TVS_7.

4.3.3.8. Finis. Because there are no more ratios available to impute a value for TVS, the final decision of the edit is to accept TVS, as the best possible value. This value is flagged appropriately as being imputed. The edit then resumes ratio testing at the point where the ratio failure (ratio 5) interrupted it.

APPENDIX A. DETERMINATION OF DEFECTIVE COMPONENT

When a ratio fails, it is assumed that one and only one component is defective. (Thus, a ratio does not fail if either neither component is defective or both components are defective.) An adjusted probability measure is computed for each component. Each measure is based on an a priori probability of the component being defective, on whether the other test ratios that contain this component are passing or failing at this time, on the a priori probabilities of the other items that are contained in those other ratios, and on edit flags previously assigned to the component. Analysis of historical edit and analyst acceptance rates has been used to determine the a priori probabilities. Should both components actually be defective, it is likely that the component not initially found defective will be detected in a subsequent ratio failure.

To illustrate these points more precisely, suppose that ratio $R = x/y$ is a failed ratio and suppose further that $R_1, R_2, \ldots, R_{n_x}$ are all the ratios containing component x and that $T_1, T_2, \ldots, T_{n_y}$ are all the ratios containing component y. Clearly, $R = R_i = T_j$ for some i and j. Then we wish to compute the following probabilities:

$$P(x) = P(x \text{ is defective} \mid R_i \text{ passes or fails}; i = 1, 2, \ldots, n_x),$$

$$P(y) = P(y \text{ is defective} \mid T_j \text{ passes or fails}; j = 1, 2, \ldots, n_y).$$

It can be shown (see Appendix B) for $P(x)$ and, similarly, for $P(y)$ that, if the failures of items are independent, then

$$P(x) = \frac{1}{1 + f(x)},$$

where

$$f(x) = \frac{q_x}{p_x} \prod_{i=1}^{n_x} \frac{s_i q_{B_i} + r_i p_{B_i}}{s_i p_{B_i} + r_i q_{B_i}}$$

and where

p_x is the a priori or historical probability of x being defective,
$q_x = 1 - p_x$,
n_x is the number of ratios containing component x,
p_{B_i} the a priori probability of component other than x in ratio R_i being defective,
$q_{B_i} = 1 - p_{B_i}$,

$$r_i = \begin{cases} 0 & \text{if ratio } R_i \text{ lies within its tolerance limits, i.e., } R_i \text{ does not fail,} \\ 1 & \text{if ratio } R_i \text{ does not lie within its tolerance limits, i.e., } R_i \text{ fails,} \end{cases}$$

and

$s_i = 1 - r_i$.

We consider $f(x)$ to be a reliability measure of component x. To compute it, the mechanics of the preceding equation imply the following procedure:

(1) Compute the ratio q_x/p_x.
(2) Test the n_x ratios containing component x sequentially.
 (a) If ratio R_i passes, then q_{B_i}/p_{B_i} becomes the next term in the product.
 (b) If ratio R_i fails, then p_{B_i}/q_{B_i} becomes the next term in the product.

The composite product of the values in (1) and (2) is $f(x)$.

It should be noted that the probabilities $P(x)$ and $P(y)$ are computed independently of one another, and it is not to be expected that they necessarily add to 1. These are conditional probabilities, and as the equations for $f(x)$ and $f(y)$ imply, they are dependent solely on the pass–fail patterns of the ratios containing the component and on the a priori probabilities of the other items in those ratios. The pass–fail pattern for one component is not affected by the pattern for the other. If, for example, y is in a ratio with a component other than x, then the passage or failure of that ratio affects only the probability measure $P(y)$; it provides no additional information for $P(x)$.

The a priori probabilities of an item being defective are all less than .5. Thus for a given ratio its passage causes a larger contribution ($q_{B_i}/p_{B_i} > 1$) to $f(x)$ than it would if it had failed. This fact lends meaning to the concept of $f(x)$ as a reliability measure for component x in that the more ratios that are satisfied, the larger is the value of $f(x)$.

Once the value $f(x)$ is computed, then $P(x)$ is found and finally $Q(x) = 1 - P(x)$ is computed. The value $Q(x)$ represents the probability that component x is not defective, given the pass–fail pattern of the ratios containing it. Similarly, the value $Q(y)$ is computed for component y. A final adjustment is then made to these probability measures prior to their comparison. This adjustment is based on the flagging system developed for the GS edit program. Each item has a flag assigned to it that indicates, for example, that the item is blank, has been previously imputed, or is rounded. A value is assigned each flag, and this value is averaged with the $Q(x)$ value to produce the final adjusted probability measure. For example, if a component has previously been imputed (say, to replace a "defective" reported value), the value associated with the impute flag increases the reliability measure more than for any other flag since the program is assumed to have imputed the best possible value. This is not to say that the imputed component will always be accepted but only that it is less likely to be found as the defective component.

In the unlikely event that the final adjusted probability measures are equal for the components of a failed ratio, a tie-breaking scheme in the form of a failure score for each component has been devised. A failure count *fc* has been assigned each ratio, indicating the relative importance we attach to it, so that if the ratio fails, this value is counted. The cumulative failure count over all ratios containing the component is the failure score $F(x)$ for that component, and the component with the highest score is judged to be defective. Not all ratios are considered equal with regard to their contribution to the failure score. In particular, the cumulative failure count for the ratios involving a set of detail items and their total should be less than the failure count for all other ratios containing the total. This is so because if the total is adjusted so that the other total-to-total ratios containing it now pass, we want this to reduce the overall failure score for the total. The detail-to-total ratios may initially have passed (that is, a consistent error occurred for both detail and total), but with the correction to the total, they now fail. However, their combined failure count is less than the failure count of the ratios containing the total, and therefore the overall failure score for the total is reduced as desired. When the failure score is thus reduced for a total, but not to zero, the edit will attempt to adjust the detail in order to reduce the failure score to zero.

Ratios with zero or blank components are specially tested to determine whether or not they are acceptable. The reason for testing them uniquely derives from the fact that we want to allow more flexibility in accepting or rejecting a zero value than would occur under the usual ratio testing. For example, under normal ratio testing, a ratio with zero in the numerator would always pass (if the lower limit is zero) or will always fail (if the lower limit is nonzero). So we would either always accept zero for the component or we would never accept it. Neither these extremes is particularly reflective of the real world. Inventories, for example, are legitimately zero for many establishments. However, the likelihood that an establishment has zero inventories decreases as the level of its shipments increases.

APPENDIX B. DERIVATION OF $P(x|R_c, c = 1, \ldots, n)$

The model on which the derivation is based uses a definition of failure of a ratio in terms of defectiveness of its components that differs from the definition given earlier that a ratio fails when the ratio is outside specified tolerance limits. The model also makes the assumption that different components are independent with respect to whether or not they are defective.

Evaluation of models and the conclusions to which they lead, especially when definitions used in the model differ from those in the process modeled, is a normal part of the application of any model. So also is the development of improved models both judgmentally, to bring the process and its model in close agreement, and empirically, as a result of evaluation.

B.1. Notation and Definitions

Let (A, B_i) be the components of the ratio R_i, $i = 1, \ldots, n$; A is one of the components (X, Y) of R, the ratio that fails, and B_i is the component that occurs with A in R_i, $i = 1, \ldots, n$. (R is one of the R_i.)

To discuss failure and defectiveness more explicitly, 2n + 1 random variables $a', b_i', r_i', i = 1, \ldots, n$ are first defined. These are the variables in terms of which the model is specified.

The $2n + 1$ random variables are defined by

$$a', b_i', \ldots, b_n', r_i', \ldots, r_n' = 1 \tag{1a}$$

if $A, B_1, \ldots, B_n$ are defective and if $R_1, \ldots, R_n$ fail, and

$$a', b_1', \ldots, b_n', r_1', \ldots, r_n' = 0 \tag{1b}$$

if $A, B_1, \ldots, B_n$ are not defective and $R_1, \ldots, R_n$ do not fail. The apriori probabilities that a component is defective are

$$Pr\{a' = 1\} = p_A,$$

$$Pr\{b_i' = 1\} = p_{B_i}.$$

Then the probabilities that the components are not defective are

$$Pr\{a_i' = 0\} = q_A = 1 - p_A$$

and

$$Pr\{b_i' = 0\} = q_{B_i} = 1 - p_{B_i}, \qquad i = 1, \ldots n.$$

Denote $(b_1', \ldots, b_n')$ by b' and $(r_1', \ldots, r_n')$ by r'.

The probability of any specified values a, r_i of $a', r_i', i = 1, \ldots, n$ is denoted by $P_{a'r'}(a, r_1, \ldots, r_n)$ the subscripts a', r' of P indicating that in calculating the probability $P_{a'r'}$, a' and r' are the random variables. The probabilities $P_{a'b'}(a, b_1, \ldots, b_n)$ are similarly defined.

B.2. The Model

The following assumption and definition determine the particular model on which the derivation is based.

B.3. Independence Assumption

Defectiveness occurs independently in different components, i.e., for all 2^n possible values of a', b' the following equations hold.

$$P_{a'b'}(a, b) = P_{a'}(a) \prod_{i=1}^{n} P_{b_i'}(b_i). \tag{2}$$

Thus any subset of the a, b are independent. In particular

$$P_{b'}(b) = \prod_{i=1}^{n} P_{b_i'}(b_i). \tag{3}$$

In the editing process a ratio fails if its value falls outside specified tolerance limits. The following definition differs and sometimes the two definitions will give different results.

B.4. Definition of Failure

R_i fails if exactly one of A and B_i fail; R_i does not fail if either both A and B_i fail or if neither fails.

The definition of failure establishes the following relationships among the random variables a, b_i and r_i.

Lemma 1. For fixed a, b and r have a one to one relationship. If $a = 1$, then $b_i = 1 - r_i$ and if $a = 0$, then $b_i = r_i$. In general

$$b_i = a(1 - r_i) + (1 - a)r_i, \qquad i = 1, 2, \ldots, n.$$

The proof consists of substituting the values that are possible for a, b_i, r_i, according to the definition of failure in the equations of Lemma 1.

The definition of failure and the resulting one to one relationships stated in Lemma 1 are important because they permit the evaluation of the conditional probability that ratios fail given that A is or is not defective in terms of the a priori probabilities, assumed known, that the components B_i are defective.

Lemma 2. For given a', the probabilities that $r' = r$, $r'_i = r_i$ are the functions of the a priori probabilities that $b' = b$ and $b'_i = b_i$ stated below:

$$\begin{aligned} P_{r'}(r|a) &= P_{b'}(a(1 - r) + (1 - a)r|a) \\ &= \prod_{i=1}^{n} P_{b'_i}\left[a(1 - r_i) + (1 - a)r_i|a\right] \\ &= \prod_{i=1}^{n} a\left[r_i q_{B_i} + (1 - r_i)p_{B_i}\right] + (1 - a)\left[r_i p_{B_i} + (1 - r_i)q_{B_i}\right]. \end{aligned} \tag{4}$$

$$P_{r'}(r|1) = \prod_{i=1}^{n} \left[r_i q_{B_i} + (1 - r_i)p_{B_i}\right]. \tag{5}$$

$$P_{r'}(r|0) = \prod_{i=1}^{n} \left[r_i p_{B_i} + (1 - r_i)q_{B_i}\right]. \tag{6}$$

Proof. The first two equalities of (4) follow from Lemma 1 and the independence assumption. The last equality of (4) may be verified by giving a and r_i their possible values of 1 and 0. The expressions (5) and (6) are special cases of (4) obtained by giving a its possible values of 1 and 0.

The derivation of $P(A)|R_i, \ldots, R_n)$ now follows.

From the definition of conditional probability it follows that, if the denominators are positive,

$$P_{a'}(a|r) = [P_{a'r'}(a, r)]/[P_{r'}(r)]$$
$$= \left[\frac{1 + P_{a'}(1 - a)P_{r'}(r|1 - a)}{P_{a'}(a)P_{r'}(r|a)}\right]^{-1}. \tag{7}$$

Substituting from (4), (5) and (6) in (7) we then have the desired result.

Theorem. The conditional probability that a component is defective or not defective given the failure pattern of the ratios containing that component is

$$p_{a'}(1|r) = [1 + f(A)]^{-1},$$
$$p_{a'}(0|r) = \left[1 + \frac{1}{f(A)}\right]^{-1},$$

where

$$f(A) = \frac{q_A}{p_A}\prod_{i=1}^{n}\left[\frac{r_i p_{B_i} + (1 - r_r)q_{B_i}}{(1 - r_i)p_{B_i} + r_i q_{B_i}}\right].$$

REFERENCES

Clark, D. F., and Kusch, G. L. (1979). Annual survey of manufactures general statistics edit. *1979 Proceedings of the Business and Economic Statistics Section*, American Statistical Association, Washington, D.C.

Cole, S., and Altman, M. (1979). An experiment with short forms in the annual survey of manufactures. *1979 Proceedings of the Business and Economic Statistics Section*, American Statistical Association, Washington, D.C.

Greenberg, B. (1982). Using an edit system to develop editing specifications. *1982 Proceedings of the Survey Research Methods Section*, American Statistical Association, Washington, D.C.

Ogus, J., and Clark, D. F. (1971). Annual Survey of Manufactures: A Report on Methodology, Technical Paper 24, Bureau of the Census, Washington, D.C.

Waite, P. J., and Cole, S. J. (1980). Selection of a new sample panel for the annual survey of manufactures. *1980 Proceedings of the Business and Economic Statistics Section*, American Statistical Association, Washington, D.C.

U.S. Bureau of the Census (1972). 1972 Census of Manufactures: General Summary Tables, Table 8, p. 98.

CHAPTER 7

Incomplete Data in the Survey of Consumer Finances

Rodger Turner
Murray Lawes

1. SURVEY DESIGN

1.1. Purpose of the Survey

Statistics Canada has conducted the Survey of Consumer Finances to collect incomes of families and individuals on a periodic basis between 1951 and 1971 and annually since 1972. In years ending with even numbers, the survey is conducted as a supplement to the April Canadian Labour Force Survey (LFS). The sample of approximately 37,000 dwelling units is large enough to generate income distributions of reasonable accuracy for each province of Canada. The subject-matter content of the survey is restricted to basic questions on income and work experience for the full calendar year, in addition to the demographic and labour force participation questions asked by the Labour Force Survey (see questionnaires in Appendix A). In the years ending with odd numbers, when the sample of dwellings is independent, the survey is not a supplement and is generally conducted between the April and May Labour Force Surveys. The sample size of approximately 16,500 dwelling units is less than half of that of the even-numbered years, with a corresponding reduction in the detail released. For example, data for the four provinces in the Atlantic region, Newfoundland, Prince Edward Island, Nova Scotia, and New Brunswick, are combined into a single distribution, as are data for the three Prairie provinces, Manitoba, Saskatchewan, and Alberta. The questionnaire's content is substantially expanded to collect in-depth information on various income-related topics, including information on such diverse topics as the asset and

INCOMPLETE DATA
IN SAMPLE SURVEYS
Volume 1, Part II

ISBN 0-12-363901-8

indebtedness of families, health and education benefits, and work histories of Canadians. An example of an expanded questionnaire is given in Appendix B.

The major vehicles for the dissemination of income data collected by the Survey of Consumer Finances are

(1) the publication of a collection of statistical reports (annual and occasional) that present income distribution statistics classified by standard demographic and labour force variables for individuals, the economic family definition, and the census family definition;

(2) the release of micro data computer files that do not violate confidentiality requirements; and

(3) the dissemination of unpublished data based on special requests.

Examples of income distribution data usage are (a) the administration of existing or the planning of new legislation by various levels of government; (b) the study of the problems of special groups such as the poor and aged by governments and social welfare agencies; (c) the analysis of earnings of Canadians by trade unions; and (d) the formulation of marketing policies by the business sector.

Illustrative tables from "Income distributions by size in Canada 1976" (Statistics Canada, 1978) are given in Appendix C. The two tables (Tables 3 and 4) shown in Appendix C are examples of the reduction of detail released in an odd-numbered year survey. For years with even numbers, these tables are produced at the provincial level.

1.2. Sample Design

The sample design of the Survey of Consumer Finances follows that of the Labour Force Survey. A summary description is given later and a detailed description is available in *Methodology of the Canadian Labour Force Survey* (Platek and Singh, 1976).

The coverage for the Survey of Consumer Finances includes the civilian noninstitutional population of persons 15 years of age and over in the ten provinces of Canada, with the exclusion of residents of Indian reserves.

The sample of dwellings for the survey is selected on the basis of area sampling using a stratified multistage sample design. Each of the ten Canadian provinces is divided into geographically contiguous economic regions, which are further split into three types of areas: self-representing (SR), non-self-representing (NSR), and special, having independent sample selections within each type of area.

Self-representing (SR) *areas* comprise the larger cities. The minimum size of city varies from province to province, with the Atlantic provinces' cities having a minimum of 7000 persons and Ontario and Quebec's having a minimum of 24,000 persons. Cities of smaller sizes may be classified as self-representing areas if they exhibit unique socioeconomic characteristics. Each city is divided into subunits whose size varies from 1000 to 12,000, depending on the size and sampling ratio for the province. Subunits so formed are subdivided into clusters consisting of approximately one city block. The design is based on a rotating panel sample, with each panel of selected dwellings being surveyed for six consecutive months and the oldest of the six panels rotating out upon the introduction of a new panel. The clusters are randomly grouped into multiples of six groups, to facilitate this six-month household rotation scheme. One cluster per group is selected with probability proportional to size, and a systematic sample of dwellings is selected from within selected clusters.

Non-self-representing (NSR) areas consist of rural and small urban areas. These areas are initially divided into strata that are in turn divided into primary sampling units (PSUs). Two or more PSUs per stratum are selected with probability proportional to size. Once selected, PSUs are further divided into rural and urban parts and each part is clustered. A sample of clusters is selected within each PSU with probability proportional to size, and within sampled clusters, a systematic sample of dwellings is selected.

Special areas are composed of areas or establishments that possess characteristics differing from the general population and that may require special interviewing techniques. Included in the special areas are hospitals, schools, hotels, military establishments, and remote areas. Special areas make up about 1% of the sample frame.

For even-numbered years, the sample for the Survey of Consumer Finances consists of either the full Labour Force Survey sample or a subset of this (in order to maintain consistency of sample size throughout the years). Since the set of sampled dwellings associated with a particular rotation number is a proper subsample of the full sample, these subsamples for the Survey of Consumer Finances are based on rotation numbers. For the independent survey in odd-numbered years, the sample is selected from the frame using

(1) subsamples identified by rotation numbers,
(2) the sample selected from only two of the PSUs per stratum in non-self-representing areas, and
(3) the sample-size stabilization capabilities (as described in Section 1.3).

The data for the survey consist of three basic blocks of data, namely, the regular labour force data collected on Forms F03 and F05 (see Appendix A), the work experience data collected in the supplementary questions section of Form F03 (Question 50) and the income details collected on Form CF06 (see Appendix A).

The data collection procedures used for collecting the regular labour force information follow the procedures used for the Labour Force Survey; indeed, for even-numbered years when the Survey of Consumer Finances is a supplement, the parent survey is in fact the official Labour Force Survey for the month in question. Because of the rotating panel feature of the Labour Force Survey, dwellings are sampled on six successive monthly survey occasions. In self-representing unit (SRU) areas, telephone interviewing was introduced, if agreeable to the household, for the second and subsequent survey occasions, although the initial contact with the household was always on the basis of a personal interview. For all other situations, and indeed for all sampled dwellings in odd-numbered years, all contacts with the sampled households are on the basis of personal visits. The data collection procedures are modeled after those for the Labour Force Survey. It is suggested that interviewers can make up to three or four calls to sampled dwellings to establish contact with the household. For dwellings having a known postal address, an introductory letter is mailed to the dweliing. Otherwise, the letter is provided to the household during the first visit. Interviewers appropriately introduce themselves, give a brochure describing the Survey of Consumer Finances, and impress upon the potential respondent the importance of the survey and of the cooperation of the respondent.

The interviewer initially collects basic household and demographic data on Form F03 and then collects information, for household members 15 years of age or more, on the labour force activities during the reference week (usually the week prior to the interview period). This Labour Force Survey information may be collected by proxy. The supplementary work experience data for the preceding calendar year are then collected on Form F03. Finally, the income questionnaires are left (or delivered, in the event of telephone interviewing) for self-enumeration, and an appointment is made by the interviewer to pick up the completed income questionnaires. To minimize nonresponse to the income questionnaire, interviewers are generally instructed to make two return calls to the household. However, if the interviewer has failed to obtain completed questionnaires because the information is not readily available, an envelope is left for the respondent to mail the questionnaire to the head office—but only as a last resort.

The completed LFS schedules, F03 and F05, are reviewed for completeness in terms of numbers of forms of the various types per dwelling relative to the response codes and the household sizes. The Labour Force Survey data and the supplementary work experience data are data-captured in the regional offices and transmitted to the head office via telecommunication lines, whereas the income questionnaires are mailed by the regional office to the head office. (There are eight regional offices across Canada that are primarily involved in the data collection phase of the survey. The head office, located in Ottawa, carries out the survey preparation, all processing of collected data, and publication of survey estimates.) On receipt in the head office, the Labour Force Survey data are processed according to the regular Labour Force Survey procedures up to the point of obtaining a clean file of labour force data. The income questionnaires are clerically reviewed for completeness, and any explanatory notes and comments are incorporated as required. The income questionnaires are then loaded onto the data base and linked with the appropriate labour force data. The file is then submitted through the edit and imputation modules, as described in later sections of this chapter, to obtain a clean, edited data file.

A preliminary report based on partially edited data of individuals and economic family units usually appears about six months after the survey date, with the final report published about a year after the survey date. Subsequently, reports on the income of census family units and income after taxes for individuals and economic family units are issued.

1.3. Estimation

Each responding sample unit in the Survey of Consumer Finances is weighted as the product of six factors, the product of the first five being termed the *subweight* (or simple survey weight), namely,

(1) the inverse of the household sampling rate;

(2) the rural–urban factor, which compensates for the over- or under-representation of population in selected PSUs with respect to the entire NSR area in a province;

(3) the nonresponse balancing factor (defined as the number of expected households, i.e., dwellings occupied by eligible individuals, divided by the number of enumerated households) calculated at the subunit level in the SR areas and at each of the rural and urban levels within PSUs in NSR areas;

(4) the cluster weight, which is the inverse of the cluster subsampling rate and has a value of one except in clusters experiencing rapid growth, for which a further subsampling of households is carried out;

(5) the sample stabilization weight, which is the inverse of the stabilization subsampling rate and has a value of one for provincial types of area (SR, NSR, or special) except in those for which the sample has exceeded a prespecified size, resulting in a random deletion of selected dwellings; and

(6) the final factor attributable to the principle of ratio estimation, which is defined as the ratio of an external estimated total divided by the subweighted (or simple survey weighted) population within the poststrata.

It is well known that ratio estimation (use of external sources of highly correlated data) results in improved efficiency, with a by-product being the ability to adjust weights to certain control totals. The adjustment takes place by determining the ratio (i.e., the sixth factor) of the external total to the sum of the subweights for responding units. If there were no income nonresponse present, then the weighting would have compensated for all nonresponse and this factor could be considered a measure of slippage (i.e., the over- or under-representation of sample units in the survey). However, with the presence of income nonresponse and the use of subweights from only responding records, this factor also represents a compensation for income nonresponse. This balancing for income nonresponse is completed for each poststratum used in the survey. Further details on poststratification appear in Section 2.3.1. The product of the subweight and the final factor gives the final weight.

The estimation of standard errors for the Survey of Consumer Finances is patterned after the corresponding Labour Force Survey model (Platek and Singh, 1976). In NSR areas, the selected PSUs within strata are assumed to have been sampled independently. In SR areas, each selected subunit is split to form "pseudo-PSUs." In small sample years, these geographic areas are collapsed to grosser levels. Using a variance formula for ratio estimators (for example, Platek and Singh, 1976, p. 75, or Cochran, 1963, p. 173) within each of the geographical levels previously identified, the variance of an estimate is determined. Appendix C provides in Tables 3 and 4 some survey estimates with their standard errors from the 1977 Survey of Consumer Finances.

2. INCOMPLETE DATA IN THE SURVEY OF CONSUMER FINANCES

All surveys, including the Survey of Consumer Finances, are faced with the problem of incomplete data. Some of the causes of missing data are the inability to contact sampled households, the failure to supply details because of refusal or other reasons, interviewer error, data capture, and transmission and processing errors. Incomplete data for the full *block of data* may occur at any of the following levels:

1. Labour force data: Questionnaires F03 and F05 in Appendix A.
2. Work experience data: Questionnaire F03, Question 50, in Appendix A.
3. Income data: Questionnaire CF06, Appendix A.

Incomplete data may also occur as missing items *within blocks* (i.e., some, but not all, information for the block is missing) for

Block 4. Labour force items: Items on Questionnaires F03 or F05.

Block 5. Work experience items: Items on Questionnaire F03 for parts of Question 50.

Block 6. Income items: Items on Questionnaire CF06.

Each of these situations will be discussed in subsequent portions of this chapter. Sections 2.1–2.3 deal with missing blocks of data and Sections 2.4–2.6 deal with missing items within blocks. There is a general hierarchy in the amount of missing data at the block level according to interviewing and processing instructions. The nonresponse rates for the Survey of Consumer

Finances provide an indication of the magnitude of the missing-data problem at the block level. There have been numerous variations in survey operations over time, including such aspects as the degree of self-enumeration, the presence of other supplementary questionnaires imposed on households during their tenure in the Labour Force Survey sample, variation in the income supplementary topics for small sample years, changes in interviewing procedures over time (the introduction of telephone interviewing for example), etc. Because of these variations over time, a historical evaluation over time is inappropriate, and also an assessment of the relative effect in reducing nonresponse of changes in the data collection procedures is not possible. Nonresponse rates for two recent surveys are presented in Table 1.

TABLE 1

Incomplete Data for the Surveys of Consumer Finances (1976 and 1977)[a]

	1976		*1977*	
Units	*Count*	*Rate*	*Count*	*Rate*
Dwelling units				
Selected	38,257	100.0	17,066	100.0
Vacant	3,687	9.6	1,780	10.4
Occupied	34,570	100.0	15,286	100.0
Nonrespondent	3,232	9.3	1,807	11.8
No contact	2,114	6.1	667	4.4
Refusal	1,118	3.2	1,140	7.5
Respondent[b]	31,335	90.6	13,479	88.2
Sampled units within respondent dwellings[c]				
Economic Family Units				
Total	33,629	100.0	14,203	100.0
Complete response	26,871	79.9	12,846	90.4
Nonrespondent	6,758	20.1	1,357	9.6
Partial nonresponse	1,195	3.6	473	3.3
Complete nonresponse	5,563	16.5	884	6.2
Individual units (age 15+)				
Total	73,447	—	31,241	—
Total with income	60,720	100.0	25,636	100.0
Provided income (Block 3) details	48,665	80.1	23,553	91.9
Did not provide income (Block 3) details	12,055	19.9	2,083	8.1

[a] For a description of the difference between odd- and even-numbered survey years, see the discussion in Section 1.2.

[b] An occupied dwelling is classified as respondent in this table if Block 1 data were provided for all eligible members.

[c] Responses to income data are given only for Block 3. Comparable rates for Block 2 were not available, since each field of Block 2 data was initially edited separately rather than all data items for the block being edited collectively.

2.1. Labour Force Survey (Block 1) Data Missing

The Survey of Consumer Finances generally deals with the problem of missing Labour Force Survey data by a *reweighting* procedure. A block of Labour Force Survey data is considered as missing if questions 10, 13-1, 33, 50, and 80 on the F05 questionnaire (Appendix A) are all missing. This reweighting is carried out at the household level within balancing units that are defined at geographic levels. In SR areas, each subunit forms a balancing unit, whereas each PSU in NSR areas is split into two balancing units—the rural part and the urban part. Within a balancing unit, the weights on Labour Force Survey respondent records are inflated to account for the reduced sample attributable to the missing data. The adjustment factor is the ratio of the total weighted count of eligible households to the weighted count of responding households in the balancing unit. These counts are weighted by the product of the cluster and stabilization factors (as described in Section 1.3). For operational considerations, if the adjustment factor is greater than three, the affected balancing unit is collapsed with an adjacent balancing unit of like type, and the balancing unit factor calculated for the collapsed balancing unit is applied to all respondent records in the collapsed balancing unit.

2.2. Work Experience (Block 2) Data Missing

For individual records possessing the Labour Force Survey data for the reference week, the block of data dealing with the annual work experience activities may be missing. This block is considered missing if all entries are blank. With the amount of historical work experience data available on the Labour Force Survey questionnaires, it is often possible to *assign*[1] the necessary work experience details. Labour Force Survey questions supplying information on whether the respondent has ever worked or when the respondent started working for his/her current employer, etc., can be used to assign work experience data referring to the preceding calendar year. In the event that the particular data available on the Labour Force Survey questionnaires are insufficient to assign work experience data, they can often be used in conjunction with reported income details (if available) to determine an assignment. For example, if unemployment insurance benefits are reported for the preceding calendar year, then it is reasonable to assign a period of unemployment to the individual. The magnitude of wages and salaries reported in conjunction with industry/occupation descriptions for the current job, along with hours worked per week,

[1] The terms "assignment" and "imputation" as defined here are for the purpose of this chapter and may have different meanings for some readers. Thus *assignment* refers to obtaining a missing value for a record on the basis of available data for the record, whereas *imputation* is based on obtaining missing values for a record on the basis of comparable data available for another record.

is available to assign weeks worked. If these procedures do not permit the assignment of work experience data, then, within basic demographic classifications, missing values are *imputed* on the basis of observed distributions for respondent individuals. Each record for which work experience data are assigned or imputed must pass all the subsequent logic and validity edits before being accepted.

2.3. Income (Block 3) Data Missing

Income data may be missing at the individual level, for entire family units or for some but not all members of family units. This block of data is considered as missing if any of the income components, except for family allowances or old age security income, is missing (i.e., any of Questions 1–8 or 11–17, Appendix A, Questionnaire CF06). The procedures adopted to compensate for missing income data are carried out in conjunction with the ratio estimation procedure and will be presented in Section 2.3.1 for individual weighting and family weighting separately. However, in both cases, the compensation for missing income data is an *implicit reweighting* of the respondent sampled records. On an experimental basis, missing income data have been imputed for the Survey of Consumer Finances. Details of this procedure will be presented in Section 2.3.2.[2]

2.3.1. Reweighting for Compensation of Missing Income Data

For the individual file, the set of individuals for whom income data are available is weighted up to predetermined totals within poststrata defined on the basis of demographic and labour force variables. The ratio adjustment is the predetermined total for the poststrata divided by the subweighted count of individuals for whom income details are available. This adjustment implicitly reweights the sample of income-responding individuals to the subweighted total of all individuals within the poststrata, followed by a ratio estimation procedure applied to the adjusted subweighted estimate of individuals for whom income details are available, which was described earlier as slippage (i.e., the over- or underestimation of surveyed population by the simple survey estimates). For the weighting of individual records, the poststrata were chosen because they are highly correlated with income. For individuals in the labour force, the poststratum is class of worker/labour force status by sex, and for individuals not in the labour force, age by sex within each province. Hence, 11 classes within each sex and province group yield a total of 220 poststrata. The weight adjustment to accommodate the inflation for nonresponse and ratio

[2] At the time of writing, imputation was still at an experimental stage. Subsequently, the imputation was adopted with slight modifications and was used in the 1978 Survey of Consumer Finances.

estimation is carried out independently within each poststratum. The external estimated population totals for the poststrata are obtained from the Labour Force Survey and, indirectly, from census sources for the corresponding month.

Family units with partially missing income data (i.e., income details missing for at least one member of the family) and with completely missing income data are likewise compensated for by an implicit reweighting of the family units for which complete income details are available. The combined reweighting and ratio adjustment is, as before, carried out by a ratio adjustment to compensate for family nonresponse and to accommodate the ratio estimation procedure for the reduced sample of family units within each poststratum. They are, for both the economic and census family definitions, class of worker/labor force status of head by sex of head within family size (one and two or more) and province. Hence five classes within each sex of head, family size, and province group yield a total of 200 poststrata. These final factors and the resulting final weights are calculated for the heads of the family units but are applicable to the whole family unit. The predetermined population totals within poststrata are obtained from census and Labour Force Survey sources.

2.3.2. Imputation of Income Data

A procedure to impute missing data at the individual level is presently in an experimental stage for the Survey of Consumer Finances. This section briefly describes the procedure as applied to the 1977 survey, as well as refinements or enhancements for the 1978 survey.[3]

For individual income nonrespondents, a considerable amount of information, including demographic data, labour force activity during the reference week, and details of the work experience during the preceding calendar year, is available for use in dealing with the missing data. By exploiting correlations between these available data and the income data, an attempt is made to reduce the nonresponse bias in survey estimates. Imputation of income data results in a substantial increase in the amount of usable data for both the individual and family unit series. The magnitude of this increase is found in Table 1.

The method of determining imputation categories for the survey is a technique known as Automatic Interaction Detection (AID), as described by Sonquest and Morgan (1964). The technique was used to select from a list of variables those, and their appropriate cross-classifications, to be used in defining imputation categories.

The AID technique operates on data vectors consisting of a single dependent variable and one or more independent, or predictor, variables. Using the principle of least squares, AID iteratively divides the sample into subgroups defined on the basis of predictor variable splits of an existing subgroup into new

[3] See footnote 2.

subgroups, subject to sample size and percentage of explained variation criteria. At each stage of all possible splits, AID chooses the split that minimizes the residual sum of squares of the dependent variable when the group is split into two parts.

Ideally, the determination of imputation categories should be carried out on the basis of respondent records for the survey in question. However, if the survey is of a continuous nature and if the questionnaire's content and design and the operational aspects are virtually unchanged from previous surveys, then the imputation categories might well be determined on the basis of past survey data, as was the case for income data. Data for three survey years were analyzed with no substantial differences in the categories so derived. The variables used to define the categories were the same for the three survey years, with the only noticeable difference being a switching in the order of variables, which affected their order but not their levels of splits. For example, education breaking on degree obtained and its complement and occupation breaking on managers plus sales and its complement would both be present for all years but education may come before occupation one year and after for the others.

In applying the techniques, the experimentally imputed data for the survey were the complete individual income questionnaire, with the exception of five items that were calculated on the basis of individual or family data and assigned to the individual record or had blanks inserted. The income items (see Questionnaire CF-06) not to be imputed are

(1) family allowances (assigned);
(2) old age security and guaranteed income supplement benefits (assigned);
(3) capital gains or losses (blanks inserted, later changed to zeros);
(4) income tax payable (blanks inserted, later calculated values were assigned); and
(5) provincial tax credit (blank inserted, later changed to zero).

The donor files consisted of the data records of all individuals who both received income and supplied details. Imputation was to be done on individuals who had income but refused details (receivers) and whose personal and labour force characteristics were known. The dependent variable chosen was total actual income because it is the primary datum used in SCF publications.

To evaluate the imputation procedure, a series of standard statistical tests of significance, carried out on these data, revealed that income nonresponse is not independent of imputation category and, furthermore, that there is a monotone relationship between income nonresponse rates and average individual incomes. The evaluation also revealed the necessity of including a geographic component in the procedure. The method chosen from several available (including geographical categorization and geographically adjacent restrictions) was a technique of nearest-record selection. Each record on the file has a geographic identification. The proximity of records on the file can then be assessed on the basis of this geographic identification. For operational reasons, the concept of nearest-record selection for donor records was adjusted to a nearest-preceding-record selection procedure. In addition, for the 1978 survey, old age security and guaranteed income supplement benefits were imputed. The refined imputation procedures yielded satisfactory results, and the 1978 Survey of Consumer Finances was the first survey to use imputed income data.[4]

2.4. Missing Labour Force Survey Items

Information on the current and historical labour force activities of individuals are collected on the Labour Force Survey questionnaire. The questionnaire is a highly structured document and the series of questions to be completed for a respondent depends on the responses given to particular key questions.

There are four types of edits applied: (1) questionnaire sequence completion, (2) subject-matter logic, (3) validity of values, and (4) logical consistency. The edit failures are identified by the computer system. The assignments and imputations are carried out on a manual basis according to prespecified procedures. There are three approaches available to obtain any missing values. Considering the internal questionnaire logic from both the sequence of applicable questions and the subject-matter interrelationships, assignments for several of the missing values can be determined from the actual record itself. For the regular survey (and for supplements), the previous month's labour force data are available for the majority of the respondents. The high degree of serial correlations for some data items means that the previous month's response can be used as an imputation value. For years in which the Survey of Consumer Finances is an independent survey, this approach is not available. The third approach is utilized when the previous two methods fail. An imputation value is derived from a similar record. The similar record selection criterion is dependent on the missing item and involves characteristics such as age, sex, industry, and occupation, which are felt to be highly correlated with values for the missing question. The similar record is generally selected from within the same geographic sampling unit (i.e., the primary sampling unit).

2.5. Missing Work Experience Items

Edits for this block of information fall into two broad categories: those requiring assignment (either manual or computer) if a failure occurs and those

[4] See footnote 2.

present for evaluation purposes only. In all cases, a comparison is made between two or more items. The flow of editing for missing items first looks for blanks in items interrogated, then for blanks in some but not all the items, inconsistencies when all items are present, and finally information that could lead an editor to suspicions that the data present may or may not be invalid. However, in the latter instance, the step is initially used for evaluation purposes only.

Edits with manual assignment are classified as major errors (or *type 1*). A way to illustrate this phase of processing is to follow one of several work experience edit streams, namely, one of the interrelationships between weeks worked and weeks unemployed. First, if weeks worked and weeks unemployed are both blank, manually update both using demographic and income data available. If only one of the two fields is blank, ask if weeks worked is blank. If weeks unemployed is not blank but is less than 52 and if other activity while not in the labour force is not applicable (i.e., has a value of zero), then manually update weeks worked since the individual was not in labour force for an unspecified period of time. Finally, the same edit is performed interchanging weeks worked and weeks unemployed. This stream is designed to fill in blanks for these two fields by manual update after survey operations weigh all other survey information available for that individual.

Edits with computer updates (classified as intermediate of *type 2* errors) fill in fields where the missing item is more apparent than situations described for manual updates. Activation of this type of edit occurs when two or more nonblank fields are inconsistent. At least one of the fields is considered master and is left unchanged, whereas the remaining field is not the master and is changed automatically. These fields are hierarchical in nature in that, at the beginning, one field may become master for other fields and never change, whereas others do. Those subject to possible change can later become master to check subsequent fields. The edit stream analogous to this is described next.

If weeks worked is blank and weeks unemployed is not blank, assign zero to weeks worked if weeks unemployed is 52; otherwise, assign the difference between 52 and weeks unemployed to weeks worked if other activity is zero (i.e., not applicable). Next, reverse weeks worked and weeks unemployed (i.e., weeks worked is not blank and weeks unemployed is blank) and repeat the edit. Then, if weeks worked and weeks unemployed are greater than 52, make weeks unemployed the difference between 52 and weeks worked. The stream completes the interediting between weeks worked and weeks unemployed.

The other categories of edits are those designated for evaluation only. They consist of a group that may require action (i.e., minor, or *type 3* errors) and a group that is present only for counting purposes (i.e., evaluation, or *type 9* errors). Since this group is closely aligned to possible missing income item portions of survey processing, the description by example of the edit stream is left to the next section.

2.6. Missing Income Items

At the individual level, amounts of money income as derived from sixteen possible income sources (items 1–16 on Questionnaire CF06 in Appendix A), along with the amounts for three income-related items, including income tax payable (items 18–20 on Questionnaire CF06, Appendix A), are collected by the Survey of Consumer Finances. At the data-capture stage, only nonzero items are captured, ignoring the indicators (Yes/No) of the presence of an income item (see Questionnaire CF06 in Appendix A), because these indicators were used at manual editing stages. Items for which no value is reported are set to zero.

There are basically two types of edits for the income data—those that require assignment if the record does not pass the edit criterion and those that are present for evaluation purposes only.[5] Missing or inconsistent items may be identified either by an examination of the questionnaire or by the computer edit system. Prior to data entry, the questionnaires are manually examined for consistency and completeness. Here, instances of an income amount being entered in the wrong cell or a reported amount being clearly in error (for example, the indicator (Yes/No) indicates the presence of an income component that is erroneously entered opposite a different item) possibly require changes to be made as a result of information in the notes section or written in detail on the questionnaire. Subsequent to data capture and linkage with Labour Force Survey data and supplementary work experience data, the record passes through the detailed income edits. Cases of edit failures are identified, and the appropriate assignment value that is generated by the computer or the relevant content of the questionnaire is listed on a turn-around document for manual examination and possible update.

The missing values may be assigned on the basis of standard values for the income component. For example, in the event of a missing amount for the Canadian Old Age Security Pension Plan, income for an individual older than 66 years of age and immigrating earlier than 1967, an amount is assigned. For other situations, an income amount may be erroneously coded in the wrong position (for example, an individual with an age of less than 60 who reports government pension plan benefits) or an income amount may be reported under two separate income sources (for example, government pension plan income is reported correctly under "Canada or Quebec Pension Plan benefits" as well as under "Retirement pensions, superannuation, and annuities"). For these types of situations, income amount are transferred to the proper cell or eliminated as considered appropriate.

A series of minor or evaluation edits is applied to the income data. These are in place to indicate that there may be an error in the income data, but in the

[5] Edits of the second type might better be termed "evaluation counts" since they measure possible inconsistencies or unusual situations, but a data change is not mandatory.

absence of a strong logical inconsistency, corrective action is often not possible. For example, one edit states that if the number of weeks unemployed is greater than zero, then there should be an amount present for unemployment insurance benefits. However, there does not have to be an entry because the insurance is available only by applying for it and meeting certain requirements, and not everyone unemployed applies and/or meets the requirements. A second example is the case in which the amount of unemployment insurance *benefits* reported as income is approximately equal to the maximum amount of unemployment insurance *contributions*. The entry may indeed be benefits. Evaluation edits are included primarily to observe problems in the collection of income data experienced because of difficulties in the income concepts or to evaluate the edits for possible future inclusion as major or intermediate edits in the regular survey processing.

In addition to the preceding assignment procedures for missing items on the income questionnaire, there are two particular items, namely, the family allowance benefits and the income tax payable, for which special procedures have been developed to obtain an imputation value derived on the basis of data available for the individual claiming the amount as well as for the other members of the census family.

For the family allowance component, the amount of the benefit is calculated by using the age and educational status of each child in the family in conjunction with the amount payable as determined under the government family allowance program. Because of variation in the amounts payable, depending on educational status and province, and because children eligible for receipt of benefits for the previous calendar year are not in the family at survey time, exact amounts of family allowance benefits cannot always be calculated. However, on the basis of data available, a calculated amount, along with maximum and minimum amounts, is determined for the family. In fact, these three values are derived for all families whether or not the family allowance component is reported. If the item is missing, then the calculated value is the value entered for the missing item. If the component is available and falls between the upper and lower amounts, no change is made.

The procedures established for assigning missing income tax payable amounts consist essentially of simulating the income tax calculation. The calculation is particularly complex because income components and demographic information available for the individual, as well as family variables and demographic and income amounts for other family members, are required. The procedure initially calculates the amount of taxable income. This amount is determined by subtraction of personal exemptions and allowable deductions as specified under the income tax law from the total income subject to tax. Using the rates of federal income tax, any applicable refinements, and provincial income tax rates, the total income tax payable is calculated. This very briefly describes the procedure, although in fact the procedure is somewhat more complex, incorporating numerous refinements and adjustments. For

TABLE 2

Selected Counts from Revenue Canada and the Survey of Consumer Finances *1977–1976*

	Individuals who:			
Item	*Filed income tax returns*[a]	*Received income*[b]	*Paid taxes*[a]	*Paid taxes*[b]
Sample size	381,622	23,399	267,709	13,869
Estimated numbers ($ million)	12.2	14.1	8.7	8.5
Aggregate income subject to tax ($ million)	124,132	128,710	114,237	112,526
Average income subject to tax ($)	10,182	9,120	13,165	13,207
Aggregate estimated tax ($ million)	20,291	20,249	20,291	20,249
Average estimated tax ($)	1,664	1,435	2,338	2,377

[a] Revenue Canada (1978).
[b] Statistics Canada (1979).

evaluation and editing purposes, this simulation derivation of income tax payable is completed for all individuals whether or not the appropriate item was reported.

In the event that the item is missing, the calculated value is entered for the item. Note that if the edit and imputation system has substantially altered reported amounts or has resulted in the imputation of unreported items, then the reported income tax payable may differ substantially from the calculated value.

Both of the aforementioned calculations—for family allowance and for income tax payable amounts—yield satisfactory results as indicated by comparisons with respondents who supplied all details and as compared with outside sources.

To illustrate comparisons with outside sources, Table 2 comprises selected counts taken from the report "Income after tax, distributions by size in Canada 1976" (Statistics Canada, 1979). The report provides more detailed explanations with respect to comparing the data than does the following summary. The table represents an attempt to reconcile data produced by Revenue Canada (1978) in their report "Taxation Statistics, 1978 Edition" and the 1977 Survey of Consumer Finances. Data in Table 2 have been adjusted for differences in income definitions and geographic coverage in order to render the data as comparable as possible. In interpreting the numbers, it must be emphasized that all income recipients do not necessarily have to file a tax return.

3. SUMMARY

The procedures for dealing with missing-data problems as presented in this chapter are applicable to the most recently published survey (SCF78). The imputation of income details, currently at an experimental stage, will be further evaluated and refinements to the procedures carried out as necessary with a view to its implementation as a regular survey feature.

The procedures for dealing with missing data are relatively dynamic in the sense that they have evolved over time (and will change in the future) as dictated by changes in the income environment of the population (as, for example, changes in the income tax law or revisions to regulations governing pension incomes) or as motivated by potential improvements in the quality of survey data. Considering current trends in the population's attitudes toward government surveys and the question of "invasion of privacy," surveys collecting data of a sensitive nature, such as income data, are particularly vulnerable to increases in nonresponse rates. Survey procedures must clearly respond to changes in the magnitude of the missing-data problem at all stages of the survey operations.

APPENDIX A. STATISTICS CANADA QUESTIONNAIRES

Questionnaire F03

Statistics Canada Statistique Canada **HOUSEHOLD RECORD DOCKET** CONFIDENTIAL when completed 1 FORM NO. 03

2 Docket No. 3 Survey date 4 Assignment No. 5 Designated Interviewer no. No change 1 or Your interviewer no.

6 P.S.U. Group Cluster Rot. No. 7 Listing Mult. 8 Type of dwelling *Enter code*

9 *Record time of every call on this household.*

Mon.	:	:	:	:	:
Tues.	:	:	:	:	:
Wed.	:	:	:	:	:
Thur.	:	:	:	:	:
Fri.	:	:	:	:	:
Sat.	:	:	:	:	:

10 Listing address

11 Mailing address

12 *INTERVIEWER CHECK ITEM*
- *If first interview at this dwelling or new household since last interview* 1 ○ *go to 13*
- *Otherwise* 2 ○ *go to 15*

13 IN WHICH OFFICIAL LANGUAGE DO YOU PREFER TO BE INTERVIEWED?
English 1 ○ French 2 ○ Either 3 ○

14 *INTERVIEWER CHECK ITEM* *In which OFFICIAL LANGUAGE was the interview conducted?*
English 1 ○ French 2 ○ Neither 3 ○

15 *INTERVIEWER CHECK ITEM*
If unable to make contact with a responsible household member, enter code in 41. For Code B, also specify reason in NOTES.

16 Telephone no. ___ – ___
or No telephone 1 ○
or Telephone no. refused 2 ○ } *go to 18*

17 *INTERVIEWER CHECK ITEM*
- *If "T" in* 1 ☐ *request permission to interview by phone*

Permission granted 2 ○
Permission denied 3 ○ } *Go to 18*
- *Otherwise* 4 ○ *go to 18*

18 *Determine and record the best time to call on this household.*

20 *INTERVIEWER CHECK ITEM*
- *If first interview at this dwelling or new household since last interview* 1 ○ *go to 21*
- *If a subsequent interview with the same household* 2 ○ *go to 25*

21 WHAT ARE THE NAMES OF ALL PERSONS NOW LIVING OR STAYING AT THIS ADDRESS WHO HAVE NO USUAL PLACE OF RESIDENCE ELSEWHERE?
Enter names in 32

22 ARE THERE ANY PERSONS AWAY FROM THIS HOUSEHOLD ATTENDING SCHOOL, VISITING, TRAVELLING OR IN HOSPITAL WHO USUALLY LIVE HERE?
Yes 1 ○ *Enter names in 32* No 2 ○

23 DOES ANYONE ELSE LIVE AT THIS ADDRESS SUCH AS OTHER RELATIVES, ROOMERS, BOARDERS, OR EMPLOYEES?
Yes 1 ○ *Enter names in 32* No 2 ○

24 *Enter answers for 33 through 40 for each person recorded in 32, then go to 41.*

25 DO THE FOLLOWING PERSONS STILL LIVE OR STAY HERE?
- *Read names of all persons recorded in 32 and*
- *Enter appropriate code in 40*

26 DOES ANYONE ELSE NOW LIVE OR STAY AT THIS ADDRESS SUCH AS OTHER RELATIVES, ROOMERS, BOARDERS, ETC.?
Yes 1 ○ *Enter answers for 32 through 40 for all new household members.* No 2 ○ *Go to 41*

31 HRD Pg	Ln	32 Names of household members	33 Age	34 Sex	35 MS	36 Fam Id.	37 R to H	38 Educ. 1	2	40 Membership
	1	given name / surname								
	2	given name / surname								
	3	given name / surname								
	4	given name / surname								
	5	given name / surname								
	6	given name / surname								
	7	given name / surname								
	8	given name / surname								

50 Answers to supplementary questions

Ln	a	b	c	d	e	f	g	h	i	j	k	l	m
1													
2													
3													
4													
5													
6													
7													
8													

41 Household Response
month
response

42 IS THIS DWELLING OWNED BY A MEMBER OF THIS HOUSEHOLD?
Yes 1 ○ *Go directly to FORMS NO. 05 & 06.*
No 2 ○ *Complete FORM NO. 04*

43 Forms Control

form	04	05	06
printed			
completed			

44 NOTES

Ln	Retain	Item no.	*See over for additional NOTES.* ○
1	1 ○		
2	2 ○		
3	3 ○		
4	4 ○		

Questionnaire F05

Statistics Canada Statistique Canada

LABOUR FORCE SURVEY QUESTIONNAIRE

CONFIDENTIAL when completed

Docket No. **2** Survey date **3** Mo Yr Assignment No. **4**

HRD page - line No. **5** Given name **6** Surname **7**

1 FORM NO. **05**

10 LAST WEEK, DID . . . DO ANY WORK AT A JOB OR BUSINESS?
Yes 1 ◯ No 2 ◯ Go to 30
PERMANENTLY unable to work 3 ◯ Go to 50

11 DID . . . HAVE MORE THAN ONE JOB LAST WEEK?
Yes 1 ◯ No 2 ◯ Go to 13

12 WAS THIS A RESULT OF CHANGING EMPLOYERS LAST WEEK?
Yes 1 ◯ No 2 ◯

13 HOW MANY HOURS PER WEEK DOES . . . USUALLY WORK AT HIS/HER:
(Main) JOB? ☐☐ If total 30 or more go to 15
Other jobs? ☐☐

14 WHAT IS THE REASON . . . USUALLY WORKS LESS THAN 30 HOURS PER WEEK?
☐ Enter code

15 LAST WEEK, HOW MANY HOURS OF OVERTIME OR EXTRA HOURS DID . . . WORK?
(Include paid and unpaid time at all jobs) ☐☐ If none enter 00

16 LAST WEEK, HOW MANY HOURS DID . . . LOSE OR TAKE OFF FROM WORK FOR ANY REASON SUCH AS ILLNESS, HOLIDAY, OR LAYOFF? (From all jobs)
☐☐ If none enter 00 and go to 18

17 WHAT WAS THE MAIN REASON FOR LOSING THESE HOURS?
☐ Enter code

18 HOW MANY HOURS DID . . . ACTUALLY WORK LAST WEEK AT HIS/HER:
(Main) JOB? ☐☐ Other jobs? ☐☐

19 IN THE PAST 4 WEEKS, HAS . . . LOOKED FOR ANOTHER JOB?
Yes 1 ◯ No 2 ◯ Go to 71

20 WHAT HAS . . . DONE IN THE PAST 4 WEEKS TO FIND ANOTHER JOB?
☐ ☐ ☐ Enter code(s) and go to 71

30 LAST WEEK, DID . . . HAVE A JOB OR BUSINESS AT WHICH HE/SHE DID NOT WORK?
Yes 1 ◯ Go to 33 No 2 ◯

31 LAST WEEK, DID . . . HAVE A JOB TO START AT A DEFINITE DATE IN THE FUTURE?
Yes 1 ◯ No 2 ◯ Go to 50

32 COUNTING FROM THE END OF LAST WEEK, IN HOW MANY WEEKS WILL . . . START TO WORK AT HIS/HER NEW JOB?
☐☐ Go to 50

33 WHY WAS . . . ABSENT FROM WORK LAST WEEK?
☐ Enter code and if code 6 go to 32

34 DID . . . HAVE MORE THAN ONE JOB LAST WEEK?
Yes 1 ◯ No 2 ◯

35 HOW MANY HOURS PER WEEK DOES . . . USUALLY WORK AT HIS/HER:
(Main) JOB? ☐☐ If total 30 or more go to 37
Other jobs? ☐☐

36 WHAT IS THE REASON . . . USUALLY WORKS LESS THAN 30 HOURS PER WEEK?
☐ Enter code

37 UP TO THE END OF LAST WEEK, HOW MANY WEEKS HAS . . . BEEN CONTINUOUSLY ABSENT FROM WORK?
☐☐

38 IS . . . GETTING ANY WAGES OR SALARY FROM HIS/HER EMPLOYER FOR ANY TIME OFF LAST WEEK?
Yes 1 ◯ No 2 ◯

39 *INTERVIEWER CHECK ITEM*
• If code 5 (layoff) in 33 1 ◯ go to 56
• Otherwise 2 ◯ go to 40

40 IN THE PAST 4 WEEKS, HAS . . . LOOKED FOR ANOTHER JOB?
Yes 1 ◯ No 2 ◯ Go to 71

41 WHAT HAS . . . DONE IN THE PAST 4 WEEKS TO FIND ANOTHER JOB?
☐ ☐ ☐ Enter code(s) and go to 71

DESCRIPTION OF MAIN JOB OR BUSINESS

71 HAS . . . CHANGED EMPLOYERS SINCE LAST MONTH?
Yes 1 ◯ No 2 ◯

72 FOR WHOM DID . . . WORK? (Name of business, government dept. or agency, or person)
▲ No change ☐ 1
or ▶

73 WHEN DID . . . START WORKING FOR THIS EMPLOYER?
☐☐ Mo. Yr. No change ☐ 1 or ☐☐ Mo. Yr. If month unknown enter – – in month

74 WHAT KIND OF BUSINESS, INDUSTRY OR SERVICE WAS THIS? (Give full description: e.g., paper-box manufacturing, retail shoe store, municipal government)
▲ No change ☐ 1
or ▶

75 WHAT KIND OF WORK WAS . . . DOING? (Give full description: e.g., posting invoices, selling shoes, teaching primary school)
▲ No change ☐ 1
or ▶

76 Class of worker:
Main job . . . . . ☐ No change ☐ 1 or ☐ Enter code

77 Other job . . . . . ☐ No change ☐ 1 or ☐ Enter code go to 80

NOTES — See over for additional NOTES ◯

Item no.	
99	
99	
99	
99	

50 HAS . . . EVER WORKED AT A JOB OR BUSINESS?
Yes 1 ◯ No 2 ◯ Go to 55

51 WHEN DID . . . LAST WORK?
☐☐ Mo Yr No change ☐ 1 or ☐☐ Mo. Yr
If month unknown, enter – – in month

52 *INTERVIEWER CHECK ITEM:*
(1) If 51 is before . . . . . . ☐☐ 1 ◯ go to 55
(2) If 51 is equal to or later than . . . . . . Mo. Yr. 2 ◯ go to 53

53 WAS THIS A FULL-TIME OR A PART-TIME JOB?
Full - time (30 or more hours per week) 1 ◯ Part - time (Less than 30 hours per week) 2 ◯

54 WHAT WAS THE MAIN REASON WHY . . . LEFT THAT JOB?
☐ Enter code

55 *INTERVIEWER CHECK ITEM:*
• If "perm. unable to work" in 10 1 ◯ go to 80
• Otherwise 2 ◯ go to 56

56 IN THE PAST 6 MONTHS HAS . . . LOOKED FOR WORK?
Yes 1 ◯ No 2 ◯ Go to 70

57 • IN THE PAST 4 WEEKS WHAT HAS . . . DONE TO FIND WORK? Mark all methods reported
Nothing ☐ 1 Go to 63
• IN THE PAST 4 WEEKS HAS . . . DONE ANYTHING ELSE TO FIND WORK? Mark all other methods reported
For each method given ask:
• WHEN DID . . . LAST ______? (Repeat method)

Checked with:	Method used	No. of weeks ago (excl. svy. week)
PUBLIC employment AGENCY	2 ◯	☐
PRIVATE employment AGENCY	3 ◯	☐
UNION	4 ◯	☐
EMPLOYERS directly	5 ◯	☐
FRIENDS or relatives	6 ◯	☐
Placed or answered ADS	7 ◯	☐
LOOKED at job ADS	8 ◯	☐
OTHER Specify in NOTES	9 ◯	☐

58 WHAT WAS . . . DOING IMMEDIATELY BEFORE HE/SHE STARTED TO LOOK FOR WORK? FOR EXAMPLE, WORKING, KEEPING HOUSE, GOING TO SCHOOL OR SOMETHING ELSE.
☐ Enter code

59 UP TO THE END OF LAST WEEK, HOW MANY WEEKS HAS . . . BEEN LOOKING FOR WORK? DO NOT COUNT WEEKS WHEN HE/SHE ALSO WORKED.
☐☐

60 HAS . . . BEEN LOOKING FOR A JOB TO LAST FOR LESS THAN 6 MONTHS, OR MORE THAN 6 MONTHS?
Less than 6 months (incl. 6 mos.) 1 ◯ More than 6 months 2 ◯

61 HAS . . . BEEN LOOKING FOR A FULL-TIME OR A PART-TIME JOB?
Full - time 1 ◯ Part - time 2 ◯

62 *INTERVIEWER CHECK ITEM:*
• If "1 week ago" for any method in 57 1 ◯ go to 64
• Otherwise 2 ◯ go to 63

63 WAS THERE ANY REASON WHY . . . DID NOT LOOK FOR WORK LAST WEEK?
☐ Enter code

64 WAS THERE ANY REASON WHY . . . COULD NOT TAKE A JOB LAST WEEK?
☐ Enter code

70 *INTERVIEWER CHECK ITEM:*
• If "No" (never worked) in 50 2 ◯ go to 80
• If upper circle in 52 is marked 3 ◯ go to 80
• Otherwise 4 ◯ go to 72

EDUCATIONAL ACTIVITY

80 LAST WEEK, WAS . . . ATTENDING A SCHOOL, COLLEGE, OR UNIVERSITY?
Yes 1 ◯ No 2 ◯ Go to 90

81 WAS . . . ENROLLED AS A FULL-TIME OR A PART-TIME STUDENT?
Full-time 1 ◯ Part-time 2 ◯

82 WHAT KIND OF SCHOOL WAS THIS?
☐ Enter code

INFORMATION SOURCE

90 HRD page-line No. of person providing the above information.
Last interview ☐☐ This interview ☐☐

91 Was this information provided over the telephone?
Yes 1 ◯ No 2 ◯

Questionnaire CF06

FORM NO. CF **06**

Name of respondent

Français au verso

Statistics Canada Statistique Canada

1977 INCOME QUESTIONNAIRE

To be completed by persons 15 years of age and over

Pick-up date:

Authority - Statistics Act, Chapter 15, Statutes of Canada 1970-71-72. This act prohibits the disclosure by Statistics Canada of any particulars which could reveal the identity of an individual.

1 R.O.	2 Docket No.	3 Survey date (Mo. Yr.)	4 Assignment No.	5 HRD page-line No.

Please refer to the GUIDE for instructions

DURING THE TWELVE MONTHS ENDING DECEMBER 31, 1977, DID YOU RECEIVE ANY INCOME FROM THE FOLLOWING SOURCES? IF "YES", PLEASE MARK "X" IN THE "YES" CIRCLE AND ENTER AMOUNT RECEIVED. IF "NO", PLEASE MARK "X" IN THE "NO" CIRCLE AND PROCEED. (TO INDICATE A LOSS, ENTER THE AMOUNT AND WRITE "LOSS" ABOVE.)

2 0 1

I.C. 25

			Dollars	Cents
1. WAGES and SALARIES before deductions	Yes ○ No ○	01		
2. MILITARY PAY and ALLOWANCES	Yes ○ No ○	02		
3. Net income from NON-FARM SELF-EMPLOYMENT	Yes ○ No ○	03		
4. Net income from FARM SELF-EMPLOYMENT	Yes ○ No ○	04		
5. Gross income from ROOMERS and BOARDERS	Yes ○ No ○	05		
6. INTEREST on bonds, deposits and savings certificates	Yes ○ No ○	06		
7. DIVIDENDS, actual amount received (not taxable amount)	Yes ○ No ○	07		
8. OTHER INVESTMENT INCOME (net rents from real estate, etc.)	Yes ○ No ○	08		
9. FAMILY ALLOWANCES	Yes ○ No ○	09		
10. OLD AGE SECURITY PENSION and GUARANTEED INCOME SUPPLEMENT from federal government only, provincial income supplements should be reported in question 13	Yes ○ No ○	10		
11. CANADA or QUEBEC PENSION PLAN BENEFITS	Yes ○ No ○	11		
12. UNEMPLOYMENT INSURANCE BENEFITS, total benefits before tax deductions	Yes ○ No ○	12		
13. SOCIAL ASSISTANCE and PROVINCIAL INCOME SUPPLEMENTS	Yes ○ No ○	13		
14. OTHER INCOME from GOVERNMENT SOURCES, **PLEASE SPECIFY** ____ Provincial tax credits should be reported in question 20	Yes ○ No ○	14		
15. RETIREMENT PENSIONS, SUPERANNUATION and ANNUITIES	Yes ○ No ○	15		
16. OTHER MONEY INCOME, **PLEASE SPECIFY** ____	Yes ○ No ○	16		
17. **TOTAL INCOME**, sum of entries in questions 1 to 16	Yes ○ No ○	17		
18. **Taxable** portion of CAPITAL GAINS or allowable capital losses	Yes ○ No ○	18		
19. INCOME TAX (federal and provincial) — **total payable** on 1977 income and capital gains	Yes ○ No ○	19		
20. PROVINCIAL TAX CREDIT applicable only to residents of ONTARIO, MANITOBA, ALBERTA and BRITISH COLUMBIA	Yes ○ No ○	20		
21. Remarks: ____		21		
		22		

APPENDIX B. SURVEY OF CONSUMER FINANCES' 1976 INCOME, ASSET, AND DEBT QUESTIONNAIRE

Français au verso

Statistics Canada Statistique Canada

SURVEY OF CONSUMER FINANCES

1976 INCOME, ASSET and DEBT QUESTIONNAIRE

A column to be completed for each person 15 years of age and over

Authority – Statistics Act, Chapter 15, Statutes of Canada 1970-71-72.

This act prohibits the disclosure by Statistics Canada of any particulars that can be related to any identifiable individual person.

CF 06

P.S.U. Group Cluster Rotation No. Listing Mult. Fam. Id.

SECTION 1. PERSONAL INCOME 2 0 1

Page/Line Number				
During the twelve months ending December 31, 1976, what was your income from the following sources?				
1- 1. WAGES and SALARIES before deductions	01	01	01	01
1- 2. MILITARY PAY and ALLOWANCES	02	02	02	02
1- 3. NET income from NON-FARM SELF-EMPLOYMENT	03	03	03	03
1- 4. NET income from FARM SELF-EMPLOYMENT	04	04	04	04
1- 5. Gross income from ROOMERS and BOARDERS	05	05	05	05
1- 6. INTEREST on bonds, deposits and savings certificates	06	06	06	06
1- 7. DIVIDENDS, actual amount received (not taxable amount)	07	07	07	07
1- 8. OTHER INVESTMENT INCOME (net rents from real estate, etc.)	08	08	08	08
1- 9. FAMILY ALLOWANCES	09	09	09	09
1-10. OLD AGE SECURITY PENSION and GUARANTEED INCOME SUPPLEMENT from federal government only, provincial income supplements should be reported in 1-13	10	10	10	10
1-11. CANADA or QUEBEC PENSION PLAN BENEFITS	11	11	11	11
1-12. UNEMPLOYMENT INSURANCE BENEFITS total benefits before tax deductions	12	12	12	12
1-13. SOCIAL ASSISTANCE and PROVINCIAL INCOME SUPPLEMENTS	13	13	13	13
1-14. OTHER INCOME from GOVERNMENT SOURCES, please specify in remarks below. Provincial tax credits should be reported in 1-20	14	14	14	14
1-15. RETIREMENT PENSIONS, SUPERANNUATION and ANNUITIES	15	15	15	15
1-16. OTHER MONEY INCOME, please specify in remarks below	16	16	16	16
1-17. **TOTAL INCOME**, sum of entries in 1-1 to 1-16	17	17	17	17
If no income in 1976, please check (√) →	☐	☐	☐	☐
1-18. Taxable portion of CAPITAL GAINS or allowable capital losses	18	18	18	18
If no taxable capital gains or allowable losses, please check (√) →	☐	☐	☐	☐
1-19. INCOME TAX (federal and provincial) – total payable on 1976 income and capital gains	19	19	19	19
If no income tax payable on 1976 income, please check (√) →	☐	☐	☐	☐
1-20. PROVINCIAL TAX CREDIT applicable only to residents of ONTARIO, MANITOBA, ALBERTA and BRITISH COLUMBIA	20	20	20	20
If no provincial tax credit, please check (√) →	☐	☐	☐	☐
Office use only	21	21	21	21
	22	22	22	22
	23	23	23	23

1-21. REMARKS

Pg./Ln.	Item No.

SECTION 2. HOUSING

2	0	2

2- 1. **INTERVIEWER CHECK ITEM** (√)
Form 03
- If "Yes" in item 42 ... ☐ Go to 2-2
- Otherwise ... ☐ Go to Section 3

2- 2. **INTERVIEWER CHECK ITEM** (√)
Form 03
- If code 0, 3 or 5 in item 8. ... ☐ Go to 2-3
- Otherwise ... ☐ Go to 2-5

2- 3. How many dwellings are there in this building? ... [03]

2- 4. Is this dwelling a condominium unit? ... [04] 1☐ Yes 2☐ No

2- 5. What share of this property is owned by your family ... % [05]

2- 6. Is part of this property used for purposes other than your residence? (√) ... [06] 1☐ Yes Go to 2-7 2☐ No Go to 2-9

2- 7. A part of this property is: (√) [07]
- rented out ... 1☐
- used for business ... 2☐
- other, please specify in remarks below ... 3☐

2- 8. What percentage of this property is occupied by your household as residence? ... % [08]

2- 9. In what year was this property acquired? ... [09]

2-10. What was the purchase price? ... $ [10]

2-11. What is the present market value? ... $ [11]

2-12. Is there a mortgage on this property today? (√) ... [12] 1☐ Yes Go to 2-13 2☐ No Go to Section 3

	1st mortgage	2nd mortgage
2-13. What is the principal outstanding today? ... $	[13]	[33]

2-14. **INTERVIEWER CHECK ITEM** (√)
- If amount is given in item 2-13 ... ☐ Go to Section 3
- Otherwise ... ☐ Go to 2-15

	1st mortgage	2 nd mortgage
2-15. What is the contractual term of your present mortgage? (√)	[15]	[35]
1-2 years	1☐	1☐
3-4 years	2☐	2☐
5 years	3☐	3☐
over 5 years	4☐	4☐
2-16. Is your present mortgage renegotiable? (√)	[16] 1☐ Yes 2☐ No	[36] 1☐ Yes 2☐ No
2-17. When did you take out your present mortgage?	[17] Mo. Yr.	[37] Mo. Yr.
2-18. What was the principal outstanding when you took out the present mortgage? ... $	[18]	[38]
2-19. What is the original amortization period of the present mortgage in years?	[19]	[39]
2-20. What is the current interest rate on the mortgage?	[20] % per	[40] % per
2-21. How often are mortgage payments made? (√)	[21]	[41]
monthly	1☐	1☐
other, please specify in remarks below	2☐	2☐
2-22. What is the amount of the regular payment? (Include only principal and interest.) ... $	[22]	[42]
2-23. What is the amount of any extra payment(s) made on principal since you took out the present mortgage? ... $	[23]	[43]

2-24. REMARKS

Item No.

2	0	3

SECTION 3. PERSONAL ASSETS

In addition, Statistics Canada would also like to obtain details on other types of assets held by Canadians. Where assets are held jointly with other persons, state your own share only. If assets relate partly to business and partly to personal use, include the full amounts here if you consider them to be primarily of a personal nature. Where such assets relate primarily to business, please include their value in Section 5 below.

Page/Line Number	☐☐	☐☐	☐☐	☐☐
What is the value of assets you now own?				
CHARTERED BANK DEPOSITS:				
3-1. Current and personal chequing accounts	01	01	01	01
3-2. Savings accounts and certificates	02	02	02	02
OTHER DEPOSITS AND SAVINGS CERTIFICATES:				
3-3. Credit unions and caisses populaires	03	03	03	03
3-4. Trust companies	04	04	04	04
3-5. Other, please specify in remarks below	05	05	05	05
3-6. CASH ON HAND	06	06	06	06
BONDS AND STOCKS: Report bonds and debentures at face value, stocks at market value.				
Canada Savings Bonds:				
3-7. 1976 issue **being paid for** (face value contracted for)	07	07	07	07
3-8. **Fully paid** – purchased in 1974 and after	08	08	08	08
3-9. – purchased before 1974	09	09	09	09
3-10. – total (only if no answer in previous two.)	10	10	10	10
3-11. Face value of matured uncashed coupons (If no matured uncashed coupons, enter 'O')	11	11	11	11
3-12. **INTERVIEWER CHECK ITEM** (√)				
• If amount given in item 3-11	☐ Go to 3-16	☐ Go to 3-16	☐ Go to 3-16	☐ Go to 3-16
• Otherwise	☐ Go to 3-13	☐ Go to 3-13	☐ Go to 3-13	☐ Go to 3-13
3-13. Matured coupons cashed from bonds – issued in 1974 and after (√)	13 1☐ None 2☐ Some 3☐ All	13 1☐ None 2☐ Some 3☐ All	13 1☐ None 2☐ Some 3☐ All	13 1☐ None 2☐ Some 3☐ All
3-14. – issued before 1974 (√)	14 1☐ None 2☐ Some 3☐ All	14 1☐ None 2☐ Some 3☐ All	14 1☐ None 2☐ Some 3☐ All	14 1☐ None 2☐ Some 3☐ All
3-15. – if not sure of issue dates, indicate in general (√)	15 1☐ None 2☐ Some 3☐ All	15 1☐ None 2☐ Some 3☐ All	15 1☐ None 2☐ Some 3☐ All	15 1☐ None 2☐ Some 3☐ All
3-16. Other Government of Canada bonds	16	16	16	16
3-17. All other bonds	17	17	17	17
3-18. Publicly traded stocks and mutual fund shares	18	18	18	18
3-19. Shares in investment clubs, etc.	19	19	19	19
3-20. MORTGAGES	20	20	20	20
3-21. OTHER LOANS TO PERSONS AND BUSINESSES	21	21	21	21
REGISTERED SAVINGS PLANS: Amount held (include total principal and accrued interest)				
3-22. Registered Home Ownership Savings Plan	22	22	22	22
3-23. Registered Retirement Savings Plan	23	23	23	23
If none of the above assets to report, please check (√) →	☐	☐	☐	☐

3-24. **REMARKS**

Pg./Ln.	Item No.	
☐☐	☐☐	
☐☐	☐☐	

SECTION 3. PERSONAL ASSETS – concluded

2	0	3

Item	Col 1	Col 2	Col 3	Col 4
Page/Line Number	☐☐	☐☐	☐☐	☐☐
PENSION PLANS: Do not include CPP/QPP or pension plans from which you are already receiving benefits. Please explain unusual situations in remarks below.				
3-25. Are you covered by a pension plan connected with your present or past work? (√)	25 1 ☐ Yes Go to 3-26 2 ☐ No Go to 3-28	25 1 ☐ Yes Go to 3-26 2 ☐ No Go to 3-28	25 1 ☐ Yes Go to 3-26 2 ☐ No Go to 3-28	25 1 ☐ Yes Go to 3-26 2 ☐ No Go to 3-28
3-26. Did you pay any pension plan premiums in 1976? (√)	26 1 ☐ Yes Go to 3-27 2 ☐ No Go to 3-28	26 1 ☐ Yes Go to 3-27 2 ☐ No Go to 3-28	26 1 ☐ Yes Go to 3-27 2 ☐ No Go to 3-28	26 1 ☐ Yes Go to 3-27 2 ☐ No Go to 3-28
3-27. Premiums paid in 1976	27	27	27	27
LIFE INSURANCE, ALL POLICIES EXCEPT GROUP AND TERM: Include only policies with loan or cash value; i.e., those with savings feature.				
3-28. Do you have any life insurance policies as described above? (√)	28 1 ☐ Yes Go to 3-29 2 ☐ No Go to 3-32	28 1 ☐ Yes Go to 3-29 2 ☐ No Go to 3-32	28 1 ☐ Yes Go to 3-29 2 ☐ No Go to 3-32	28 1 ☐ Yes Go to 3-29 2 ☐ No Go to 3-32
3-29. Face value	29	29	29	29
3-30. Current cash surrender or loan value	30	30	30	30
3-31. Premiums paid in 1976	31	31	31	31
3-32. OTHER FINANCIAL ASSETS, please specify in remarks below	32	32	32	32
REAL ESTATE:				
Vacation home				
3-33. – market value	33	33	33	33
3-34. – mortgage debt outstanding	34	34	34	34
Other real estate				
3-35. – market value	35	35	35	35
3-36. – mortgage debt outstanding	36	36	36	36
PASSENGER CARS:				
3-37. Do you own a passenger car? (√)	37 1 ☐ Yes Go to 3-38 2 ☐ No Go to 3-44	37 1 ☐ Yes Go to 3-38 2 ☐ No Go to 3-44	37 1 ☐ Yes Go to 3-38 2 ☐ No Go to 3-44	37 1 ☐ Yes Go to 3-38 2 ☐ No Go to 3-44
CAR-1 3-38. Estimated market value	38	38	38	38
CAR-1 3-39. Model year	39	39	39	39
CAR-1 3-40. Model				
CAR-1 Office use only	40	40	40	40
CAR-2 3-41. Estimated market value	41	41	41	41
CAR-2 3-42. Model year	42	42	42	42
CAR-2 3-43. Model				
CAR-2 Office use only	43	43	43	43
3-44. MISCELLANEOUS ASSETS, please specify in remarks below	44	44	44	44
If none of the above assets to report, please check (√) →	☐	☐	☐	☐
Office use only T.A.	45	45	45	45
Office use only A.C.	46	46	46	46
Office use only	47	47	47	47

Pg./Ln.	Item No.	3-45. REMARKS
☐☐	☐☐	
☐☐	☐☐	
☐☐	☐☐	
☐☐	☐☐	

2	0	4

SECTION 4. PERSONAL DEBTS AND MONEY OWED

Statistics Canada would like to complete its study of the financial position of Canadians by obtaining some information, in Sections 4 and 5, on personal debts and business investments. Where debts relate jointly to business and personal use, please report the amounts in Section 4 if you consider them to be primarily of a personal nature. Where you consider such debts to relate primarily to business, please include them in Section 5.

Page/Line Number	☐☐	☐☐	☐☐	☐☐
What do you owe on the following at present?				
CHARGE ACCOUNTS AND INSTALMENT DEBT:				
4- 1. Credit cards issued by banks	01	01	01	01
4- 2. Other universal credit cards issued by institutions other than banks	02	02	02	02
4- 3. Charge accounts and instalment debt	03	03	03	03
LOANS FROM CHARTERED BANKS:				
4- 4. Secured by stocks and bonds	04	04	04	04
4- 5. Secured by household goods	05	05	05	05
4- 6. Student loans	06	06	06	06
4- 7. Other bank loans	07	07	07	07
LOANS FROM OTHER INSTITUTIONS:				
4- 8. Sales finance and consumer loan companies	08	08	08	08
4- 9. Credit unions and caisses populaires	09	09	09	09
4-10. Other institutional loans, please specify in remarks below	10	10	10	10
4-11. OTHER DEBTS AND LOANS, please specify in remarks below	11	11	11	11
If no debts to report, please check (√) →	☐	☐	☐	☐
Office use only T.D.	12	12	12	12
D.C.	13	13	13	13
	14	14	14	14

4-12. REMARKS

Pg./Ln.	Item No.	
☐☐	☐☐	
☐☐	☐☐	
☐☐	☐☐	
☐☐	☐☐	

2	0	5

SECTION 5. PROFESSIONAL AND BUSINESS INTERESTS

This section relates to your professional practice or business only if it takes the form of a sole proprietorship, a partnership or a private corporation, including farming, fishing and other self-employment activities. Assets should be recorded net of accumulated depreciation allowance and together with liabilities, should be recorded as at the end of the last fiscal year.

Page/Line Number	☐☐	☐☐	☐☐	☐☐
5- 1. Are you at present engaged in a business or profession as defined above? (√)	01 1 ☐ Yes Go to 5-2 2 ☐ No End	01 1 ☐ Yes Go to 5-2 2 ☐ No End	01 1 ☐ Yes Go to 5-2 2 ☐ No End	01 1 ☐ Yes Go to 5-2 2 ☐ No End
ACTIVITY – 1				
5- 2. Type of activity				
Office use only	02	02	02	02
5- 3. Legal status (√) Sole propr.	1 ☐	1 ☐	1 ☐	1 ☐
Partnership	2 ☐	2 ☐	2 ☐	2 ☐
Private corp.	03 3 ☐	03 3 ☐	03 3 ☐	03 3 ☐
5- 4. What is the value of your equity in this business?	04	04	04	04
5- 5. **INTERVIEWER CHECK ITEM (√)**				
• If amount is given in item 5-4	☐ Go to 5-10	☐ Go to 5-10	☐ Go to 5-10	☐ Go to 5-10
• Otherwise	☐ Go to 5-6	☐ Go to 5-6	☐ Go to 5-6	☐ Go to 5-6
5- 6. Assets – market value	06	06	06	06
5- 7. book value	07	07	07	07
5- 8. Liabilities – book value	08	08	08	08
5- 9. Share of business owned %	09	09	09	09
ACTIVITY – 2				
5-10. Type of activity				
Office use only	10	10	10	10
5-11. Legal status (√) Sole propr.	1 ☐	1 ☐	1 ☐	1 ☐
Partnership	2 ☐	2 ☐	2 ☐	2 ☐
Private corp.	11 3 ☐	11 3 ☐	11 3 ☐	11 3 ☐
5-12. What is the value of your equity in this business?	12	12	12	12
5-13. **INTERVIEWER CHECK ITEM (√)**				
• If amount is given in item 5-12	☐ End	☐ End	☐ End	☐ End
• Otherwise	☐ Go to 5-14	☐ Go to 5-14	☐ Go to 5-14	☐ Go to 5-14
5-14. Assets – market value	14	14	14	14
5-15. book value	15	15	15	15
5-16. Liabilities – book value	16	16	16	16
5-17. Share of business owned %	17	17	17	17
Office use only	18	18	18	18

5-18. REMARKS

APPENDIX C. INCOME DISTRIBUTIONS BY SIZE IN CANADA

TABLE 3

Percentage Distribution of Families and Unattached Individuals by Income Groups and Regions (1974)[a]

INCOME GROUP TRANCHE DE REVENU	CANADA	ATLANTIC PROVINCES PROVINCES DE L'ATLANTIQUE	QUEBEC	ONTARIO	PRAIRIE PROVINCES PROVINCES DES PRAIRIES	BRITISH COLUMBIA COLOMBIE-BRITANNIQUE
			PER CENT POURCENTAGE			
UNDER $3,000/MOINS DE $3,000............	8.2	9.2	8.9	7.5	8.9	7.3
$ 3,000-$ 4,999..........................	9.1	9.4	7.9	8.4	11.8	9.8
5,000- 6,999..........................	8.4	11.8	8.5	7.2	8.9	8.5
7,000- 8,999..........................	7.3	9.4	7.6	6.9	7.0	7.2
9,000- 10,999..........................	7.5	9.0	8.0	7.3	7.5	5.8
11,000- 11,999..........................	3.6	4.4	4.2	3.2	3.6	2.6
12,000- 12,999..........................	3.5	4.0	4.0	3.1	3.6	3.4
13,000- 13,999..........................	3.3	3.1	3.6	3.4	3.2	3.1
14,000- 14,999..........................	3.7	4.4	4.0	3.7	3.0	3.7
15,000- 15,999..........................	3.4	4.0	3.7	2.9	3.7	3.6
16,000- 16,999..........................	3.9	3.1	3.7	4.6	3.3	3.7
17,000- 17,999..........................	3.1	3.9	3.3	3.1	2.3	3.3
18,000- 19,999..........................	6.0	5.4	6.0	6.3	5.8	5.6
20,000- 21,999..........................	5.5	4.1	5.0	5.9	5.4	6.7
22,000- 24,999..........................	6.5	5.4	6.1	7.4	5.7	6.6
25,000- 29,999..........................	7.4	4.4	6.3	8.8	7.0	7.9
30,000- 34,999..........................	4.0	2.6	3.8	4.3	3.8	4.9
35,000 AND OVER/ET PLUS................	5.5	2.3	5.3	6.0	5.7	6.4
TOTALS/TOTAL..........................	100.0	100.0	100.0	100.0	100.0	100.0
AVERAGE INCOME/REVENU MOYEN............$	15,849	13,049	15,369	16,879	15,144	16,706
MEDIAN INCOME/REVENU MEDIAN............$	13,716	11,272	13,257	14,801	12,650	14,618
SAMPLE SIZE/TAILLE DE L'ECHANTILLON.....	12,734	2,302	2,818	3,664	2,715	1,235
STANDARD ERROR OF AVERAGE INCOME/ ERREUR TYPE DU REVENU MOYEN............$	155	229	292	318	319	294

[a] Statistics Canada (1978).

TABLE 4

Percentage Distribution of Individuals by Income Groups, Regions, and Metropolitan and Nonmetropolitan Areas (1976)[a]

INCOME GROUP TRANCHE DE REVENU	CANADA	ATLANTIC PROVINCES PROVINCES DE L'ATLANTIQUE	QUEBEC	ONTARIO	PRAIRIE PROVINCES PROVINCES DES PRAIRIES	BRITISH COLUMBIA COLOMBIE-BRITANNIQUE
			PER CENT POURCENTAGE			
ALL AREAS/ENSEMBLE DES REGIONS						
UNDER $1,000/MOINS DE $1,000	9.2	9.3	7.8	9.7	10.6	8.5
$ 1,000-$ 1,999	8.0	8.2	7.5	8.3	8.0	7.8
2,000- 2,999	8.9	13.5	10.2	7.5	8.8	6.9
3,000- 3,999	7.2	8.2	7.0	6.7	7.8	8.0
4,000- 4,999	5.6	6.4	6.0	5.2	5.6	5.8
5,000- 5,999	5.3	5.8	5.8	5.0	5.7	4.1
6,000- 6,999	5.1	6.1	5.4	4.9	4.8	4.4
7,000- 7,999	4.8	5.3	5.6	4.2	5.0	4.1
8,000- 8,999	4.4	5.3	5.3	4.1	3.6	4.0
9,000- 9,999	4.3	4.0	4.8	4.6	3.7	3.0
10,000- 10,999	4.4	4.8	4.8	4.4	3.8	4.3
11,000- 11,999	3.6	3.3	4.0	3.6	3.3	3.8
12,000- 14,999	10.1	8.0	10.2	10.7	9.5	10.6
15,000- 19,999	10.5	7.4	8.8	12.2	9.9	12.1
20,000- 24,999	4.3	2.8	3.2	4.6	4.7	6.3
25,000 AND OVER/ET PLUS	4.3	1.6	3.6	4.4	5.3	6.2
TOTALS/TOTAL	100.0	100.0	100.0	100.0	100.0	100.0
AVERAGE INCOME/REVENU MOYEN $	9,265	7,394	8,814	9,663	9,197	10,592
MEDIAN INCOME/REVENU MEDIAN $	7,151	5,773	7,063	7,642	6,734	8,076
SAMPLE SIZE/TAILLE DE L'ECHANTILLON	23,399	4,407	5,232	6,777	4,845	2,138
STANDARD ERROR OF AVERAGE INCOME/ ERREUR TYPE DU REVENU MOYEN $	89	160	198	156	194	203
METROPOLITAN AREAS/ REGIONS METROPOLITAINES						
UNDER $1,000/MOINS DE $1,000	8.2	8.0	6.5	8.9	9.3	8.4
$ 1,000-$ 1,999	7.8	6.9	7.5	8.1	7.4	8.2
2,000- 2,999	7.8	10.7	9.7	6.8	8.0	5.7
3,000- 3,999	6.8	7.8	7.2	6.1	6.4	8.1
4,000- 4,999	5.6	6.4	5.9	5.1	5.6	6.0
5,000- 5,999	5.3	4.8	6.0	5.0	6.0	3.9
6,000- 6,999	4.9	5.6	5.5	4.7	4.9	3.8
7,000- 7,999	4.8	5.3	5.4	4.2	5.3	4.3
8,000- 8,999	4.4	5.7	5.0	4.2	4.0	3.8
9,000- 9,999	4.5	4.3	4.8	4.8	4.0	3.1
10,000- 10,999	4.7	5.7	4.9	4.8	4.0	4.7
11,000- 11,999	3.7	3.9	3.9	3.6	3.3	4.2
12,000- 14,999	10.6	9.5	10.3	11.2	10.1	10.3
15,000- 19,999	11.2	9.4	9.2	12.7	10.4	12.4
20,000- 24,999	4.7	3.5	3.8	4.9	5.6	5.9
25,000 AND OVER/ET PLUS	5.0	2.6	4.4	4.9	5.7	7.1
TOTALS/TOTAL	100.0	100.0	100.0	100.0	100.0	100.0
AVERAGE INCOME/REVENU MOYEN $	9,866	8,391	9,277	10,146	9,740	10,989
MEDIAN INCOME/REVENU MEDIAN $	7,772	6,961	7,310	8,245	7,452	8,400
SAMPLE SIZE/TAILLE DE L'ECHANTILLON	12,771	1,115	3,542	5,057	1,956	1,101
STANDARD ERROR OF AVERAGE INCOME/ ERREUR TYPE DU REVENU MOYEN $	121	312	274	186	294	222
NON-METROPOLITAN AREAS/ REGIONS NON METROPOLITAINES						
UNDER $1,000/MOINS DE $1,000	11.1	9.9	10.5	12.2	12.4	8.9
$ 1,000-$ 1,999	8.4	8.8	7.6	9.0	8.8	7.0
2,000- 2,999	10.9	14.8	11.3	9.4	9.8	9.6
3,000- 3,999	8.1	8.4	6.7	6.4	9.5	7.6
4,000- 4,999	5.8	6.3	6.0	5.5	5.6	5.4
5,000- 5,999	5.3	6.3	5.3	5.0	5.3	4.6
6,000- 6,999	5.4	6.4	5.1	5.6	4.7	5.6
7,000- 7,999	4.8	5.3	6.0	4.0	4.6	3.7
8,000- 8,999	4.5	5.2	5.9	3.8	3.0	4.6
9,000- 9,999	3.9	3.9	4.8	3.9	3.3	2.8
10,000- 10,999	3.9	4.4	4.6	3.4	3.6	3.2
11,000- 11,999	3.5	3.0	4.3	3.5	3.2	3.0
12,000- 14,999	9.2	7.3	10.1	9.1	8.6	11.3
15,000- 19,999	9.0	6.4	7.8	10.6	9.2	11.5
20,000- 24,999	3.4	2.5	2.0	3.8	3.5	7.1
25,000 AND OVER/ET PLUS	2.9	1.1	1.8	2.9	4.8	4.2
TOTALS/TOTAL	100.0	100.0	100.0	100.0	100.0	100.0
AVERAGE INCOME/REVENU MOYEN $	8,101	6,904	7,802	8,215	8,484	9,731
MEDIAN INCOME/REVENU MEDIAN $	6,073	5,268	6,496	6,095	5,728	7,367
SAMPLE SIZE/TAILLE DE L'ECHANTILLON	10,628	3,292	1,690	1,720	2,889	1,037
STANDARD ERROR OF AVERAGE INCOME/ ERREUR TYPE DU REVENU MOYEN $	116	184	198	282	226	425

[a] Centres with a population of 30,000 and over are classified as metropolitan areas and the rest of the country as nonmetroplitan. (From Statistics Canada, 1978.)

TABLE 5

Percentage Distribution and Estimated Standard Error for the Percentage Distribution of Individuals by Income Group (1976)

Income group ($)	*Percentage*	*Standard error*
Under 1,000	9.2	.22
1,000–1,999	8.0	.25
2,000–2,999	8.9	.21
3,000–3,999	7.2	.24
4,000–4,999	5.6	.17
5,000–5,999	5.3	.17
6,000–6,999	5.1	.19
7,000–7,999	4.8	.21
8,000–8,999	4.4	.18
9,000–9,999	4.3	.16
10,000–10,999	4.4	.19
11,000–11,999	3.6	.19
12,000–14,999	10.1	.19
15,000–19,999	10.5	.37
20,000–24,999	4.3	.19
25,000 and over	4.3	.19
Total	100.0	—
Average income	9,265	89

TABLE 6

Percentage Distribution and Estimated Standard Error for the Percentage Distribution of Economic Families and Unattached Individuals by Income Groups (1976)

Income group ($)	*Percentage*	*Standard error*
Under 3,000	7.7	.35
3,000–4,999	8.7	.33
5,000–6,999	8.3	.33
7,000–8,999	7.3	.28
9,000–10,999	7.4	.32
11,000–11,999	3.6	.21
12,000–12,999	3.6	.23
13,000–13,999	3.4	.21
14,000–14,999	3.8	.19
15,000–15,999	3.5	.19
16,000–16,999	4.0	.31
17,000–17,999	3.1	.18
18,000–18,999	6.1	.25
20,000–21,999	5.6	.22
22,000–24,999	6.7	.31
25,000–29,999	7.6	.29
30,000–34,999	4.1	.17
35,000 and over	5.7	.23
Total	100.0	—
Average income	16,095	154

ACKNOWLEDGMENTS

The authors acknowledge the assistance and support of Mrs. G. Oja, Mrs. M. Levine, and Mr. M. Meere of the Consumer Income and Expenditure Division and Mr. R. Platek of the Census and Household Surveys Methods Division in the preparation of this chapter.

REFERENCES

Cochran, W. G. (1963). *Sampling Techniques.* New York: Wiley.

Platek, R., and Singh, M. P. (1976). Methodology of the Canadian Labour Force Survey. Statistics Canada, Census and Household Surveys Methods Division.

Revenue Canada (1978). Taxation Statistics, 1978 Edition. Catalogue Rv44-1978, Ottawa.

Sonquist, J. N., and Morgan, J. A. (1964). *The Detection of Interaction Effects.* Monograph No. 35, Survey Research Centre, Institute for Social Research, University of Michigan.

Statistics Canada (1978). Income distributions by size in Canada 1976. Catalogue 13-207, Ottawa.

Statistics Canada (1979). Income after tax, distributions by size in Canada 1976. Catalogue 13-210, Ottawa.

CHAPTER 8

An Empirical Investigation of Some Item Nonresponse Adjustment Procedures*

M. Haseeb Rizvi

1. INTRODUCTION TO THE NATIONAL LONGITUDINAL STUDY (NLS)

1.1. Objectives and Population Description

The National Longitudinal Study (NLS) of the High School Class of 1972 is a large-scale, long-term survey effort supported by the National Center for Educational Statistics (NCES), Office of the Assistant Secretary for Education in the Department of Health, Education, and Welfare (DHEW).

* This case study is based on the Third Follow-up National Longitudinal Study of the High School Class of 1972 and is adapted from: Brenda G. Cox and Ralph E. Folsom, Jr., An Empirical Investigation of Alternate Item Nonresponse Adjustment Procedures. NLS Sponsored Reports Series No. RTI/0884/52-071, National Center for Education Statistics, Washington, D.C., 1979. (This report is now available from the Educational Resources Information Center, Princeton, New Jersey.) Appendixes D and E have been adapted from Jay R. Levinsohn and Katherine C. McAdams (ed.), NLS Third Follow-up Survey Final Methodological Report. Project No. 224-884, Center for Educational Research and Evaluation, Research Triangle Institute, North Carolina, 1978. Reports 1 and 2 listed above were prepared for the National Center for Education Statistics, Education Division, under Contract No. OEC-0-73-6666 with the U.S. Department of Health, Education, and Welfare. Contractors undertaking such projects are encouraged to express freely their professional judgment. These reports, therefore, do not necessarily represent positions or policies of the Education Division, and no official endorsement should be inferred. Great appreciation is recorded here for the very helpful reviews by Brenda G. Cox, Richard A. Platek, and Donald B. Rubin of an earlier draft of this chapter.

INCOMPLETE DATA
IN SAMPLE SURVEYS
Volume 1, Part II

ISBN 0-12-363901-8

The primary purpose of the NLS is observation of the educational and vocational activities, plans, aspirations, and attitudes of young people after they leave high school and investigation of the relationships of this information to their prior educational experiences and their personal and biographical characteristics. Ultimately, it is hoped the study will allow a better understanding of the development of students as they pass through the American educational system and of the complex factors associated with individual educational and career outcomes. Thus broadly stated, NLS is designed to provide statistics on a national sample of students as they move out of the American high school system into the critical years of early adulthood. Data have been gathered from several sources, coded and edited for analysis purposes, and stored on magnetic computer tapes for future access. The current tapes contain base-year (1972) survey data, collected by the Educational Testing Service, integrated with first follow-up (1973), second follow-up (1974), and third follow-up (1976) survey data, collected by the Research Triangle Institute (RTI). This tape package is augmented periodically as data from subsequent follow-up surveys become available.

The merged NLS data file represents a rich and complex source of information for potential use by a broad spectrum of educators, researchers, policy analysts, and decision makers. The need to maintain a faithful record of the original raw data (including respondent errors and inconsistencies) and the need to provide a straightforward set of data for the typical researcher were both taken into account in editing the data file; however, the former was given somewhat greater importance. It was felt that editing, recoding, imputation, or other data transformation procedures should be minimized so as to limit tacit assumptions about the respondents or subjective interpretation of the data that would be required for these procedures. Thus the data transformation procedures that were used were directed toward making the data available in a consistent and useful format that would preserve as accurately as possible the original responses of the study participants.

The file processing that was performed involved extensive verification, cleaning, and supplementary coding of the original data. Editing transformations were limited to verifying that respondents followed the written instructions and routing patterns in the questionnaire. If either type of instruction were violated, then supplementary codes were inserted in the data to indicate the location and nature of the violation. The extent and details of the editing are discussed in the NLS Users Manual (Levinsohn, Henderson, Riccobono, and Moore, 1978).

Since little summarization of the data has been done, the analyst must exercise care in using the data file. The NLS data base is large and complex. The routing patterns in the instruments and the fact that data have been collected at different times require that the analyst view the data base holistically.

Many of the individual items may not be suited to independent analysis. Moreover, it may be necessary to consider the interdependent nature of individual items and to study the pattern of response to those items.

1.2. Sampling Plan

Following an extensive period of planning, which included the design and field test of survey instrumentation and procedures, a full-scale survey was initiated in the spring of 1972. The sample design called for a deeply stratified national probability sample of 1200 schools with 18 seniors per school, school size permitting. The resulting base-year sample of 19,144 students from 1009 high schools provided base-year data on up to three data collection forms—a Test Battery (TB), a Student Record Information Form (SRIF), and a Student Questionnaire (SQ). The key form, the SQ, was completed by 16,683 seniors.

The first follow-up survey began in October 1973 and ended in April 1974. Added to the base-year sample were 4450 1972 high school seniors from 257 additional schools that were unable to participate earlier. After a few deletions from the base-year sample, this addition brought the total first follow-up sample to 23,451 potential respondents. First follow-up forms were mailed to 22,654 students. There were 21,350 sample members who completed a First Follow-Up Questionnaire, 69% by mail and 31% by personal interview. Of the 16,683 seniors who completed a base-year Student Questionnaire, 15,635 took part in the first follow-up survey—a sample retention rate of 93.7%. Participants were asked where they were in October 1973 and what they were doing with regard to work, education, and/or training. Similar information was requested for the same time period in 1972 to facilitate tracing of progress since leaving high school and to define the factors that affect that progress.

The second follow-up survey began in October 1974 and was completed in April 1975, with forms sent to 22,364 potential respondents. There were 20,872 sample members who completed a Second Follow-Up Questionnaire, 72% by mail and 28% by personal interview. Of the 21,350 persons who completed a First Follow-Up Questionnaire, 20,194 (94.6%) also responded to the second follow-up survey.

The third follow-up survey began in October 1976 and ended in May 1977. Questionnaires were mailed to the last-known addresses of the sample members whose addresses appeared sufficient and correct and who had not been removed from active status by prior refusal, reported death, or other reason. Some 20,092 sample members completed a Third Follow-Up Questionnaire, 80% by mail and 20% by personal interview. The overall response rate was approximately 92% of the initial 21,807 mailouts. The retention rate of second follow-up respondents was 93.9%; the retention rate of the sample members who completed the third follow-up survey among those who had completed all three previous student instruments was 94.7%.

1.3. Data Collection Procedures

In the NLS follow-up studies, a planned sequence of reminder postcards, additional questionnaire mailings, reminder mailgrams, and personal interviews contributed to the high instrument response rates. Efforts for maximizing participation and response, with some assessment of these efforts, are described in Appendix D. As a supplement to these procedures, to reduce the level of nonresponse, adjustments may be used in an attempt to reduce possible bias that was due to nonresponse. The NLS uses a weighting-class procedure, based on classifier variables obtained from the students' high school records via the SRIF, to adjust sampling weights for instrument nonresponse; Appendix E discusses these weight calculations.

A different type of nonresponse, especially in a mail survey such as NLS, is item nonresponse. A large number of the returned questionnaires had one or more blank items. The level of item nonresponse depends on the type of question and the information being solicited. For example, categorical questions typically have a smaller rate of nonresponse than quantitative questions; the two items in NLS that requested information on family income had the largest rate of nonresponse, with approximately 12% of the incoming questionnaires having missing responses. Associated with the problem of item nonresponse is that of inconsistent or invalid responses and violations of routing patterns (often referred to as "skip patterns"). This is an obvious source of bias that must also be considered before the data are analyzed. For categorical questions, inconsistencies may represent a greater source of bias than nonresponse.

1.4. Data Processing

The first step in treating all of the preceding cases of item nonresponse and inconsistencies is to check the questionnaire to determine whether or not logical imputations, based on the responses to other questions, can be made for the missing or inconsistent items. If a logical imputation cannot be made, the next step is to contact those who failed to answer or incorrectly answered a question. The NLS designated certain items as critical questions. Those individuals who did not give a valid response to a critical question were telephoned, and the missing item was completed or the inconsistency resolved. Obviously, obtaining the correct data is the best solution to the problem of invalid responses. In view of the large number of items on the survey instrument and the frequency of item nonresponse and inconsistencies, this procedure was not implemented for all items. An attempt to resolve inconsistencies and obtain missing information for every item on the instrument by contacting the individual would have been too expensive. Hence logical editing rules were needed to resolve or discard the inconsistent responses. Item nonresponse adjustment procedures were also considered in order to adjust for the bias that could arise because of differences between characteristics of individuals who responded or failed to respond to an item.

1.5. Brief Description of the Present Investigation

Presently, when computing survey estimates, NLS makes no adjustment for possible bias effects that are due to item nonresponse. Thus the sample means and proportions are the weighted respondent averages, where the weights reflect the sample selection weight for the individual after adjustments are made within weighting classes for instrument nonresponse. This procedure, which does not adjust or impute for missing responses to an item but which instead uses the estimated respondent mean, will henceforth be referred to as the "no imputation procedure."

A general review of two possible item nonresponse imputation and adjustment procedures suited to NLS is given in Section 2; these are a hot-deck imputation procedure and a weighting-class imputation procedure. An empirical investigation of these two procedures appears in subsequent sections. This imputation study was conducted using NLS data in such a manner that the bias, variance, and mean square error of the resulting imputation-based estimators could be evaluated and compared with the bias, variance, and mean square error of the estimates produced by the no imputation procedure. The bias and mean square error were determined by the data set, with missing responses secured by telephone follow-up.

2. ITEM NONRESPONSE IMPUTATION AND ADJUSTMENT PROCEDURES

When an individual fails to respond to an item, there is still some information available about him from the completed questionnaire. For NLS, classifier variables such as race, sex, high school curriculum, high school grades, and parents' education are known about the nonrespondent; so also are his responses to other items on the questionnaire and prior questionnaires. The procedures discussed next divide the sample members into categories based on information that is available for the two groups: those who respond to an item and those who fail to respond. The assumption made here is that the responses of individuals within the same poststratum cell are relatively homogeneous.

2.1. The Hot-Deck Imputation Procedure

To use a hot-deck procedure for imputation, the individuals completing the questionnaire must be divided into categories. An initial value is determined for each category based on previous data. As the new data are processed, the category to which each individual belongs is determined. If the questionnaire being processed is complete, then that individual's responses replace the responses stored in the relevant category of the hot deck. Thus new responses are supplied for each cell of the hot deck as they appear in the data file. When a questionnaire is encountered with a missing item, the response in the same cell of the hot deck is imputed for the missing response. When all questionnaires have been processed and the missing data imputed, the means and variances are again computed in the usual manner.

The hot deck is one of the most commonly used item nonresponse imputation procedures, so the quality of its imputation needs to be assessed. One reason why the hot deck is so commonly used is the ease with which it can be implemented; another is its flexibility. However, a study by Bailar and Bailar (1978) demonstrates that the hot-deck procedure increases the variance of sample means (compared with the variance of sample means computed when missing values are ignored). The magnitude of this increase in variance and the bias reduction resulting from imputation are studied in this investigation. For the hot-deck technique, the variances are analytically intractable and are, therefore, estimated by using some form of pseudoreplication such as balanced repeated replication (BRR). In this investigation, BRR estimates of variances were computed and compared with results obtained when imputation was ignored. It should be emphasized that there is no probability mechanism attached to the assignment of missing values in the hot deck and that the same individual's responses may be used repeatedly to supply missing information.

2.2. The Weighting-Class Imputation Procedure

In the present analysis of NLS data, a weighting-class adjustment procedure is used to adjust for instrument nonresponse (i.e., when a questionnaire is not obtained for a sample member). Basically, the weighting-class adjustment procedure assigns sample members to various classes based on information available for both respondents and nonrespondents. Within these classes, each individual is assigned an adjusted sampling weight. Respondents within weighting class l have their sampling weight (which is the inverse of their probability of selection) multiplied by the weight adjustment factor $\mathrm{WS}(l)/\mathrm{WR}(l)$ to produce the nonresponse adjusted sampling weight, where $\mathrm{WS}(l)$ is the sum of the sampling weights for all sample individuals in weighting class l and $\mathrm{WR}(l)$ the sum of the sampling weights of the respondents in weighting class l. Nonrespondents are assigned adjusted sampling weights of zero. Sample estimates are then obtained by using these weights adjusted for nonresponse.

Such a weighting-class adjustment procedure may also be used for item nonresponse and adapted as an imputation technique. For quantitative data, an imputation technique equivalent to the usual weighting-class adjustment procedure would be to assign the average value of the responses in a weighting class to all nonrespondents in that weighting class. For qualitative data, in which categories of responses are reported, this technique cannot be used. For example, a yes–no-type question might be coded 1 for yes and 0 for no, but to impute the average (say, .2) to nonrespondents is not reasonable and also does not allow the usual tabulation of the data. To impute categorical responses, a technique was developed for this investigation that is analogous to weighting-class adjustments. For each item, this "weighting-class" imputation procedure first determines the weighted response option proportions from all responding members of the various weighting classes. For instance, the proportion of all respondents in weighting class l who make response k could be denoted by $\rho(k|l)$. Next, the sum of the sample weights for the nonrespondents who belong to weighting class l is found and denoted by $\mathrm{WN}(l)$. The response k is then randomly imputed to nonrespondents in weighting class l so that their sample weights sum to $\mathrm{WN}(lk) = \rho(k|l)\mathrm{WN}(l)$. The nonrespondents in class l to which the response k is to be imputed is determined in the following manner. First, list the nonrespondents with their sample weights in random order. Go down the list, summing weights until the sum equals $\mathrm{WN}(lk)$. Impute the response k to the corresponding sample members. Continue this procedure until all nonrespondents in weighting-class l have had a response imputed for the missing item. The estimated proportions making each response would then be determined in the usual way. Estimates of response-level proportions will then be the same as those that would have resulted had a weighting-class adjustment procedure been used, except for weight accumulations that do not break precisely at the place desired. To the extent that this categorical imputation technique produces the same estimated response proportions as a weighting-class adjustment procedure, variance approximations for the weighting-class adjustment procedure should be appropriate for the imputation-based procedure. This is thus different from the hot-deck procedure for which an analytical expression for the variance is generally not available.

2.3. The Creation of Weighting Classes for Use in Imputation

Before implementing a weighting-class imputation procedure, the sample must be partitioned into classes. The hot-deck imputation procedure also performs well if the sample is divided into classes before imputation occurs. In both instances, characteristics must be identified that define weighting classes that vary with respect to response rates and survey estimates. For this investigation, the weighting classes were based on the student's race, sex, high school grades, high school curriculum, and parents' education. (Weight calculations are given in Appendix E.) These weighting classes were initially created to adjust for total questionnaire nonresponse, but they would also be applicable in this investigation because of the large number and diversity of items studied. However, an analyst using only a few items from the NLS data file may want to construct special weighting classes before imputing for missing values. The remainder of this section discusses procedures for forming the weighting classes.

In constructing weighting classes, the overall goal is to form classes for which the responses within classes are homogeneous (and between classes are heterogeneous) and for which the response rate varies. Further, the characteristics used to define the weighting classes must be known for both respondents and nonrespondents. The choice of survey characteristics should be based on the following considerations:

(1) Usually a coarser division of several variables is more advantageous than the finer divisions of one.

(2) Smaller cells may be combined while forming cells; symmetry is not necessary.

(3) Different criteria may be used for different subgroups; for instance, males may be partitioned with respect to characteristics different from those for females.

(4) The classifier variables should not be closely related to each other. If two variables are highly correlated, then either will describe approximately the same amount of variation.

In short, the weighting classes that are formed should be as different as possible with respect to outcome variables but should be as internally homogeneous as possible. Cluster analysis may help in choosing an appropriate set of classifier variables and determining how each variable chosen should be used to subdivide the units into classes; a particular clustering technique in this regard is Automatic Interaction Detection (AID) analysis (Hartigan, 1975). The AID algorithm operates by successively dichotomizing the sample according to the factor level (classifier variable) that minimizes the within-weighting-class sum of squares for the dependent variable. The result is a "tree" of clusters (weighting classes) having similar dependent variable values in which the clusters are defined by the levels of the factors selected in the computing algorithm.

3. THE DESIGN OF THE EMPIRICAL INVESTIGATION

The two techniques discussed in the previous section assume that the respondents and nonrespondents within each weighting class have responses that are similarly distributed. The extent to which respondents and nonrespondents differ will influence the amount of bias in the resulting estimators.

3.1. Construction of the Experimental Data Set

To assess the bias associated with each of these nonresponse adjustment procedures, a data set was needed in which some item nonresponse occurred but for which the missing responses were subsequently obtained. The data set with the missing item responses could then be analyzed, using the weighted-class imputation technique, the hot-deck imputation technique, and finally the no imputation technique. These results, in turn, could be compared with those obtained when the missing responses were added to the data set via telephone follow-up.

A data set was constructed from the NLS Third Follow-Up (TFU) Survey by taking the following considerations into account. Certain items on the questionnaire were designated critical items by the NLS staff. When an incoming questionnaire had a missing response or an inconsistent set of responses for one or more of these critical items, the questionnaire was marked as having failed edit and the individual involved was telephoned and the missing response(s) added by the telephone interviewer. The data records for individuals whose questionnaire's failed edit contained the responses to these critical items but did not indicate which responses were obtained by telephone editing or what the original responses were.

In order to obtain this information on the responses before telephone resolution, the questionnaires were reexamined by data editors and the original responses to the selected critical items recorded. In all, a total of 10,850 questionnaires failed edit. For reasons of economy, a subsample of 5854 was selected for reexamination. The following 20 items (given in Appendix A) were chosen to be representative of the types of items on the NLS instrument and were examined on each of the selected questionnaire: TQ1, TQ9, TQ10, TQ12, TQ15, TQ16, TQ29, TQ33, TQ51, TQ52, TQ66, TQ89, TQ90, TQ101, TQ102, TQ118, TQ129, TQ131, TQ136, and TQ141. Here TQ preceding the item number refers to the Third Follow-up Questionnaire.

Except for TQ15, TQ16, TQ89, and TQ141, the other selected items have categorical responses, with TQ1, TQ9, TQ131, and TQ136 allowing the student to choose multiple response options. Item TQ10 was the lead-in question to a routing pattern, with TQ12 to be answered by the unemployed and TQ15, TQ16, and TQ29 to be answered by the employed. Another major routing pattern was controlled by TQ51, which directed those not attending school during 1974–1976 to skip to item TQ98. Other items found within routing patterns were TQ33, which had as its lead-in item TQ32, TQ102 with lead-in item TQ101, and TQ131 and TQ136 with lead-in item TQ129. Finally, the four continuous items, TQ15, TQ16, TQ89, and TQ141, requested hours worked, weekly salary, college expenses, and annual income. These questions were more sensitive and historically these types of questions exhibit higher rates of item nonresponse.

The subsample of TFU questionnaires that failed edit was examined, and a working file was constructed that contained the student identification code and additional codes to indicate whether or not each of the 20 items failed edit, and if so, what the original responses were. The working file was then merged with the NLS Third Follow-Up File to create the two data files that were used in this investigation. The first file, referred to as the "data file of telephone corrected and completed information," which was abstracted from the NLS Third Follow-Up File, contains the data records for those students who passed edit, combined with the subsample of those who failed edit. This master file contains the responses to the items after missing and inconsistent responses were corrected by telephone follow-up. The second file, referred to as the "pretelephoning file," contains the responses to the 20 questions before telephoning was used to correct the data set. Since only a sample of the fail-edit mail questionnaires was included in the investigation, the sampling weights on the data records corresponding to these individuals were adjusted so that the sum of the weights of the subsampled mail questionnaires that failed edit for each weighting class equaled the sum of the weights of all the mail questionnaires that failed edit from that weighting class. No weight adjustments were needed for the data records corresponding to those questionnaires that passed edit or those that were completed by personal interview since all of these questionnaires were included in the study.

3.2. Response Error Rates in the Experimental Data Set

For each of the 20 selected items, responses before telephone follow-up were noted. These results were based on the subsample of 5854 drawn from the 10,850 questionnaires that failed edit. Adding the 9235 questionnaires that passed edit (and hence had complete, consistent responses to the critical items) and adjusting for the sampling of fail-edit questionnaires yield the estimated response error rates (for the full NLS sample) presented in Table 1. Since the experimental data set had the sampling weights of fail-edit questionnaires adjusted to account for the subsampling, these error rates also apply to the experimental data set. Except for two multiple-response option questions (TQ1 and TQ9) and four financial questions corresponding to columns (a) and (b) of TQ89 and TQ141 (TQ89HA, TQ89HB, TQ141FA, TQ141FB), 95% of the questionnaires contained a response for an item that was consistent with other responses on the questionnaire. The highest rates of missing or blank responses were found for the income items, TQ141FA and TQ141FB, with about 13% nonresponse. Items TQ1 and TQ9 had the highest inconsistency rates; that is, the responses to TQ1 and TQ9 were most frequently in conflict with other questionnaire items. The "other" category in Table 1 is composed of those persons who failed an item but could not be contacted for telephone resolution.

From Table 1, one can see that the original responses to the 20 critical items for the full NLS sample give a relatively small rate of item nonresponse and a somewhat larger rate of inconsistent responses. Since the experimental data set contained only half of the fail-edit questionnaires, with weight adjustments to account for this subsampling, these rates also apply (with respect to their effect on survey estimates) to the experimental data set as well. However, in the control process of imputing data, the actual number of nonrespondents to each of the 20 items becomes important. This information is presented in the last column of Table 1. Since the experimental data set contains 15,089 records, the (small) number of nonrespondents is obviously of little practical importance, except for the income items TQ141FA and TQ141FB.

TABLE 1

Classification of Original Responses to the 20 Selected Critical Items

	Original response (%)				*Number of missing responses*
Item	*Consistent*	*Blank*	*Inconsistent*	*Other*	
Discrete					
TQ1	92.4	.2	6.7	.7	16
TQ9	87.1	.3	11.7	1.0	32
TQ10	96.0	.9	2.7	.4	96
TQ12	97.3	.5	2.0	.3	51
TQ29	99.0	.5	.3	.2	59
TQ33	95.1	.5	4.0	.4	54
TQ51	97.5	.9	1.2	.4	98
TQ52	96.1	.9	2.6	.4	95
TQ66	94.5	1.1	4.0	.5	114
TQ90	97.3	1.4	.9	.4	150
TQ101	98.5	1.0	.2	.3	104
TQ102	99.1	.2	.5	.2	25
TQ118	97.1	2.4	.1	.4	258
TQ129	97.2	.5	.1	.1	57
TQ131	99.5	.2	.2	.1	23
TQ136	99.6	.2	.1	.1	22
Continuous					
TQ15	98.6	.6	.5	.3	66
TQ16	97.3	1.8	.6	.4	192
TQ89HA	92.9	2.6	3.8	.7	286
TQ89HB	92.5	2.8	4.0	.8	303
TQ141FA	82.9	12.9	2.4	1.9	974
TQ141FB	82.5	12.9	2.7	1.9	1030

3.3. Estimation of the Bias Associated with Survey Estimates

By using actual data that contained item nonresponse for which the answers were subsequently obtained by telephone follow-up, the bias could easily be estimated for the two imputation procedures and the no imputation method as the difference between the estimates obtained using the imputation (or no imputation) procedure on the pretelephone data set and the estimates obtained using the data set that had been corrected by telephone interviewing. Since the pretelephone data set contained inconsistent responses as well as missing responses, the bias that was obtained reflected these two response error sources and hence contained nonresponse bias and measurement error bias attributable to inconsistent data. It was possible to estimate the inconsistent response bias. The magnitude of this response bias establishes an upper limit on the effectiveness of the imputation procedures since they were designed primarily to reduce the nonresponse or missing-data component of the overall bias.

3.4. Estimation of the Variance Associated with Survey Estimates

The variance of the sample means and proportions was estimated using a variation of the balanced repeated replication (BRR) technique proposed by McCarthy (1966) for estimating the variance of complex survey statistics. The BRR utilizes a balanced set of half-samples to compute the sampling variance of these statistics, where the variability of the replicated estimates approximates the variance of the full-sample statistic. In this investigation, the item nonresponse imputation procedures were separately applied to the balanced half-samples to obtain BRR variance estimates that reflect the variability induced by the imputation procedures.

As mentioned earlier, most users of imputation procedures ignore the fact that imputation has occurred when computing the variance of survey estimates because computing variance approximations that allow for the fact that imputation occurred is both difficult and costly. The pseudoreplication methods can provide valid estimates of the variance induced by imputation, but the standard software packages for these procedures do not allow one to employ the imputation procedure on each individual half-sample before computing the half-sample estimates. Consequently, the imputation procedure must be applied independently to the half-samples in order to estimate the variance induced by the imputation procedure. Thus some software development would be necessary to modify existing packages or create one's own before the variance of imputation-based statistics could be estimated. Among the objectives of this investigation was the determination of the effect of imputation on the variance of survey estimates and the underestimation effect caused by ignoring imputation in computing variance estimates. A discussion of the theory underlying the BRR procedure is given in Appendix B, along with a justification for using the procedure in the manner specified for computing the variance of imputation-based statistics.

To estimate the variance when imputation was ignored, STDERR (an RTI software package) was used. This package utilizes a Taylor series linearization to estimate the variance of complex survey estimates (see Appendix C for specific variance estimation formulas). For the ratio estimates of means and proportions used exclusively in this investigation, the Taylor series expansion of the variance results in the usual estimate for the variance of a ratio estimate as given in standard sampling texts (e.g., Cochran, 1977, pp. 153–154). As a side issue of this investigation, the STDERR and BRR variance estimates were computed for the survey estimates obtained when no imputation was used. In this situation, the two procedures were measuring the same quantity so that the two variance estimates could be compared.

3.5. Estimation of the Mean Square Error Associated with Survey Estimates

As in Section 3.3, the bias induced by item nonresponse and inconsistencies is defined to be the expected value of the difference between the imputation-based (or no imputation) estimate $\hat{\mu}$ and the corresponding estimate $\hat{\mu}_{\mathrm{T}}$ obtained when the telephone-corrected and consistent follow-up data are used. The magnitude of this bias is estimated by

$$\mathrm{bias}(\hat{\mu}) = \hat{\mu} - \hat{\mu}_{\mathrm{T}}.$$

The mean square error (MSE) of the statistic $\hat{\mu}$ is defined to be the expected value of the squared difference of the estimate from the value obtained using the posttelephone data and is estimated as

$$\mathrm{MSE}(\hat{\mu}) = [\mathrm{bias}(\hat{\mu})]^2 + \mathrm{var}(\hat{\mu}).$$

Note that this estimate will be biased to the extent that the correlation between $\hat{\mu}$ and $\hat{\mu}_{\mathrm{T}}$ is different from unity since

$$E[\mathrm{bias}(\hat{\mu})]^2 = [\mathrm{bias}(\hat{\mu})]^2 + \mathrm{var}(\hat{\mu}) + \mathrm{var}(\hat{\mu}_{\mathrm{T}}) - 2\,\mathrm{cov}(\hat{\mu}, \hat{\mu}_{\mathrm{T}}).$$

Since $\hat{\mu}$ and $\hat{\mu}_{\mathrm{T}}$ are estimated from largely the same data set (except for moderate nonresponse), one would expect their correlation to be close to unity. With this anticipated high correlation, the joint contribution of the extra terms

$$\mathrm{var}(\hat{\mu}) + \mathrm{var}(\hat{\mu}_{\mathrm{T}}) - 2\,\mathrm{cov}(\hat{\mu}, \hat{\mu}_{\mathrm{T}})$$

should be negligible.

4. RESULTS OF THE EMPIRICAL INVESTIGATION

The items studied in this investigation are similar in format to other items on the instrument; however, the frequency of response errors may be greater among the study items since the anticipated worst cases were selected. The original responses before telephone editing were used to form the data base for the investigation, with the responses after telephone resolution used to judge the quality of the data set and the performance of the imputation procedures. The effect of response errors in the original data and the efficacy of the two nonresponse adjustment procedures—hot deck and weighting class—are discussed next.

4.1. Response Error Bias in the Experimental Data Set

Population values that are estimated from sample surveys are subject to two kinds of error. The first kind is variable error (which is random); this includes the sampling error attributable to the random selection of individuals rather than to the complete census of all individuals and also the variable measurement error attributable to the natural fluctuations in questionnaire responses and data transcriptions. The second kind of error is bias, which is a systematic error that can result from the estimation procedure, nonresponse, or nonsampling errors inherent in the measurement process. One may model the total error associated with using the sample estimate $\hat{\theta}$ to predict the population parameter θ by

$$\hat{\theta} - \theta = [\hat{\theta} - E(\hat{\theta})] + [E(\hat{\theta}) - \theta],$$

where $E(\hat{\theta})$ is the expected value of the statistic $\hat{\theta}$ over a conceptual sequence of repeated samplings and repeated interviewing–transcribing trials for a given sample. The first term represents the variable error and the second term represents the bias. The mean square error (MSE), defined as the expectation of the squared total error, is given by

$$\text{MSE}(\hat{\theta}) = E[\hat{\theta} - \theta]^2 = E[\hat{\theta} - E(\hat{\theta})]^2 + [E(\hat{\theta}) - \theta]^2 = \text{var}(\hat{\theta}) + [\text{bias } \hat{\theta}]^2.$$

The first term of the MSE is the sampling variance of the estimate $\hat{\theta}$ and the second term is the square of the bias of the estimate.

The ultimate problem of a sample designer is to construct a sample survey that minimizes the mean square error of the estimate. This means that concern must be given to the protocol for collecting the data and the design of the survey instrument as well as the number of individuals to be surveyed. Similarly, in utilizing the survey data, the investigator must also be concerned with the total error. After the data are collected, however, the variable or sampling portion of the error is fixed. In this case, the investigator should be concerned with the extent of nonresponse, both instrument and item nonresponse, and with the amount of systematic measurement or response error that is evident in the data. Various imputation procedures are available to reduce the bias caused by nonresponse, and logical editing and/or deletion may be used to resolve inconsistencies that are symptomatic of response or measurement errors in the data.

In most surveys the bias of sample estimates is not known. However, in this empirical investigation, the bias in survey estimates that is due to using data containing item nonresponse and items whose responses are logically inconsistent with one another could be determined since telephone-corrected and completed data were also available. The experimental data set that was constructed from the original responses to the 20 selected critical items had a relatively low rate of item nonresponse and a somewhat larger rate of inconsistencies. But in general, the quality of the data appears to be relatively good, even before the telephone editing. The relative bias (RB) of an estimate is defined for this study as the bias divided by the true value of the estimate ($\bar{Y}$-TRUE) obtained when the telephone-edited and corrected data were used. Of the 51 estimates computed for the entire population, using the experimental data set of 14 discrete items with no imputation or editing procedure (NI), 42 have moderate to low relative biases of less than 5% (see Table 2). The average relative bias over all 51 estimates is 2.25%. The estimates of the mean value for the entire population for the six continuous questions has an average relative bias of 0.8%, with all estimates having relative biases of less than 5% (see Table 2).

The relative root mean square error (R $\sqrt{\text{MSE}}$), which is defined to be the square root of the mean square error of the statistic estimating θ divided by the value of the estimate obtained when the telephone-edited and corrected data were used, is a better measure of the quality of the data base. However, the estimates that are based on the experimental data set do not reflect the quality of the NLS data directly because the variances for the estimates that use the experimental data set are larger, since the experimental data set was only about three-fourths the size of the full data set. This was so because only one-half of the fail-edit mail questionnaires was included in the data set. The weights were adjusted to account for the subsampling, so that the means and proportions are valid estimates; however, the variances of the estimates will be larger than what would be obtained had the full data set been used. Thus the relative root mean square errors of the estimates will also be larger than what would be obtained if the full data set had been used. For the estimates of proportions, using the entire population as the domain, 34 of the 51 estimates have relative root mean square errors of less than 5% and 46 have less than 10%. The average relative root mean square error for the 51 proportions estimated from the 14 discrete items is 4.64%. All of the estimated means for the entire population, resulting from the six continuous items, have relative root mean square errors that are less than 5%, with an average value of 1.29% (see Table 2).

TABLE 2

Comparison of No Imputation Estimates When Inconsistent Data Are Removed (NIC) and When Retained (NI)[a]

ITEM	RESPONSE	SAMPLE SIZE	$\bar{Y}$-TRUE	ME BIAS	ME RB%	NI RB%	NIC RB%	NI BR	NIC BR	NI R$\sqrt{MSE}$%	NIC R$\sqrt{MSE}$%
Part 1. Proportions estimated for the total population											
TQ1A	1	15089	72.29	-1.74	-2.40	-2.38	.96	-4.21	1.52	2.45	1.14
TQ1C	1	15089	17.16	- .07	- .40	- .31	-8.97	- .14	-4.40	2.21	9.19
TQ1D	1	15089	4.12	.30	7.28	7.06	-8.37	1.67	-1.70	8.23	9.70
TQ1G	1	15089	9.19	-1.25	-13.59	-13.67	-13.37	-5.06	-4.90	13.93	13.64
TQ9A	1	15089	67.82	-4.49	-6.62	-6.69	- .98	-9.05	-1.13	6.73	1.30
TQ9C	1	15089	32.15	- .56	-1.74	-1.63	-9.72	-1.03	-6.81	2.27	9.83
TQ9D	1	15089	3.96	.01	.25	.53	-18.47	.11	-4.43	4.67	18.93
TQ9G	1	15089	7.00	-1.57	-22.41	-22.54	-22.84	-5.21	-4.67	22.96	23.36
TQ10	1	15089	61.22	.06	.09	.09	1.99	.10	2.28	.87	2.18
TQ10	2	15089	13.06	- .05	- .38	- .24	-9.53	- .09	-3.83	2.66	9.85
TQ10	3	15089	1.45	- .06	-4.13	-5.97	-14.93	- .77	-1.82	9.76	17.02
TQ10	4	15089	24.27	.05	.20	.26	.98	.16	.64	1.62	1.81
TQ12	1	3644	25.66	1.38	5.37	5.03	-12.77	1.41	-3.38	6.16	13.32
TQ12	2	3644	7.20	- .01	- .13	.52	5.38	.07	.72	7.05	9.17
TQ12	3	3644	67.14	-1.37	-2.04	-1.98	4.30	-1.18	2.39	2.59	4.66
TQ29	1	11439	91.40	.01	.01	.03	.06	.09	.15	.39	.40
TQ29	2	11439	8.60	- .01	- .11	- .39	- .64	- .09	- .15	4.22	4.32
TQ33	1	4234	17.56	1.26	7.17	6.79	-22.97	1.51	-4.52	8.14	23.53
TQ33	2	4234	3.57	.31	8.68	9.81	5.95	1.33	.70	12.26	10.33
TQ33	3	4234	78.87	-1.57	-1.99	-1.95	4.84	-1.90	3.71	2.20	5.01
TQ51	1	15089	47.13	.42	.89	.48	- .23	.36	- .18	1.40	1.31
TQ51	2	15089	52.87	- .42	- .79	- .43	.21	- .36	.18	1.25	1.17
TQ52	1	7579	49.75	.24	.48	.61	- .18	.45	- .12	1.47	1.47
TQ52	2	7579	50.25	- .24	- .47	- .60	.18	- .45	.12	1.46	1.46
TQ66	1	7579	20.29	.29	1.42	1.32	-1.17	.77	- .59	2.16	2.29
TQ66	2	7579	32.03	1.08	3.37	4.15	3.73	2.32	2.15	4.52	4.11
TQ66	3	7579	47.68	-1.38	-2.89	-3.35	-2.00	-2.26	-1.50	3.66	2.40
TQ90	1	7579	65.64	.12	.18	.19	.81	.16	.73	1.16	1.38
TQ90	2	7579	4.67	.26	5.57	5.73	- .64	.77	- .09	9.33	6.58
TQ90	3	7579	5.79	.02	.34	.93	-1.15	.14	- .18	6.59	6.36
TQ90	4	7579	23.90	- .40	-1.67	-1.87	-1.84	- .52	- .52	4.03	3.99

TQ101	1	15089	84.17	- .01	- .01	- .12	- .15	- .30	- .38	.42	.43
TQ101	2	15089	15.83	.01	.06	.66	.82	.30	.38	2.26	2.30
TQ102	1	2199	66.64	- .53	- .79	- .46	.34	- .23	.16	2.04	2.08
TQ102	2	2199	33.36	.53	1.58	.92	- .69	.23	- .16	4.09	4.15
TQ118	1	15089	92.68	.00	.00	- .09	- .05	- .42	- .21	.25	.24
TQ118	2	15089	.90	.01	1.11	1.98	-2.23	.19	- .20	10.33	11.01
TQ118	3	15089	6.42	- .01	- .15	1.16	1.04	.29	.26	4.06	4.00
TQ129	1	15089	9.78	.02	.20	- .17	- .55	- .04	- .14	3.92	3.97
TQ129	2	15089	45.95	.00	.00	- .06	- .02	- .07	- .02	.91	.91
TQ129	3	15089	4.02	- .01	- .24	.07	.02	.01	.00	5.08	5.07
TQ129	4	15089	40.25	- .01	- .02	.10	.16	.15	.22	.70	.73
TQ131A	1	6336	73.00	.11	.15	.07	.03	.08	.03	.91	.93
TQ131C	1	6336	7.87	- .02	- .25	- .03	.19	.00	.03	6.38	6.44
TQ131D	1	6336	4.30	.00	.00	.08	.32	.00	.02	11.53	11.54
TQ131F	1	6336	26.45	- .02	- .07	.16	.26	.04	.07	3.38	3.41
TQ136BOX	1	7010	18.19	.02	.10	.00	.06	.00	.01	3.45	3.45
TQ136A	1	5743	71.91	.00	.00	.00	- .01	.00	- .01	1.06	1.06
TQ136C	1	5743	9.09	.00	.00	.37	.39	.06	.06	5.71	5.72
TQ136D	1	5743	3.89	.00	.00	- .19	- .16	- .02	- .02	7.76	7.76
TQ136F	1	5743	21.81	.01	.04	- .39	- .43	- .11	- .12	3.52	3.51

Part 2. Means estimated for the total population

IQ15	11445	38.75	.00	.00	.04	.14	.12	.43	.33	.37
IQ16	11445	160.26	- .21	- .13	.14	.27	.22	.44	.63	.68
IQ89HA	7579	2102.30	7.68	.36	1.20	.60	.87	.45	1.81	1.45
IQ89HB	7579	2173.09	40.05	1.84	2.38	.58	1.50	.44	2.86	1.41
IQ141FA	15089	7039.68	-17.34	- .24	- .44	- .30	- .44	- .29	1.08	1.06
IQ141FB	15089	8704.41	-38.25	- .43	- .57	- .41	- .66	- .53	1.04	.89

[a] Terms used in table.

SAMPLE SIZE number of sample members eligible to respond to a particular item for the domain under consideration.
$\bar{Y}$-TRUE the estimate obtained using the telephone corrected and completed data.
ME measurement error caused by the use of data containing logical inconsistencies.
RB% the relative bias defined to be the bias divided by the value of $\bar{Y}$-TRUE, expressed as a percentage.
BR the bias ratio defined as the bias divided by the standard error of the estimate.
R$\sqrt{MSE}$% the relative root mean square error defined as the square root of the mean square error divided by the value of $\bar{Y}$-TRUE and expressed as a percentage.

Yet another measure of the quality of the survey estimates is the bias ratio (BR), defined as the bias of an estimate divided by its standard deviation. The bias ratios presented in this study are again not the same as those that would have resulted had the full data set been used. In this case, the bias ratios should be smaller than those for the full data set since the denominator, or standard deviation of the estimates, is larger because of subsampling. Considering that the bias ratios obtained by using the experimental data set would be less than those obtained by using the full data set, the quality of the estimates with respect to the bias ratios is not as good. Because the large sample size results in small variances, even a moderate bias can have a large effect. Looking at total population estimates, 17 of the 51 estimated proportions and three of the six estimated means have absolute bias ratios greater than 0.5, with 13 proportions having bias ratios greater than 1.0 (see Table 2). The large bias ratios are mainly associated with TQ1 and TQ9 and with items within routing patterns.

To determine the effect on the bias of inconsistencies, which were occurring at a high rate, a data set was constructed that contained the original inconsistent data but had all nonresponse replaced by the response obtained in the telephone interview. The difference between the estimates obtained using this data and the estimates obtained using the data set with both inconsistencies and nonresponse corrected is the bias due to the presence of inconsistent responses, which will be referred to as the "measurement error (ME) bias." Note that the estimates obtained from the experimental data set, using no imputation or editing (NI), contain both this measurement error bias as well as nonresponse bias. Referring to Table 2, which gives both of these bias terms for the proportions estimated for the 14 discrete items, one can see that the measurement error caused by inconsistent (and incorrect) responses is the most important component of the bias in the no imputation estimates. This is to be expected since the item nonresponse rate for discrete items is less than 2% in the experimental data set and the associated nonresponse bias would therefore be small as well.

A general conclusion that can be made for the discrete items is that nonresponse is not an important factor in the bias of the no imputation estimates and any imputation procedure that merely replaces missing values will not compensate for the most important source of bias, namely, measurement error that is due to inconsistent and incorrect responses. In fact, in many instances the nonresponse bias and the measurement error bias have opposite effects, so that the total bias in the no imputation estimates is less than the measurement error bias. One easily implemented approach to the problem of inconsistent data might be to remove the inconsistent responses to various items and code the responses as missing. This procedure was used on the experimental data set and estimates were obtained using no imputation on this new data set, which had only consistent responses but many more missing responses. It was found that, in general, the bias of the no imputation estimates when inconsistent data are removed (NIC, for no imputation, consistent data) is larger than that of the no imputation estimates obtained using the inconsistent data (NI). A probable explanation for this peculiar behavior is that a data item may be inconsistent with another data item and yet be correct, so that, by discarding all responses to an item that are inconsistent with responses to other items on the same instrument, information is lost and nonresponse bias becomes a more serious problem.

Another concern in evaluating the importance of response error biases is the effect on the bias of domain estimates. Table 3 presents the results for selected domains of interest.

4.2. Comparison of the Performance of Hot-Deck and Weighting-Class Estimators with That of No Imputation Estimators

For discrete items in general and for continuous items with routing patterns, logical inconsistencies occurring in the data can have a serious effect on the bias. For instance, if all missing data were replaced by the true value, using either an imputation procedure or telephone follow-up, the measurement error bias caused by the failure of an individual to interpret or respond in a correct manner to the questions would still be present and could cause serious problems in interpreting the estimates obtained from some of the items. The focus of this investigation was to investigate nonresponse bias, so the potential benefits of logical editing to correct inconsistent data were not investigated. However, the hot-deck and weighting-class imputation procedures did force the response within a routing pattern to agree with the lead-in question. When an inconsistent response was encountered within a routing pattern, the entire set of responses to the items within the routing pattern was replaced. Thus the performance of the two imputation procedures depends on three factors: the amount of nonresponse bias in the estimates, the extent to which the imputation procedure reduces the nonresponse bias, and the degree to which replacing inconsistent responses within a routing pattern improves the quality of the estimates. These issues will be discussed separately for Case I: Discrete Items and Case II: Continuous Items.

4.2.1. Case I: Discrete Items

The weighting-class imputation procedure devised for use with discrete items almost uniformly produced estimates whose total error is greater than that of the hot-deck and no imputation estimates. Part of the reason why the weighting-class imputation technique performed so poorly was that the number of nonrespondents to an item within each weighting class was so small (often as few as one or two individuals) that it was impossible to have the proportion

TABLE 3

Average Quality of the Data for Selected Domains[a]

	Discrete items (average)				*Continuous items (average)*			
Domain	*ME* $\|RB\%\|$	*NI* $\|RB\%\|$	*NI* $\|BR\|$	*NI* $R\sqrt{MSE}\%$	*ME* $\|RB\%\|$	*NI* $\|RB\%\|$	*NI* $\|BR\|$	*NI* $R\sqrt{MSE}\%$
Total	2.11	2.25	.91	4.64	.50	.80	.64	1.29
Sex								
Male	1.98	2.08	.65	5.90	.72	1.01	.50	1.87
Female	2.62	2.78	.64	6.96	.29	.43	.26	1.59
Ability								
Low	2.63	2.98	.43	8.83	.69	1.42	.37	3.26
Middle	2.72	2.83	.64	7.07	.86	.79	.35	1.97
High	2.31	2.71	.59	8.62	.55	.61	.32	1.80
SES								
Low	2.41	2.50	.49	7.02	.78	1.41	.54	2.50
Middle	2.51	2.74	.66	6.20	.32	.58	.38	1.38
High	2.49	2.73	.65	8.18	.72	.78	.37	1.97
Race								
Black	3.41	3.74	.49	10.25	1.16	2.01	.62	4.02
White	2.00	2.16	.74	5.23	.47	.72	.55	1.30
Hispanic	4.97	3.03	.35	12.10	.29	9.36	.48	17.28
Other	2.74	3.03	.35	12.10	1.04	1.19	.28	3.51
Region								
Northeast	2.64	2.86	.51	9.19	.88	1.28	.49	2.54
South	2.16	2.43	.69	7.11	.38	.38	.26	1.50
North Central	2.39	2.74	.59	6.23	1.12	1.50	.66	2.60
West	2.32	2.53	.51	8.09	.35	.51	.23	2.30
Race × Ability								
Black								
Low	4.35	4.96	.42	14.76	3.75	4.72	.67	8.15
Middle	3.15	3.69	.35	21.99	.53	1.27	.22	6.45
White								
Low	2.42	2.73	.26	12.13	.79	.69	.21	3.71
Middle	2.86	3.00	.58	7.82	.82	.72	.31	2.01
High	2.31	2.77	.57	8.64	.57	.63	.33	1.79
Other								
Low	4.58	5.06	.25	22.76	1.46	2.85	.33	7.98
Middle	3.24	3.68	.20	22.12	2.67	2.86	.28	7.25
High	3.22	3.54	.22	31.08	.59	1.21	.20	7.08

[a] The domain estimates for blacks of high ability and cross tabulations involving Hispanics were not included in this table since there were too few in the sample to compute valid variance estimates. For the rest of the domains, the average of the absolute value of the relative biases expressed as a percentage (|RB%|) is given for the measurement error due to inconsistencies in the data (ME) and for the total error of the no imputation (NI) estimates. For the no imputation estimates, the average of the absolute values of the bias ratios (|BR|) and the average of the relative root mean square errors ($R\sqrt{MSE}$) are also given for each domain. Note that SES is socioeconomic status.

of nonrespondents assigned each response equal to the proportion of respondents who gave the response. Because of its obvious lack of efficacy in reducing the total error of the discrete items, the weighing-class imputation procedure is not analyzed further for comparing its estimates with no imputation estimates.

In general, the hot-deck imputation procedure does appear to have some effect in reducing the bias of survey estimates. Table 4 gives the average (over all 51 estimated proportions) of the absolute relative bias, the absolute bias ratio, and the relative root mean square errors for selected domains of interest. For almost all of these domains, the average absolute relative bias and the average absolute bias ratio are less for the hot-deck estimates than for the no imputation estimates. For instance, the average absolute relative bias of the estimates for the total population is 1.92% as compared with 2.25% when no imputation is used. Similarly, the average absolute bias ratio is 0.83 for the hot-deck estimates, as compared with .91 for the imputation estimates. The bias ratio would be expected to be less since the variance of the hot-deck estimates is usually greater than that of the no imputation estimates, reflecting the greater variability of the hot-deck procedure. However, the fact that the average relative bias is smaller does indicate that over all of the 51 estimates the hot-deck procedure generally reduces the bias. The domains in which the hot-deck estimates are not better on the average tend to be small domains such as the Hispanic or Other categories of race. For those domains in which the hot-deck estimates on the average performed better, the average relative bias reduction is less than 0.5%. Much of the bias reduction resulting from the use of the hot-deck technique is associated with a corresponding increase in the variance so that the difference between the average values of the relative root mean square error for the two procedures is usually small (0.1% or less). Nine of the twenty-five domains have the average relative root mean square error less for the hot-deck estimates, including the estimates based on the total population, where the average of the relative root mean square errors is 4.53%, as opposed to 4.64% for the no imputation estimates.

Another way of comparing the hot-deck procedure to the no imputation procedure is to count the number out of the 51 proportions in which the hot-deck estimate has a smaller absolute relative bias or a smaller relative root mean square error. Table 4 presents counts out of a total of 51 estimates for which the hot-deck estimates have smaller absolute relative biases (the column is labeled as Number $\mathrm{HD}|\mathrm{RB}| < \mathrm{NI}|\mathrm{RB}|$) and the counts of estimates that have smaller relative root mean square errors (the column is labeled Number $\mathrm{HD\ R}\sqrt{\mathrm{MSE}} < \mathrm{NI\ R}\sqrt{\mathrm{MSE}}$). For the entire population, 34 of the 51 hot-deck estimates have smaller absolute relative biases than do those for the corresponding no imputation estimate, whereas only 22 of the hot-deck estimates have smaller relative root mean square errors.

It is interesting to note that for only 8 of the 24 domains more than half of the hot-deck estimates were better (in the sense of relative bias) than the no

TABLE 4

Comparison of the Average Performance of Hot-Deck and Weighting-Class Estimates with That of the No Imputation Estimates for Discrete Items[a]

Domain	Average NI $\lvert RB\%\rvert$	Average HD $\lvert RB\%\rvert$	Average WC $\lvert RB\%\rvert$	Number HD $\lvert RB\rvert$ < NI $\lvert RB\rvert$	Number WC $\lvert RB\rvert$ < NI $\lvert RB\rvert$	Average NI $\lvert BR\rvert$	Average HD $\lvert BR\rvert$	Average WC $\lvert BR\rvert$	Average NI $R\sqrt{MSE}\%$	Average HD $R\sqrt{MSE}\%$	Average WC $R\sqrt{MSE}\%$	Number HD $R\sqrt{MSE}\%$ < NI $R\sqrt{MSE}\%$	Number WC $R\sqrt{MSE}\%$ < NI $R\sqrt{MSE}\%$
Total	2.25	1.92	3.12	34	15	.91	.83	1.06	4.64	4.53	5.14	22	14
Sex													
Male	2.08	1.82	2.67	29	18	.65	.58	.70	5.90	5.88	6.24	17	14
Female	2.78	2.59	5.15	25	19	.64	.62	.81	6.96	6.89	8.78	26	13
Ability													
Low	2.98	2.82	5.56	23	14	.43	.44	.64	8.83	8.92	10.93	23	16
Middle	2.83	2.45	3.49	25	14	.64	.57	.70	7.07	7.26	7.49	21	17
High	2.71	2.39	3.06	26	18	.59	.54	.60	8.62	8.46	9.06	32	18
SES													
Low	2.50	2.41	3.72	24	13	.49	.47	.61	7.02	7.27	8.17	15	12
Middle	2.74	2.70	3.92	20	18	.66	.64	.81	6.28	6.39	6.96	20	16
High	2.73	2.40	3.11	35	17	.65	.58	.69	8.18	8.03	8.51	30	24
Race													
Black	3.74	3.29	4.99	29	20	.49	.43	.56	10.25	10.26	10.98	17	12
White	2.16	1.89	2.83	24	12	.74	.70	.88	5.23	5.13	5.53	22	14
Hispanic	6.18	14.51	29.69	11	12	.23	.33	.35	37.92	47.02	52.39	9	10
Other	3.03	3.08	4.86	22	20	.35	.35	.42	12.10	12.39	13.78	21	16

Region													
Northeast	2.86	2.40	3.99	27	12	.51	.45	.59	9.19	9.20	9.98	20	12
South	2.43	2.39	2.73	25	18	.69	.68	.74	7.11	7 04	7.31	30	19
North Central	2.74	2.61	3.45	20	19	.59	.55	.67	6.23	6.25	6.73	15	11
West	2.53	2.32	4.55	23	17	.51	.48	.64	8.09	8.20	9.79	22	13
Race × Ability													
Black													
Low	4.96	4.36	5.95	21	21	.42	.38	.47	14.76	14.52	15.76	23	14
Middle	3.69	4.36	5.47	18	8	.34	.40	.42	21.99	22.59	26.72	16	7
White													
Low	2.73	2.38	4.78	25	11	.26	.29	.43	12.13	12.18	14.11	21	12
Middle	3.00	2.57	3.48	28	15	.58	.52	.61	7.82	7.74	8.21	22	20
High	2.77	2.35	3.06	31	20	.57	.50	.57	8.64	8.82	9.40	31	17
Other													
Low	5.06	7.34	12.03	16	19	.25	.34	.40	22.76	24.48	29.78	24	18
Middle	3.68	3.30	4.38	12	7	.20	.20	.24	22.12	22.91	23.68	7	8
High	3.54	4.17	4.21	15	17	.22	.25	.25	31.08	31.50	31.50	12	10

[a] The domain estimates for blacks of high ability and cross tabulations involving Hispanics were not included in this table since there were too few in the sample to compute valid variance estimates. For the rest of the domains, the average of the absolute values of the relative biases expressed as a percentage ($|RB\%|$), the average of the absolute values of the bias ratios ($|CR|$), and the average of the relative root mean square errors ($R\sqrt{MSE}$) are given. The acronym "Number HD $|RB| < $ NI $|RB|$" refers to the number of estimates for which the absolute relative bias is less for hot deck (HD) than for no imputation (NI) out of the total of 51 estimates. Similarly, "Number HD $R\sqrt{MSE} <$ NI $R\sqrt{MSE}$" refers to the number of estimates for which the relative root mean square error of the hot deck (HD) estimate is less than that of the no imputation (NI) estimate. The acronym "Number WC$|RB| <$ NI$|RB|$" has a similar definition with respect to weighting class (WC) and no imputation estimates. Note that SES is socioeconomic status.

imputation estimates. Since for 20 of the 25 domains the hot-deck estimates do better than the no imputation estimates on the average with respect to relative bias, one can hypothesize that the hot-deck procedure reduces bias for those items for which there is the most nonresponse and/or that the process by which the hot deck removes inconsistent responses within routing patterns and replaces them with consistent responses effectively reduces the measurement error associated with inconsistent responses within routing patterns. If one examines the average of the absolute relative biases and relative root mean square errors on an item-by-item basis (see Table 5), one sees that the hot-deck procedure achieves most of its gains on the questions within routing patterns. Of the within-routing-pattern items TQ12, TQ33, TQ52, TQ66, TQ90, TQ102, TQ131, and TQ136, items TQ90 and TQ136 are the only ones in which the hot-deck estimates do not have a smaller absolute relative bias. Again, much of this bias reduction is compensated for by an associated increase in the variance, so that the hot-deck estimates for TQ66, TQ102, and TQ131 have larger relative root mean squares than do the no imputation estimates.

TABLE 5

A Comparison of the Hot-Deck (HD) and No Imputation (NI) Estimates on an Item-by-Item Basis

	Average of the absolute relative biases		*Average of the relative root mean square errors*	
Item	*NI*	*HD*	*NI*	*HD*
TQ1	5.86	5.84	6.71	6.70
TQ9	7.85	7.87	9.16	9.22
TQ10	1.65	1.79	3.73	3.93
TQ12	2.52	1.67	5.27	5.23
TQ29	.22	.53	2.32	2.30
TQ33	6.19	1.74	7.54	5.08
TQ51	.46	.66	1.34	1.42
TQ52	.62	.31	1.47	1.14
TQ66	2.95	2.93	3.45	3.66
TQ90	2.19	2.58	5.29	5.38
TQ101	.39	.10	1.35	1.31
TQ102	.69	.67	3.07	3.15
TQ118	1.08	.14	4.89	4.91
TQ129	.10	.12	2.66	2.67
TQ131	.09	.17	5.56	5.61
TQ136	.19	.13	4.30	4.32

4.2.2. Case II: Continuous Items

The average values (over all six continuous items) of the absolute relative bias, the absolute bias ratio, and the relative root mean square error are given in Table 6 for the selected domains of interest. The hot-deck and no imputation estimates are more or less comparable with respect to the absolute relative bias, with the hot-deck estimates having a slightly smaller average relative bias for many of the domains, including the total. The weighting-class estimates have slightly larger average relative biases than both the no imputation and the hot-deck estimates for many domains. The average bias ratios for the no imputation estimates are, in general, larger than the corresponding ratios for both the hot-deck and weighting-class estimates. This reflects the greater variability of the hot-deck and weighting-class procedures, which results in a larger denominator for the ratio. The last measure of quality of the data, the average value of the relative root mean square errors, is smallest for the no imputation estimates for 16 of the 25 domains, including the total. It is interesting to note that, although the weighting-class estimates do not exhibit the bias reduction potential of the hot-deck estimates, the average relative root mean square error is less for the weighting-class estimates than it is for the hot-deck estimates for 23 of the 25 domains studied. In comparison with no imputation estimates, the weighting-class estimates do better with respect to average relative bias and average relative root mean square error for the smaller domains.

The six continuous items over which averages were taken were quite diverse and the imputation procedures were different for those that were within routing patterns. Of these, the last two continuous items included in the instrument were TQ141FA and TQ141FB, which requested the total income of the individual and his/her spouse in 1975 and 1976, respectively. These two items were the only continuous items included in this study that did not fall within a routing pattern. Whereas the interpretation of the results for the other four continuous items within routing patterns is not particularly clear, the results for these two items clearly show the weighting-class estimates to be superior to the no imputation and hot-deck estimates with respect to their relative bias as well as their relative root mean square error. Also, the weighting-class estimates for all six items are better than the hot-deck estimates with respect to the average relative root mean square error, suggesting that the weighting-class imputation procedure is more suitable as an imputation device when means are being estimated. For the items within routing patterns, one might hypothesize that the weighting-class estimates would have been better or comparable to the no imputation estimates if the procedure had been implemented differently for the routing patterns. Basically, the procedures forced all data within the pattern to agree by deleting all the responses to the items within a routing pattern when one or more responses were in logical disagreement with the lead-in question. A check of this procedure for two very small weighting classes in the data base revealed that many good data were being discarded by this procedure. Further,

TABLE 6

Comparison of the Average Performance of Hot-Deck and Weighting-Class Estimates with That of the No Imputation Estimates for Continuous Items[a]

Domain	Average *NI* $\lvert RB\% \rvert$	Average *HD* $\lvert RB\% \rvert$	Average *WC* $\lvert RB\% \rvert$	Average *NI* $\lvert BR \rvert$	Average *HD* $\lvert BR \rvert$	Average *WC* $\lvert BR \rvert$	Average *NI* $R\sqrt{MSE}\%$	Average *HD* $R\sqrt{MSE}\%$	Average *WC* $R\sqrt{MSE}\%$
Total	.80	.76	.86	.64	.46	.64	1.29	1.46	1.40
Sex									
Male	1.01	.94	1.15	.50	.39	.53	1.87	2.05	2.03
Female	.43	.65	.61	.26	.33	.42	1.59	1.84	1.69
Ability									
Low	1.42	2.02	1.92	.37	.36	.55	3.26	5.41	3.38
Middle	.79	.90	1.18	.35	.32	.41	1.97	2.02	2.19
High	.61	.61	.53	.32	.31	.30	1.80	1.84	1.72
SES									
Low	1.41	1.55	1.50	.54	.59	.57	2.50	2.82	2.53
Middle	.58	.52	.85	.38	.26	.54	1.38	1.79	1.57
High	.78	.76	.88	.37	.34	.43	1.97	2.14	2.00
Race									
Black	2.01	1.98	2.04	.62	.40	.41	4.02	4.43	4.06
White	.72	.61	.72	.55	.33	.50	1.30	1.49	1.36
Hispanic	9.36	7.75	8.90	.48	.40	.65	17.28	17.12	14.96
Other	1.19	1.97	1.43	.28	.54	.33	3.51	3.84	3.58

Region									
Northeast	1.28	.74	.87	.49	.28	.35	2.54	2.48	2.33
South	.38	.65	.54	.26	.38	.34	1.50	1.67	1.51
North Central	1.50	2.06	1.76	.66	.74	.68	2.60	3.11	2.80
West	.51	.35	.73	.23	.08	.33	2.30	3.33	2.23
Race × Ability									
Black									
Low	4.72	5.42	4.80	.67	.67	.71	8.15	8.99	7.77
Middle	1.27	.92	2.20	.22	.14	.39	6.45	6.61	6.47
White									
Low	.69	1.35	.69	.21	.31	.30	3.71	7.49	3.41
Middle	.72	.79	1.09	.31	.25	.35	2.01	2.09	2.17
High	.63	.58	.48	.33	.29	.24	1.79	1.83	1.71
Other									
Low	2.85	3.75	2.47	.33	.52	.30	7.98	8.16	7.20
Middle	2.86	2.78	2.64	.28	.30	.28	7.25	7.26	7.14
High	1.21	1.65	1.54	.20	.29	.28	7.08	7.74	7.41

[a] The domain estimates for blacks of high ability and cross tabulations involving Hispanics were not included in this table since there were too few in the sample to compute valid variance estimates. For the rest of the domain, the average of the absolute values of the relative biases (|RB%|), the average of the absolute values of the bias ratios (|**BR**|), and the average of the relative root mean square errors are given for the no imputation (NI), hot-deck (HD), and weighting-class (WC) estimates. Note that SES is socioeconomic status.

the quality of the weighting-class estimates of the means for continuous items within routing patterns is expected to suffer because of the poor quality of the adaptation of the weighting-class procedure used for discrete data. When a routing pattern had a missing response for the lead-in question (which was always an item with categorical or discrete responses), the response to the lead-in question was first imputed and then responses to the items within the routing pattern were imputed based on the imputed response to the lead-in item. Since the weighting-class procedure as applied to discrete items is more biased, one would expect that this bias for the responses to lead-in items would also have an effect on the bias of the estimates for continuous items within the routing pattern.

4.3. Comparison of Balanced Repeated Replication and Taylor Series Linearization Variance Estimates

To determine the effect of ignoring imputation in computing variance estimates, two sets of variance estimates are computed for each set of survey estimates. First, the Taylor series linearization estimate for the variance of a ratio is computed using STDERR, an RTI package (see Appendix C for variance estimation formulas). This package calculates the standard approximation for the variance of a ratio in terms of the variance–covariance matrix of the totals of the numerator and denominator. Second, an estimate of the variance of the ratio is obtained using the balanced repeated replication (BRR) technique, which allows one to account for the variability induced by the imputation of missing data (see Appendix B).

For the no imputation estimates obtained using the full experimental data set (NI estimates) and the experimental data set with all inconsistent data removed (NIC estimates), the two variance estimates should be equal since no imputation occurs and both data sets measure the same variability. To compare the two sets of variance estimates, the ratio of the standard deviation (SD ratio) of the estimate obtained using STDERR over the standard deviation obtained using BRR is computed. An examination of the SD ratios for the 51 discrete estimates and for the six continuous estimates indicates that the ratio is essentially one except for sampling variation; Table 7 presents the average SD ratio for both the discrete and continuous items for seven specific domains. The SD ratio is more variable for the continuous items, reflecting the fact that only six estimates are averaged for each domain. The largest SD ratio in Table 7 is 1.29, and this corresponds to individuals of low socioeconomic status for the continuous NIC data.

It is interesting that almost all of the average values of the SD ratios exceed one. Since STDERR, unlike the BRR procedure, allows the use of finite correction factors, at the school level, that can be significantly less than unity, one would expect the STDERR estimate to be smaller than the BRR estimate and

TABLE 7

The Average Value of the Ratio of the STDERR Standard Deviation to the BRR Standard Deviation (SD Ratio) for the No Imputation Estimates When Inconsistent Data Are Retained (NI) and When Removed (NIC)

	Average of the SD ratio			
	Discrete items		*Continuous items*	
Domain	*NI*	*NIC*	*NI*	*NIC*
Total population	1.03	1.01	1.01	1.01
Males	1.09	1.07	.94	1.01
Individuals of high ability	1.07	1.07	1.06	1.18
Individuals with low socioeconomic status	1.09	1.07	1.16	1.29
Blacks	1.03	1.03	1.01	1.05
Individuals from the South	1.04	1.03	1.18	1.13
Blacks of average ability	1.00	1.00	.97	.95

hence to obtain SD ratios that are less than one. The fact that the converse is the case indicates that the variability of the estimates is affected by within-schools rather than between-schools variation since in that case the use of finite correction factors at the school level would have no important effect.

4.4. The Effect of Ignoring Imputation on Variance Estimation

Typically, most users of the hot-deck and other imputation procedures ignore the fact that missing values have been imputed and use a standard software package that cannot distinguish between imputed and naturally occurring data in computing the variance estimates. One additional goal of this investigation is to determine how the variance estimate is affected if the actual imputation of missing responses is ignored. The variances of survey estimates are computed using STDERR, but they do not reflect the added variability induced by the use of the imputation procedure. To obtain estimates of the variances of survey statistics that accounted for the added variability induced by imputation, a BRR procedure is used in which the imputation procedure is applied to each independent half-sample (see Appendix B). The comparison of these two variance estimates provides some insight into the effect on the variance when imputation is ignored. Table 8 gives, for seven specific domains, the average of the SD ratios obtained using STDERR and BRR for the 51 discrete estimates when the hot-deck procedure is used and the SD ratios for the six continuous estimates obtained using both weighting-class and hot-deck procedures. For discrete items, one can see that the average SD ratio is essentially one. Since the nonresponse rates for these items are very low, the imputation does not seem to have a substantial impact on the variance. For continuous items, larger rates of nonresponse meant on the average that more missing responses were imputed. The average SD ratios for continuous items were less than one in many instances. However, should TQ141FA and TQ141FB be excluded, the nonresponse rates would be fairly low and little imputation would occur, and the two sets of standard deviations would then be expected to be approximately equal.

One can obtain a better understanding of the effect of ignoring imputation by comparing the standard errors for the 25 domains studied in this investigation when the hot-deck and weighting-class procedures are applied to TQ141FA, an income item that has a 13% nonresponse rate (see Table 9). The standard deviation estimated by STDERR, using the usual formula for the variance of a ratio, does not account for the variability induced by the imputation procedures. The differential nonresponse rates within domains lead to varying values for the SD ratios, but they are usually less than one. Some of the ratios are inexplicably large, such as the ratio of 1.47 for blacks of low ability that is obtained when the weighting-class procedure is used. Note that the standard deviations obtained using STDERR and BRR for the weighting-class estimates are more nearly equal than are those obtained using the hot-deck procedure; thus the weighting-class procedure probably induces less variation through imputation than does the hot-deck procedure.

TABLE 8

Average of the Ratios of the STDERR Standard Deviation Estimates to the BRR Standard Deviation Estimates for the Hot-Deck and Weighting-Class Procedures for Selected Domains

	Discrete estimates	*Continuous estimates*	
Domain	*Hot deck*	*Hot deck*	*Weighting class*
Total	1.02	.86	.94
Males	1.08	.80	.87
Individuals of high ability	1.06	1.03	1.05
Individuals with low socioeconomic status	1.05	.91	1.13
Blacks	.97	.83	.95
South	1.07	1.08	1.13
Blacks of average ability	.96	.92	1.13

TABLE 9

The Ratio of the STDERR Standard Deviation Estimates to the BRR Estimates for TQ141FA[a]

	Hot deck			*Weighting class*		
Domain	*STDERR SD*	*BRR SD*	*SD ratio*	*STDERR SD*	*BRR SD*	*SD ratio*
Total	67.93	84.55	.80	63.61	70.11	.91
Sex						
Male	79.90	95.41	.84	76.01	95.18	.80
Female	101.89	172.30	.59	95.47	138.95	.69
Ability						
Low	175.90	223.51	.79	149.76	150.34	1.00
Middle	104.45	93.23	1.12	106.37	77.07	1.38
High	122.64	115.58	1.06	111.48	106.99	1.04
SES						
Low	124.76	167.62	.74	106.42	122.43	.87
Middle	88.52	92.87	.95	82.66	81.72	1.01
High	106.48	153.47	.69	111.23	110.06	1.01
Race						
Black	157.84	183.67	.86	147.76	156.84	.94
White	73.38	82.33	.89	69.51	64.46	1.08
Hispanic	853.14	965.70	.88	674.42	980.63	.69
Other	280.55	298.39	.94	268.93	263.13	1.02
Region						
Northeast	144.72	178.95	.81	148.26	158.11	.94
South	133.94	118.67	1.13	121.88	84.12	1.45
North Central	124.51	211.63	.59	106.64	145.55	.73
West	120.38	147.77	.81	119.72	118.44	1.01
Race × Ability						
Black						
Low	250.55	247.68	1.01	207.14	140.71	1.47
Middle	362.80	417.60	.87	353.30	382.02	.92
White						
Low	207.43	247.73	.84	160.00	145.49	1.10
Middle	112.69	97.11	1.16	115.96	69.06	1.68
High	124.96	99.68	1.25	113.68	98.19	1.16
Other						
Low	633.72	644.23	.98	624.49	638.94	.98
Middle	381.12	432.27	.88	356.86	376.07	.95
High	562.22	569.54	.99	556.75	515.27	1.08

[a] The black of high ability domain and the cross classifications of Hispanics by ability were omitted from this table since these domains had such a small sample size that valid variance estimates could not be obtained. Note that SES is socioeconomic status.

To summarize, when the nonresponse rate is large, the variance of sample means is underestimated if imputation is ignored in computing variance estimates. This underestimation effect will vary from domain to domain just as the nonresponse rate varies from domain to domain. The weighting-class imputation procedure results in a lesser underestimation effect for the continuous items than does the hot-deck procedure, because the imputation procedure itself contributes less variability than that of the hot deck and hence ignoring the variability added by weighting-class imputation has less of an effect than it does when the hot-deck procedure is used.

5. CONCLUDING REMARKS

In the experimental data set that contained the original responses to the 20 selected critical items (from the third follow-up instrument of the NLS) before editing and telephone resolution, inconsistent data represented the most serious source of error in the survey estimates, with the discrete items, especially TQ1, TQ9, and the items nested within routing patterns most susceptible to the resulting bias. The effect of this source of measurement error bias is reflected in the relative bias that, for many of the estimates for the discrete data, is over 5%. Because of the large sample size, the variance of the estimates is small, so that even a moderate amount of bias relative to the variance becomes important; this effect is measured by the bias ratio, which for many estimates has values of .5 or more. The most appropriate overall measure of the quality of the estimates is the relative mean square error. However, this measure is quite large for a few of the estimates. For any analysis of the NLS data, some attention should be given to the problem of resolving logical contradictions contained within an individual's responses to the items on the survey instrument. Although some minor editing has been done with flags indicating violations of routing patterns, the majority of the editing has been left to the user. The present investigation points out that merely discarding the inconsistent responses and coding these items as blank is not a desirable solution. The user should design edit checks and choose procedures for the imputation or correction of missing and faulty data that are best suited for the items under study and the analysis at hand.

For the discrete items investigated in this study, the hot-deck procedure does reduce the overall bias of the estimates, but much of the improvement may have been related to the editing procedure that removed inconsistent data within routing patterns and replaced these responses with data from an individual in the same weighting class who responded in a consistent manner. Except for this minor editing, the imputation procedures of this study were not designed to correct for the inconsistencies within an individual's responses.

Procedures designed to do both logical editing and imputation of responses for missing and logically erroneous data, e.g., CAN-EDIT of Statistics Canada (Hill, 1978), may be effective in reducing the bias resulting from missing or faulty data for discrete items.

The weighting-class procedure devised for discrete items in this study did not work well in practice. In retrospect, the distribution of responses could be estimated for each item within weighting classes and a response assigned randomly for each missing value so that the probability of each of the assigned response options equals the proportion of respondents from the same weighting class who circled that response option. Such a procedure could compete with the hot-deck procedure in reducing the bias.

For the continuous items, the nonresponse rate is higher but the effect of bias that is due to nonresponse is not great even for TQ141FA and TQ141FB, for which the nonresponse rate is 13%. The inconsistencies in the data give rise to the bias in the estimates for the continuous items as well, especially for those within routing patterns. The weighting-class procedure is most effective in reducing the bias for the continuous estimates, but the reduction in the relative root mean square error is less than .5% in general.

For both the discrete and continuous data, on the whole, the bias that is due to nonresponse is insignificant; however, the bias resulting from the failure of individuals to interpret the questions correctly is large in relation to the bias due to nonresponse, and some attention should be given to this source of error.

It should be emphasized that the present study focuses exclusively on the estimation of univariate means and proportions. For more complex analyses such as regression, factor analysis, and correlation studies involving multivariate statistics, the conclusions of this study concerning the effects of imputation may not be appropriate.

APPENDIX A. THIRD FOLLOW-UP QUESTIONNAIRE—SELECTED ITEMS

National Center for Education Statistics
Education Division
Department of Health, Education, and Welfare
Washington D.C. 20202

GENERAL INSTRUCTIONS

This questionnaire is divided into the following seven sections:

A. General Information
B. Work Experience
C. Education and Training
D. Military Service
E. Family Status
F. Experiences and Opinions
G. Background Information

Start by answering questions in Section A. You will need to answer the first question in each section, but you may not need to answer all the questions in every section. You may be able to skip most of some sections. We have designed the questionnaire with special instructions in red beside responses which allow you to skip one or more questions. Follow these instructions when they apply to you.

Read carefully each question you answer. It is important that you follow the directions for responding, which are:

- **(Circle one.)**
- **(Circle as many as apply.)**
- **(Circle one number on each line.)**

Sometimes you are asked to fill in a blank—in these cases, simply write your response on the line provided.

Where you are asked to circle a number, make a heavy circle. Here is an example:

Why did you leave high school?	**(Circle one number on each line.)**	
	My Reasons	**NOT My Reasons**
Graduated	(1)	2
Entered college	1	(2)
Went to work	(1)	2

Many questions ask what you were doing during a specific time period; for example, "What were you doing during the **first week of October 1976?**" Because it has been two years since we last heard from you, we also ask some questions about what you were doing in **1975.** As you go through the questionnaire, please watch for these time references and make sure you are thinking about the correct time period for each question.

This questionnaire is authorized by law 20 USC 1221e-1.

The Federal Privacy Act of 1974 requires that each survey respondent be informed of the following:

(1) Solicitation of information about the respondent as detailed in the questionnaire is authorized by Section 415 of the General Education Provisions Act as amended (20 USC 1226b).

(2) Disclosure of this information by the respondent is subject to no penalty for not providing all or any part of the requested information.

(3) The purpose for which this information is to be used is to provide statistics on a national sample of students as they move out of the American high school system into the critical years of early adulthood and relate these statistics to postsecondary educational costs and financial aid and other factors on the educational, work, and career choices of young adults.

(4) The routine uses of these data will be statistical in nature as detailed in 9 in Appendix B of the Departmental Regulations (45 CRF 56) published in the *Federal Register*, Vol. 40, No. 196, October 8, 1975.

When you complete this questionnaire, please place it in the post-paid, addressed envelope provided and mail it to:

OPERATION FOLLOW-UP
Research Triangle Institute
Post Office Box 12036
Research Triangle Park, North Carolina 27709

THANK YOU FOR YOUR COOPERATION

A
START

SECTION A: GENERAL INFORMATION

1. **What were you doing the first week of October 1976?**

(Circle as many as apply.)

Working for pay at a full-time or part-time job 1
Enrolled in graduate or professional school 2
Taking academic courses at a two- or four-year college 3
Taking vocational or technical courses at any kind of school or college (for example, vocational, trade, business, or other career training school) 4
On active duty in the Armed Forces (or service academy) 5
Homemaker 6
Temporary layoff from work, looking for work, or waiting to report to work 7
Other (describe: ______________________________) 8

2. **How would you describe your living quarters as of the first week of October 1976?**

(Circle one.)

Private house or mobile home 1
Private apartment 2
Dormitory or apartment operated by a school or college 3
Fraternity or sorority house 4
Rooming or boarding house 5
Military service barracks, on board ship, etc. 6
Other (describe: ______________________________) 7

3. **With whom did you live as of the first week of October 1976?**

(Circle one.)

By myself 1
With my parents 2
With my husband or wife 3
With parents and husband or wife 4
With other relatives 5
With person(s) not related to me 6

4. **Which of the following best describes the location of the place where you lived in the first week of October 1976?**

(Circle one.)

In a rural or farming community 1
In a small city or town of fewer than 50,000 people that is not a suburb of a larger place 2
In a medium-sized city (50,000-100,000 people) 3
In a suburb of a medium-sized city 4
In a large city (100,000-500,000 people) 5
In a suburb of a large city 6
In a very large city (over 500,000 people) 7
In a suburb of a very large city 8
A military base or station 9

A
CONTINUED

5. Is this the SAME city or community where you lived in October 1974?

Yes ....1 *GO TO Q. 8*
No..........................2 *CONTINUE WITH Q. 6*

6. How far is this from where you lived in October 1974?

(Circle one.)

Less than 50 miles 1
50 to 99 miles 2
100 to 199 miles 3
200 to 499 miles 4
500 miles or more 5

7. What was the main reason you moved to the place where you live now?

(Circle one.)

To find or take a job 1
Was transferred...2
Other job-related reason 3
To go to school 4
To follow my parents or spouse to a new location 5
To follow another relative or friend to a new location 6
Wanted a better place to live 7
Other (specify: ____________________________________) 8

8. Which of the following items do you have the use of as your own because you (or your spouse) have bought them or have been given them, or because they belong to your parents, roommates, dormitory, apartment building, etc.?

(Circle one number on each line.)

	Have As My Own	Have But Don't Own	Don't Have Use Of
a. Daily newspaper	1	2	3
b. Dictionary	1	2	3
c. Encyclopedia or other reference books	1	2	3
d. Magazines	1	2	3
e. Record player	1	2	3
f. Tape recorder or cassette player	1	2	3
g. Color television	1	2	3
h. Typewriter	1	2	3
i. Electric dishwasher	1	2	3
j. Two or more cars or trucks that run	1	2	3

9. Now please think back a year to Fall 1975. What were you doing in October 1975?

(Circle as many as apply.)

Working for pay at a full-time or part-time job 1
Enrolled in graduate or professional school 2
Taking academic courses at a two- or four-year college 3
Taking vocational or technical courses at any kind of school or college (for example, vocational, trade, business, or other career training school) 4
On active duty in the Armed Forces (or service academy)........5
Homemaker 6
Temporary layoff from work, looking for work, or waiting to report to work 7
Other (describe: ____________________________________) 8

START

SECTION B: WORK EXPERIENCE

In this section, we would like to find out about the jobs you have held in the two-year period from October 1974 through October 1976. Include full-time jobs, part-time jobs, apprenticeships, on-the-job training, military service and so on.

We are interested in learning about the kinds of jobs you have held, the hours you worked and your income from these jobs, the level of your job satisfaction, and the relation of your training and education to your work experience. This information will help us better understand the movement of young people into the world of work and the reasons for changes in job situations.

JOBS HELD IN OCTOBER 1976

10. Did you hold a job of any kind during the first week of October 1976?

(Circle one.)

Yes, working full-time (35 hours or more per week)	1	*GO TO Q. 13, next page*
Yes, working part-time (34 hours or fewer per week)	2	
Yes, but on temporary layoff from work or waiting to report to work	3	
No	4	*CONTINUE WITH Q. 11*

11. What were the reasons you were not working during the first week of October 1976?

(Circle one number on each line.)

	My Reasons	NOT My Reasons
a. Did not want to work	1	2
b. Was full-time homemaker	1	2
c. Going to school	1	2
d. Not enough job openings available	1	2
e. Required work experience I did not have	1	2
f. Jobs available offered little opportunity for career development	1	2
g. Health problems or physical handicap	1	2
h. Could not arrange child care	1	2
i. Other family responsibilities (including pregnancy)	1	2
j. Not educationally qualified for types of work available	1	2
k. There were jobs but none where I could use my training	1	2
l. Spouse preferred that I didn't work	1	2
m. Other (specify: ____________________)	1	2

12. Were you looking for work during the first week of October 1976?

(Circle one.)

Yes	1	*GO TO Q. 32, p. 7*
No, but DID look for work sometime during the month of September 1976	2	
No, and did NOT look for work at any time during the month of September 1976	3	

B
CONTINUED

13. **Please describe below the job you held during the first week of October 1976. (If you held more than one job at that time, describe the one at which you worked the most hours.)**

a. For whom did you work? (Name of company, business organization, or other employer)
(Write in): ______________________________
b. What kind of business or industry was this? (For example, retail shoe store, restaurant, etc.)
(Write in): ______________________________
c. What kind of job or occupation did you have in this business or industry? (For example, salesperson, waitress, secretary, etc.)
(Write in): ______________________________
d. What were your most frequent activities or duties on this job? (For example, selling shoes, waiting on tables, typing and filing, etc.)
(Write in): ______________________________
e. Were you:

(Circle one.)

An employee of a PRIVATE company, bank, business, school, or individual working for wages, salary, or commissions? 1
A GOVERNMENT employee (Federal, State, county, or local institution or school)? 2
Self-employed in your OWN business, professional practice, or farm? 3
Working WITHOUT PAY in family business or farm? 4

f. When did you start working at this job? __________ (month) __________ (year)
g. Are you currently working at this job?
Yes 1
No .... 2 Date left: __________ (month) __________ (year)

14. **How did you find this job?**

(Circle as many as apply.)

a. School or college placement service 1
b. Professional periodicals or organizations 2
c. Civil Service applications 3
d. Public employment service 4
e. Private employment agency 5
f. Community action or welfare groups 6
g. Newspaper, TV, or radio ads 7
h. Direct application to employers 8
i. Registration with a union 9
j. Relatives 10
k. Friends 11
l. Other (specify: ______________________) 12

15. **How many hours did you usually work at this job in an average week?**

__________ Hours per week

16. **In an average week, approximately how much did you earn at this job? (Report your gross earnings before deductions. If not paid by the week, please estimate.)**

$ __________ per week
(Earnings before deductions)

B
CONTINUED

25. Were you hired for this job because your employer knew you had been trained in a school or college to do this kind of work?

Yes 1
No..........................2
Don't know 3

26. Did the school at which you received your training for this job refer you to this job?

Yes 1
No..........................2

27. Do you expect to be working in October 1977?

No..............1 } *GO TO Q. 29*
Don't know ....2 }
Yes ....3 *CONTINUE WITH Q. 28* ⟶

28. Do you plan to work at the SAME KIND OF WORK?

Yes 1
No 2
Don't know..........3

29. Were you working at a second job in the first week of October 1976 at the SAME TIME as you held the job you described above?

No..........................1 *GO TO Q. 32*
Yes ....2 *CONTINUE WITH Q. 30*

30. How many hours did you usually work at this job in an average week?

_____________ Hours per week

31. In an average week, approximately how much did you earn at this job? (Report your gross earnings before deductions. If not paid by the week, please estimate.)

$_____________ per week
(Earnings before deductions)

JOBS HELD IN OCTOBER 1975

32. Now please think back to Fall 1975. Did you hold a job of any kind during the month of October 1975?

(Circle one.)

Yes, working full-time (35 hours or more per week) ....1 }
Yes, working part-time (34 hours or fewer per week) ....2 } *GO TO Q. 34, next page*
Yes, but on temporary layoff from work or waiting to report to work ....3 }
No ....4 *CONTINUE WITH Q. 33*

33. Were you looking for work during October 1975?

(Circle one.)

Yes ....1 }
No, but DID look for work sometime during the month of September 1975 ....2 } *GO TO Q. 41, p. 9*
No, and did NOT look for work at any time during the month of September 1975 ....3 }

C
START

SECTION C: EDUCATION AND TRAINING

This section asks information about your training and education. We would like to find out about the schools you have attended during the last two years, from October 1974 to October 1976. This information, combined with information you have given us in earlier follow-ups, will help to give us a complete picture of your educational experiences since high school. (Persons in the military service should also answer the questions in this section.)

EDUCATIONAL PROGRESS AND PLANS

47. Since high school, had you earned any certificate, license, diploma or degree of any kind prior to October 1976?

No ....1 *GO TO Q. 49*
Yes ....2 *CONTINUE WITH Q. 48*

48. What kind of certificate, license, diploma or degree have you earned?

	(Circle as many as apply.)	Date Received Month	Year	Area of Certificate, License, or Degree (For Example, Real Estate License, Shorthand Certificate, Degree in History)
A certificate	1	______	19__	______
A license	2	______	19__	______
A 2-year or 3-year vocational degree or diploma	3	______	19__	______
A 2-year academic degree	4	______	19__	______
A 4-year or 5-year college Bachelor's degree	5	______	19__	______
A Master's degree or equivalent	6	______	19__	______
Other (specify: ______)	7	______	19__	______

49. a. As of the first week of October 1976, what was your highest level of education or training? (Column A)

b. As things stand now, how far in school do you think you actually will get? (Column B)

		A. Had in October 1976 (Circle one.)	B. Plan to get (Circle one.)
Finished high school		1	1
Vocational trade or business school	Less than two years	2	2
	Two years or more	3	3
College program	Less than two years of college	4	4
	Two or more years of college (including two-year degree)	5	5
	Finished college (four- or five-year degree)	6	6
	Master's degree or equivalent	7	7
	Ph.D., or advanced professional degree	8	8

C
CONTINUED

50. With regard to your education and training during the last year you were in school, how satisfied as a whole were you with the following?

(Circle one number on each line.)

	Very Satisfied	Somewhat Satisfied	Neutral or No Opinion	Somewhat Dissatisfied	Very Dissatisfied
a. The ability, knowledge, and personal qualities of most teachers	1	2	3	4	5
b. The social life	1	2	3	4	5
c. Development of my work skills	1	2	3	4	5
d. My intellectual growth	1	2	3	4	5
e. Counseling or job placement	1	2	3	4	5
f. The buildings, library, equipment, etc.	1	2	3	4	5
g. Cultural activities, music, art, drama, etc.	1	2	3	4	5
h. The intellectual life of the school	1	2	3	4	5
i. Course curriculum	1	2	3	4	5
j. The quality of instruction	1	2	3	4	5
k. Sports and recreation facilities	1	2	3	4	5

51. During the two-year period from October 1974 through October 1976 were you enrolled in or did you take classes at any school like a college or university, graduate or professional school, service academy or school, business school, trade school, technical institute, vocational school, community college, and so forth?

No .... 1 *GO TO Q. 98, p. 22*
Yes .... 2 *CONTINUE WITH Q. 52*

SCHOOL ATTENDANCE IN OCTOBER 1976

52. Did you attend school in the first week of October 1976?

No .... 1 *GO TO Q. 66, p. 15*
Yes .... 2 *CONTINUE WITH Q. 53*

53. What is the exact name and location of the school you were attending in the first week of October 1976? (Please print and do not abbreviate.)

School Name: ________________________

City: ________________________ State: ____________

54. What kind of school is this?

(Circle one.)

Vocational, trade, business, or other career training school 1
Junior or community college (two-year) 2
College or university (four years or more) 3
Independent graduate or professional school (medical, dental, law, theology, etc.) 4
Other (describe: ________________________) 5

C
CONTINUED

65. During October 1976, did you work for the school you were attending?

(Circle one.)

Yes, working for pay 1
Yes, working off cost of tuition, housing, or meals 2
Yes, both of the above 3
No 4

SCHOOL ATTENDANCE IN OCTOBER 1975

66. Now please think back to Fall 1975. Were you taking classes or courses at any school during the month of October 1975?

No .... 1 *GO TO Q. 79, p. 17*
Yes, at the same school I attended in October 1976 and reported above in Q. 53 .... 2 *GO TO Q. 70*
Yes, at a school I have not yet reported .... 3 *CONTINUE WITH Q. 67*

67. What is the exact name and location of the school you were attending in October 1975? (Please print and do not abbreviate.)

School Name: ____________________

City: ____________________ State: ____________

68. What kind of school is this?

(Circle one.)

Vocational, trade, business or other career training school 1
Junior or community college (two-year) 2
College or university (four years or more) 3
Independent graduate or professional school (medical, dental, law, theology, etc.) 4
Other (describe: ____________________) 5

69. When did you first attend this school? ____________ (month) ____________ (year)

70. During October 1975, were you classified by this school as a full-time student?

Yes 1
No 2
Don't know 3

71. During October 1975, about how many hours a week did your classes meet in the subjects or courses in which you were enrolled? Include time in lectures, shop, laboratories, etc.

____________ hours per week

C

CONTINUED

85. a. Estimate how well you have done in all of your coursework or programs since high school and until October 1976. Do not include grades from graduate or professional school. (Circle one number in Column A.)

b. Estimate how well you have done in your coursework or programs only in the 2-year period from October 1974 through October 1976. Do not include grades from graduate or professional school. (Circle one number in Column B.)

	A. From High School to October 1976	B. October 1974-October 1976
Mostly A (3.75-4.00 grade point average)	1	1
About half A and half B (3.25-3.74 grade point average)	2	2
Mostly B (2.75-3.24 grade point average)	3	3
About half B and half C (2.25-2.74 grade point average)	4	4
Mostly C (1.75-2.24 grade point average)	5	5
About half C and half D (1.25-1.74 grade point average)	6	6
Mostly D or below (less than 1.25)	7	7
Have not taken any courses for which grades were given	8	8

86. Considering all of the schools you have attended since high school, do ANY of these schools or programs give credits which can be used for a 4-year college Bachelor's degree?

I don't know ....1 } *GO TO Q. 88*
No..............2 }
Yes ....3 *CONTINUE WITH Q. 87* ⟶

87. Since leaving high school, about how many credits had you earned by October 1976? (Write in.)

______Number of quarter hours
______Number of semester hours
______Number of other type of credits (specify type: ________________)

SCHOOL FINANCES FROM FALL 1974 THROUGH SUMMER 1976

The following questions ask about your school finances for the two time periods of (a) Fall 1974 through Summer 1975 and (b) Fall 1975 through Summer 1976. Please make sure you answer each question for both time periods. If you are unsure about the actual dollar amount for a particular item, give your best estimate.

88. Were you in school at any time during either of the twelve-month periods from (a) Fall 1974 through Summer 1975 or (b) Fall 1975 through Summer 1976?

(a) Fall 1974 - Summer 1975

Yes ....1 How many months? ______
No..............2

(b) Fall 1975 - Summer 1976

Yes ....1 How many months? ______
No..............2

C
CONTINUED

89. Considering the two time periods of (a) Fall 1974 through Summer 1975 and (b) Fall 1975 through Summer 1976, what is your estimate of how much it cost for you to live and go to school, regardless of who paid? Estimate the amounts and record them below. Enter a zero, "0," where you had no expenses. Do not include costs after Summer 1976. Record your expenses for the time you were in school only.

	(a) Fall 1974 - Summer 1975	(b) Fall 1975 - Summer 1976
Tuition and fees	$ ______	$ ______
Books and supplies	$ ______	$ ______
Transportation to and from class from where I lived while attending school	$ ______	$ ______
Other school-related expenses	$ ______	$ ______
Housing and meals while enrolled in school	$ ______	$ ______
All other expenses while enrolled in school: medical, dental expenses, debt payments, insurance, taxes, child care, etc.	$ ______	$ ______
HOW MUCH MONEY IS THIS IN TOTAL?	$ ______	$ ______

SCHOLARSHIPS, FELLOWSHIPS, GRANTS, AND BENEFITS

90. Considering the two time periods of (a) Fall 1974 through Summer 1975 and (b) Fall 1975 through Summer 1976, did you receive any kind of scholarship, fellowship, grant, or benefits to go to school?

No	1	*GO TO Q. 92*
Yes, Fall 1974 - Summer 1975	2	*CONTINUE WITH Q. 91*
Yes, Fall 1975 - Summer 1976	3	
Yes, both of these periods	4	

91. Estimate the amounts for each scholarship, fellowship, grant, or benefit you received, and record them below. Enter a zero, "0," where you received no financial assistance. Do not include loans.

	(a) Fall 1974 - Summer 1975	(b) Fall 1975 - Summer 1976
a. Basic Educational Opportunity Grant	$ ______	$ ______
b. Supplemental Educational Opportunity Grant	$ ______	$ ______
c. College scholarship or grant from college funds	$ ______	$ ______
d. ROTC scholarship or stipend	$ ______	$ ______
e. Nursing Scholarship Program	$ ______	$ ______
f. Social Security Benefits (for students 18-22 who are children of disabled or deceased parents)	$ ______	$ ______
g. Veterans Administration War Orphans or Survivors Benefits Program	$ ______	$ ______
h. Veterans Administration Direct Benefits (GI Bill)	$ ______	$ ______
i. State scholarship	$ ______	$ ______
j. Other scholarship or grant (write in: ______)	$ ______	$ ______
TOTAL DOLLAR VALUE	$ ______	$ ______

LOANS

92. Considering the same two periods from (a) Fall 1974 through Summer 1975 and (b) Fall 1975 through Summer 1976, did you receive a loan to go to school?

No	1	*GO TO Q. 94, next page*
Yes, Fall 1974 - Summer 1975	2	*CONTINUE WITH Q. 93*
Yes, Fall 1975 - Summer 1976	3	
Yes, both of these periods	4	

C
CONTINUED

SCHOOL FINANCES FROM FALL 1976 THROUGH SUMMER 1977

98. Are you or will you be in school at any time from Fall 1976 through Summer 1977?

No .... 1 } *GO TO Q. 101*
Don't know .... 2 }
Yes .... 3 *CONTINUE WITH Q. 99*

99. What is your estimate of how much it will cost for you to live and go to school this year, regardless of who pays? Estimate your expenses and record them below. Enter a zero, "0," where you expect no expenses.

Tuition and fees	$ ________
Books and supplies	$ ________
Transportation to and from class from where I live while attending school	$ ________
Other school-related expenses	$ ________
Housing and meals while enrolled in school	$ ________
All other expenses while enrolled in school: medical, dental expenses, debt payments, insurance, taxes, child care, etc.	$ ________
HOW MUCH MONEY IS THIS IN TOTAL?	$ ________

100. How are you meeting (or planning to meet) these expenses? Estimate the amounts you expect to receive from each source and record them below. Enter a zero, "0," where you expect no money.

	Amount will receive from each source
Grant	$ ________
Fellowship	$ ________
Scholarship	$ ________
Loan	$ ________
Teaching or research assistantship	$ ________
Job other than assistantship	$ ________
Spouse's income	$ ________
Savings	$ ________
Parents	$ ________
Other relatives or friends	$ ________
Other (specify: ____________________)	$ ________

GRADUATE OR PROFESSIONAL SCHOOL

101. Have you received a Bachelor's degree from a four-year college or university?

No .... 1 *GO TO Q. 108, p. 24*
Yes .... 2 *CONTINUE WITH Q. 102*

102. Did you formally apply for admission (fill out a form and send it in) to any graduate or professional school at any time before October 1976?

No .... 1 *GO TO Q. 104, next page*
Yes .... 2 *CONTINUE WITH Q. 103*

D

SECTION D: MILITARY SERVICE

118. Since October 1974, have you served in the Armed Forces, or a Reserve or National Guard Unit?

(Circle one.)

No .... 1 } *SKIP TO SECTION E, next page*
Yes, National Guard or Reserves but not active duty .... 2 }
Yes, active duty .... 3 *CONTINUE WITH Q. 119*

119. In which branch of the Armed Forces did you serve? (Write in): ____________

120. When did you begin active duty? ________ (month) ________ (year)

121. Have you received (or are you receiving) four or more weeks of specialized schooling while in the Armed Forces?

No .... 1 *GO TO Q. 123*
Yes .... 2 *CONTINUE WITH Q. 122*

122. What is the name of the specialized schooling program in which you spent the longest period of time? (Please print and do not abbreviate.)

Name of program: ____________

123. Specify your current primary military specialty code (Army-MOS, Air Force-AFSC, Marines-MOS, Navy-NEC). (Please print and use standard abbreviations.) Specialty Code: ________

124. What is the highest pay grade you have held?

Pay grade: ________

125. Have you taken any courses while in the Armed Forces that:

(Circle one number on each line.)

	Yes	No
Prepared you for the high school equivalency test?	1	2
Prepared you for equivalency tests that can be taken for college credit?	1	2
Were college-sponsored courses which gave college credits?	1	2

126. Are your currently on active duty?

No (Date left: ________ month ________ year) .... 1 *SKIP TO SECTION E, next page*
Yes .... 2 *CONTINUE WITH Q. 127*

127. How long do you expect to be on active duty in the Armed Forces?

(Circle one.)

For a two-year tour of duty only 1
For a three- or four-year tour of duty 2
For more than one enlistment, but less than a full career 3
For a full career (20 years minimum) 4
Have not decided 5

128. What do you plan to do when you get out of the Armed Forces?

(Circle one number on each line.)

	My Plans	NOT My Plans
Full-time or part-time work	1	2
Graduate or professional school, either full-time or part-time	1	2
College, either full-time or part-time	1	2
Technical, vocational, or business or career training school, either full-time or part-time	1	2
Registered apprenticeship or on-the-job training program	1	2
Retire	1	2
Undecided	1	2
Other (specify: ____________)	1	2

START

SECTION E: FAMILY STATUS

129. What was your marital status, as of the first week of October 1976?

(Circle one.)

Never married, but plan to be married within the next 12 months	1	*GO TO Q. 137, next page*
Never married, and don't plan to be married within the next 12 months	2	
Divorced, widowed, separated	3	*CONTINUE WITH Q. 130*
Married	4	

130. What was the date of your marriage?

____________ (month) ________ (year)

131. As of the first week of October 1976, what was your husband or wife doing?

(If you were not married in the first week of October 1976, check here ☐ and go to Q. 136, next page.)

(Circle as many as apply.)

Working for pay at a full-time or part-time job	1
Enrolled in graduate or professional school	2
Taking academic courses at a two- or four-year college	3
Taking vocational or technical courses at any kind of school or college (for example, vocational, trade, business, or other career training school)	4
On active duty in the Armed Forces (or service academy)	5
Homemaker	6
Temporary layoff from work, looking for work, or waiting to report to work	7
Other (describe: ____________)	8

132. Please describe below the job your husband or wife held during the first week of October 1976.

(If your spouse was not working, check here ☐ and go to Q. 135, next page.)

a. For whom did he/she work? (Name of company, business organization, or other employer)
(Write in): ____________

b. What kind of business or industry was this? (For example, retail shoe store, restaurant, etc.)
(Write in): ____________

c. What kind of job or occupation did he/she have in this business or industry? (For example, salesperson, waitress, secretary, etc.)
(Write in): ____________

d. What were his/her most frequent activities or duties on this job? (For example, selling shoes, waiting on tables, typing and filing, etc.)
(Write in): ____________

e. Was he/she:

(Circle one.)

An employee of a PRIVATE company, bank, business, school, or individual working for wages, salary, or commissions?	1
A GOVERNMENT employee (Federal, State, county, or local institution or school)?	2
Self-employed in his/her OWN business, professional practice, or farm?	3
Working WITHOUT PAY in family business or farm?	4

E

CONTINUED

133. How many hours did he/she usually work at this job in an average week?

_____________ Hours per week

134. In an average week, approximately how much did he/she earn at this job? (Report his/her gross earnings before deductions. If not paid by the week, please estimate.)

$_____________ per week
(Earnings before deductions)

135. As of October 1976, what was the highest level of education that your husband or wife had attained?

(Circle one.)

Some high school, or less 1
Finished high school 2
Vocational trade or business school
- Less than two years 3
- Two years or more 4

College program
- Less than two years of college 5
- Two or more years of college (including two-year degree) 6
- Finished college (four- or five-year degree) 7
- Master's degree or equivalent 8
- Ph.D., or advanced professional degree 9

136. Now please think back a year to Fall 1975. What was your husband or wife doing in October 1975?
(If you were not-married in October 1975, check here ☐ and continue with Q. 137.)

(Circle as many as apply.)

Working for pay at a full-time or part-time job 1
Enrolled in graduate or professional school 2
Taking academic courses at a two- or four-year college 3
Taking vocational or technical courses at any kind of school or college (for example, vocational, trade, business, or other career training school) 4
On active duty in the Armed Forces (or service academy) 5
Homemaker 6
Temporary layoff from work, looking for work, or waiting tc report to work 7
Other (describe: ______________________________) 8

137. Are you a twin?

Yes 1
No 2

E
CONTINUED

138. a. How many children altogether do you eventually expect to have?

(Circle one.)

0........1........2........3........4........5........6........7........8 or more

b. As of the <u>first week of October 1976,</u> how many children did you have?

(Circle one.)

0........1........2........3........4........5........6 or more

c. When do you expect to have your first (next) child?

(Circle one.)

Don't expect to have a (another) child	1
Within the next year	2
Between one and two years from now	3
Between two and three years from now	4
Between three and five years from now	5
More than five years from now	6
Don't know	7

139. Not including yourself, how many persons were dependent upon you for more than one-half of their financial support as of the first week of October 1976?

(Circle one.)

0........1........2........3........4........5........6 or more

140. As of the <u>first week of October 1976,</u> were you dependent upon your parents, spouse, or any other relatives or friends for more than one-half of your financial support?

(Circle one number on each line.)

	Yes	No
Parents	1	2
Spouse	1	2
Other relatives or friends	1	2

141. What is the best estimate of your income before taxes for (a) <u>ALL OF 1975</u> and for (b) <u>ALL OF 1976?</u> If married, include your spouse's income in the total. Do not include loans. Please make a dollar amount entry on each line. If you did not receive any income from a source, enter a zero, "0."

	(a) Amount Received 1975	(b) Amount Will Receive 1976
Your own wages, salaries, commissions, or net income from a business or farm	$ ______	$ ______
Your spouse's (husband or wife) wages, salaries, commissions, or net income from a business or farm	$ ______	$ ______
Public assistance or welfare (include spouse's)	$ ______	$ ______
Unemployment compensation (include spouse's)	$ ______	$ ______
All other income you and your spouse received (include interest, dividends, rental property income, gifts, scholarships, fellowships, etc.)	$ ______	$ ______
TOTAL INCOME FOR YOU AND YOUR SPOUSE	$ ______	$ ______

APPENDIX B. USE OF BALANCED REPEATED REPLICATION METHOD TO COMPUTE THE VARIANCE OF IMPUTATION-BASED STATISTICS

The balanced repeated replication (BRR) method is a device for estimating the precision of estimates that come from surveys with complex sample designs. In this methodological study, BRR was used to estimate the variance of survey estimates when missing values in the data set were replaced using an imputation procedure. This appendix discusses the statistical theory underlying the BRR method of estimating variances and explains why applying the imputation procedure to the individual half-samples before computing the half-sample estimates results in a variance estimate that accounts for the added variation induced by the imputation procedure. The actual implementation of BRR in this investigation is also discussed.

The balanced repeated replication (or balanced half-sample pseudoreplication as it is sometimes called) was introduced by McCarthy (1966) as a method for estimating the variance of survey estimates, including the more complex statistics such as ratio estimates and regression coefficients for which analytical expressions for the variance are not readily available in the literature. The BRR method was specifically developed for the common survey design of two replicates per strata. In this situation, a total of 2^N half-samples may be formed, with each half-sample containing one replicate from each stratum, where N is the number of strata. For linear estimators, it has been shown that if K half-samples are independently selected from the entire set of 2^N possible half-samples, then the average squared deviation of the half-sample estimates from the full-sample estimate equals the expectation of the usual variance estimate of the full-sample statistic (McCarthy, 1966). Thus the variance of the full-sample statistic $\hat{\mu}_{\mathrm{F}}$ may be estimated, using the half-sample estimates ($\hat{\mu}_{\mathrm{H}i}$), by

$$\operatorname{var}(\hat{\mu}_{\mathrm{F}}) = \sum_{i=1}^{K} (\hat{\mu}_{\mathrm{H}i} - \hat{\mu}_{\mathrm{F}})^2/K.$$

When all 2^N possible half-samples are used, this estimate of the variance equals the usual full-sample estimate of the variance; the reason for this is that the fluctuations among half-sample variance estimates arise from between-stratum contributions to the estimates that are canceled out when all 2^N possible half-samples are used. McCarthy (1966) demonstrates that one may also eliminate this between-stratum contribution to the variance by constructing a balanced set of half-samples, using orthogonal matrices such as those of Plackett and Burman (1946). The number of half-samples needed to achieve this balance will be a multiple of four and greater than or equal to the number of strata. When the sample design is based on a large number of strata (e.g., the sample design for NLS has 608 strata), it is not economically feasible to use the large number of half-samples that would be required (at least 608 for NLS). In this situation, it is possible to construct a set of K partially balanced half-samples that will yield a more precise estimate of the variance than would have been obtained had K independently selected half-samples been utilized (McCarthy, 1966).

It should be emphasized that in using the BRR technique the average of the half-sample estimates equals the full-sample estimate when a linear statistic is computed. Similarly, the BRR estimate of the variance of a linear statistic equals the usual variance estimate. However, for nonlinear statistics such as ratio estimates of means and proportions, which are the statistics computed in this study, the average of the balanced half-sample means is not strictly equal to the full-sample estimate so that no general claims of unbiasedness can be made for the BRR variance estimate. However, based on the results of some simulation studies, the BRR technique appears to yield relatively unbiased estimates of the variance, which may even be robust (Frankel, 1971).

B.1. Use of BRR to Estimate the Variance of Imputation-Based Statistics

The BRR estimators used in this study to approximate the sampling variance of imputation-based means and proportions have the same formal justification that would apply if no imputation were involved. When the primary sampling units (PSUs) are drawn from primary strata in pairs and with replacement, statistics based entirely on a half-sample composed of one PSU from each stratum are independent of the companion statistics based on the complementary half. Therefore, with independent paired selections of primary units, the average of a half-sample statistic and its complementary estimate has a variance that is estimated unbiasedly by the squared difference of the companion estimators divided by four. With the weighting-class and hot-deck imputations made independently within each half-sample and its complement, this argument continues to hold.

To improve the precision of these pseudoreplication variance approximations, the technique for generating balanced and partially balanced sets of half-samples was developed by McCarthy (1966). Thus by averaging K estimates of the form

$$\mathrm{var}(\hat{\mu}_i) = (\hat{\mu}_{\mathrm{H}i} - \hat{\mu}_{\mathrm{C}i})^2/4,$$

where $\hat{\mu}_{\mathrm{H}i}$ is the estimate obtained by using the ith half-sample and $\hat{\mu}_{\mathrm{C}i}$ the estimate obtained by using the complementary half-sample, one obtains a variance approximation that is approximately $\sqrt{K}$ times as precise as any one of the separate components. For linear sample statistics with no imputation involved, the average of a half-sample statistic and its complementary estimate is equivalent to the usual full-sample estimate based on two PSUs per stratum, so that the half-sample variance estimator previously described also applies to the full-sample estimate.

The real utility of the BRR technique lies in estimating the variance of nonlinear statistics such as means and proportions for which the average of half-sample and complement statistics is not equivalent to the corresponding full-sample ratio. Second-order Taylor series approximations suggest that averages of half-sample ratios should be subject to roughly twice the estimation bias of corresponding full-sample ratios. Therefore, for a rigorous justification of BRR variance estimates, one should report the associated half-sample average along with the pseudoreplication standard error.

The means and proportions based on full-sample ratios and full-sample imputation are, however, not reported in this study, but it should be pointed out that the differences between the full-sample and half-sample statistics are small when compared with the BRR standard errors.

B.2. Implementation of the BRR Procedure in the Present Investigation

For this investigation, a total of 608 half-samples (since there are 608 strata in the NLS sample) would have been required to achieve full balance and eliminate all between-stratum contributions to the variance, but this was impractical since the processing of results from such a large number of half-samples would have been too costly. The following system was used to construct a set of 16 partially balanced half-samples that eliminated some (but not all) of the between-stratum contribution to the variability of variance estimates.

First, a set of 15 superstrata was formed that contained approximately the same number of strata, using the following procedure. The strata were sorted first according to type of school (low versus high socioeconomic status), then size of school, type of control (public versus private), geographic region, and, finally, proximity to a college or university. The first 300 strata were regions with low socioeconomic status (SES) and the second 300 were regions with high SES. The final eight strata were composed of schools that had not been listed on the frame for the base-year sample. In creating the superstrata, seven were formed from the first 300 strata with low SES. The sorted group of low SES strata were partitioned so that the first 43 strata were designated as superstratum number 1, the second 43 strata as superstratum number 2, and so on, until the seventh superstratum contained the remaining 42 strata. The remaining eight superstrata were selected from the 300 strata with high SES and the eight strata of supplemental schools in a similar manner with the fifteenth superstratum containing the last 31 strata with high SES from the sorted file and the eight strata containing supplemental schools.

These 15 superstrata were then used in the BRR procedure to create the 16 half-samples. Since the BRR procedure is designed for two replicates per stratum, the following adaptation of the procedure was needed for this investigation. Most of the NLS strata contain two schools but a few have only one school because of nonresponse. Also, some schools did not participate in the base-year survey, so substitutions were made for those schools. Subsequently, some students from the schools that did not participate in the base-year survey were included in the follow-up surveys, so that some of the NLS strata contained three or four schools. To partition the schools into two replicates, the file was first sorted by final stratum and then by school. Within each superstratum, the schools were numbered as they occurred in the file. Within each superstratum, the even-numbered schools were regarded as the first "replicate" and the odd-numbered schools as the second "replicate." A balanced repeated replicate design for 15 strata, which required 16 half-samples, was then used, in which all of the even-numbered schools from a superstratum were included in the half-sample if the BRR design specified that the first replicate from that stratum was to be used, and the odd-numbered schools otherwise.

APPENDIX C. STDERR: STANDARD ERRORS PROGRAM FOR SAMPLE SURVEY DATA

An RTI computer program (STDERR) was used to estimate the proportions, means, and their sampling errors for the 20 selected critical items and domains. Since the sample size for each domain is not fixed, it is necessary to compute the estimated proportions $\hat{P}_d$ and means $\hat{m}_d$ as the ratios of two random variables.

For the ratio estimate of a proportion, the numerator estimated the total number of students in domain d who chose the particular response option and the denominator estimated the total number of students in the domain. Explicitly,

$$\hat{P}_d = 100 \frac{\sum_{h=1}^{L} \sum_{i=1}^{a_h} \sum_{j=1}^{b_{hi}} W_{hij} D_{hij} Y_{hij}}{\sum_{h=1}^{L} \sum_{i=1}^{a_h} \sum_{j=1}^{b_{hi}} W_{hij} D_{hij}} = 100 \frac{\hat{Y}_{(d)}}{\hat{D}_{(d)}},$$

where

L is the number of final strata,
a_h the number of sample schools in final-stratum h,
b_{hi} the number of sample students in school hi,
W_{hij} nonresponse adjusted weight for student j in school i from stratum h,

$$D_{hij} = \begin{cases} 1 & \text{if student } hij \text{ belonged to domain } d, \\ 0 & \text{otherwise,} \end{cases}$$

$$Y_{hij} = \begin{cases} 1 & \text{if student } hij \text{ answered the questionnaire item in a specified manner (e.g., yes),} \\ 0 & \text{otherwise,} \end{cases}$$

$\hat{Y}_{(d)}$ is the estimated number of students who answered the questionnaire item in a specified manner for domain d, and
$\hat{D}_{(d)}$ the estimated number of students in domain d.

The standard error of $\hat{P}_d$ is estimated by the square root of $\text{var}(\hat{P}_d)$, where (see Woodruff, 1971; Cochran, 1977)

$$\text{var}(\hat{P}_d) = (100)^2 \left\{ \sum_{h=1}^{M} a_h \sum_{i=1}^{a_h} \frac{[\hat{Z}_{hi+}(d) - \hat{Z}_{h.+}(d)]^2}{(a_h - 1)} \right.$$
$$+ \sum_{h=M+1}^{L} \frac{A_h - a_h}{A_h} a_h \sum_{i=1}^{a_h} \frac{[\hat{Z}_{hi+}(d) - \hat{Z}_{h.+}(d)]^2}{(a_h - 1)}$$
$$\left. + \sum_{h=M+1}^{L} \frac{a_h}{A_h} \sum_{i=1}^{a_h} \left(\frac{B_{hi} - b_{hi}}{B_{hi}} \right) b_{hi} \sum_{j=1}^{b_{hi}} \frac{[\hat{Z}_{hij}(d) - \hat{Z}_{hi.}(d)]^2}{(b_{hi} - 1)} \right\}$$

and

M is the total number of final strata with schools selected with probabilities proportionate to size,
A_h the total number of schools in sampling frame for final stratum h,
B_{hi} the total number of senior students enrolled in school hi,
$\hat{Z}_{hij}(d) = W_{hij}(\hat{Y}_{hij} - \hat{P}_d)/\hat{D}_{(d)}$ the weighted Taylorized deviation,
$\hat{Z}_{hi+}(d) = \sum_{j=1}^{b_{hi}} \hat{Z}_{hij}(d)$ the weighted deviation totals by school,
$\hat{Z}_{h.+}(d) = \sum_{i=1}^{a_h} \hat{Z}_{hi+}(d)/a_h$ the average school totals by stratum,
$\hat{Z}_{hi.}(d) = \hat{Z}_{hi+}(d)/b_{hi}$ the average weighted deviations by school, and

$$\hat{D}_{(d)} = \sum_{h=1}^{L} \sum_{i=1}^{a_h} \sum_{j=1}^{b_{hi}} W_{hij} D_{hij}.$$

The expression for $\text{var}(\hat{P}_d)$ assumes, without loss of generality, that the final strata have been reordered so that the first M are those in which schools are selected with unequal probabilities proportional to size (PPS). The first term in the expression for $\text{var}(\hat{P}_d)$ represents the cluster estimate of variance for the PPS strata, the second term represents between-cluster estimate of variance using the first-stage finite population correction factor for the strata where schools are selected with equal probabilities, and the third term represents the appropriate proportion of the within-school variance for the equal-probability strata. This expression gives an unbiased estimate of the variance for the equal-probability final strata and a slight overestimate of the variance for the PPS strata. Also, it should be noted that $\hat{Z}_{hij}(d)$ is the Taylorized deviation of the ratio estimate $\hat{P}_d$. Such approximations are valid for large samples.

The estimated mean $\hat{m}_d$ for domain d is also obtained as the ratio of two random variables. Explicitly,

$$\hat{m}_d = \frac{\sum_{h=1}^{L} \sum_{i=1}^{a_h} \sum_{j=1}^{b_{hi}} W_{hij} D_{hij} Y_{hij}}{\sum_{h=1}^{L} \sum_{i=1}^{a_h} \sum_{j=1}^{b_{hi}} W_{hij} D_{hij}},$$

where Y_{hij} now represents the value of the quantitative variable under consideration for student hij and the other quantities are as defined in the previous formula. The standard error of $\hat{m}_d$ is the square root of the variance of $\hat{m}_d$. The variance estimator for $\hat{m}_d$ has the same form as that presented for the sample proportion $\hat{P}_d$, with $\hat{m}_d$ replacing $\hat{P}_d$ in the definition of the Taylorized deviation $\hat{Z}_{hij}(d)$.

APPENDIX D. MAXIMIZING PARTICIPATION AND RESPONSE

The NLS population is young, highly mobile, and involved in decisions about future work and life patterns. Continued participation in the NLS depends largely on the success of the contractor in keeping in touch with, motivating, and developing rapport with individuals in the survey population. Materials used by RTI and NCES to increase response included newsletters, thank-you letters, prompting "blue fliers," mailgrams, and reminder post-cards. Additional methods involved telephone tracing of sample members whose newsletters and/or questionnaires were returned as undeliverable, prompting telephone calls to nonrespondents, and field interviews with mail nonrespondents.

Response to the mail questionnaire will become more crucial and costly each year with each survey. For this third follow-up, a $3.00 incentive payment was mailed with each questionnaire. This payment, in addition to the letter, postcard, and telephone contacts, was designed to produce a mail response rate of at least 60%; the remaining 40% were to be individually interviewed by RTI field personnel.

The target population for the third follow-up survey consisted of 22,135 sample members. Data collection activities took place from August 1976 through June 1977. Newsletters were developed and mailed to all sample members with good addresses on file, not only to encourage participation but also to use as a vehicle for updating names and addresses. When mail was returned by the postal service as undeliverable, telephone tracing procedures were employed to obtain current addresses where possible.

During October and November 1976, third follow-up questionnaires, each with a $3.00 incentive check attached, were mailed to the last known addresses of all NLS sample members. This was followed by a planned sequence of thank-you/reminder postcards, prompting postcards or mailgrams, and additional questionnaire mailings. Active mail-return efforts continued through December 1976; by early January 1977, the questionnaire return rate by mail was 70%. Questionnaires continued to arrive through the mail during the field-interview phase of the survey.

In January 1977, the names and addresses of 5060 sample members who failed to mail back their questionnaires were turned over to the RTI field staff

TABLE 10

Telephone Tracing Cases, 1976–1977

Task	*Total cases assigned*	*Cases successfully completed*		*Unable to contact*		*Removed from active file*		*Chargeable operator hours*	*Average hours per case*	*Chargeable telephone calls*	*Average calls per case*
		Number	*Percentage*	*Number*	*Percentage*	*Number*	*Percentage*				
Undeliverable newsletters	2808	2518	89.7	246[a]	8.8	44	1.6	1137	.50	5632	2.01
Undeliverable questionnaires	1426	1194	83.7	180[b]	12.6	52	3.6	722	.50	2622	1.84
Totals	4234	3712	87.7	426	10.1	96	2.3	1859	.50	8254	1.95

[a] Of these, 57 had responded to either first or second follow-up and were sent to the field for further tracing. No attempt was made to contact the 189 members who had not responded since base year.

[b] Of these, 143 had responded to either first or second follow-up and were sent to the field for further tracing. No further attempt was made to contact the 37 members who had not responded since base year.

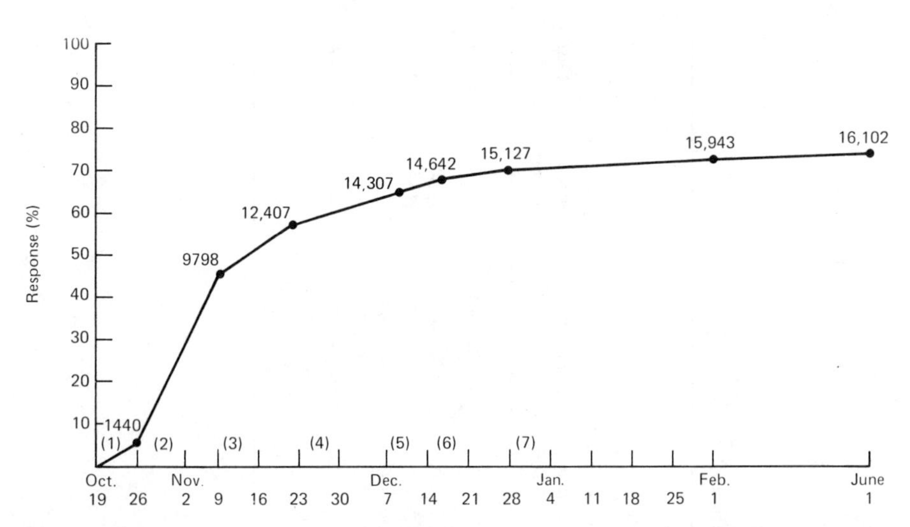

Fig. 1. *Cumulative response to all mailout types, 1976–1977. (1) Initial questionnaire mailout, (2) thank-you/reminder postcard, (3) second questionnaire mailout, (4) first prompting mailgram, (5) third questionnaire mailout, (6) second prompting postcard, and (7) final mailgram.*

for personal interview. This personal-interview phase of third follow-up data collection continued through April 1977, at which time the overall response rate had been increased to 92.1%, that is, 20,092 respondents out of 22,135 targeted sample members.

Figure 1 and Tables 10–14 summarize the results of the third follow-up for the 1976–1977 period involved.

Table 12 gives costs required to obtain response from sampled units according to level of effort expended. Thus the overall average expenditure per

TABLE 11

Overall Mail Returns and Response Rates, 1976–1977

Date	Action	*Number mailed*	*Elapsed days*	*Questionnaires received*	*Cumulative total*	*Response rate*
October 19	Initial questionnaire	21,481	—	—	—	—
October 26	Initial questionnaire	326	—	—	—	—
October 26	Thank-you/reminder postcard	21,481	7	1,440	1,440	6.60
November 9	Second questionnaire	11,982	21	8,358	9,798	44.93
November 22	First prompting mailgram	8,724	34	2,609	12,407	56.89
December 9	Third questionnaire	6,410	51	1,900	14,307	65.61
December 16	Second prompting postcard	5,668	58	335	14,642	67.14
December 28	Final mailgram	5,315	70	485	15,127	69.37
February 2	—	—	106	816	15,943	73.11
March 7	—	—	139	133	16,076	73.72
April 1	—	—	164	9	16,085	73.76
May 4	—	—	197	15	16,100	73.83
June 1	—	—	225	2	16,102	73.84

TABLE 12

Estimated Effect and Cost of the Initial Questionnaire Mailout and Various Prompting Actions

Action	*1 Number of mailings*	*2 Estimated direct cost per item*	*3 Total cost*	*4 Number of returns*	*5 Cost per return*	*6 Response rate (%) (number of returns ÷ 21,807)*	*7 Cumulative response (%) (total returns ÷ 21,807)*	*Cumulative cost for the unit*	*Cost ($)*
1. First questionnaire mailout	21,807	1.15	29,439[a]	10,882	2.71	50.0	10,832 (50)	1.35	14,691
2. Thank-you/reminder postcard	21,807	.20							
3. Second questionnaire mailout	11,982	1.15	13,779	2,841	4.85	13.0	13,723 (63)	2.50	7,103
4. First prompting mailgram	8,724	1.37	11,952	893	13.38	4.1	14,616 (67)	3.87	3,456
5. Third questionnaire mailout	6,410	1.15	7,372	348	21.18	1.6	14,964 (68.6)	5.02	1,747
6. Second prompting postcard	5,668	.20	1,134	330	3.44	1.5	15,294 (70.1)	5.22	1,723
7. Final mailgram	5,315	1.39	7,388	808[b]	9.14	3.7	16,102 (73.8)	6.61	5,341
									34,061
								6.61	37,710

[a] Total cost for items 1 and 2 is the sum of the estimated direct costs for those items × the number of mailings.
[b] The number of responses credited to the item 7 includes all responses received from January 2 through June 1.

TABLE 13

Results of Field-Interview Phase of NLS Third Follow-Up Survey

	Good address cases	*No address cases*[a]	*Total*
Interviewed			
Interviewed in person	3451	88	3539
Interviewed via phone	437	14	451
Total interviewed	3888	102	3990
Chargeable noninterviews			
Refused	451	6	457
Unable to contact	320	86	406
Total chargeable noninterviews	771	92	863
Percentage of assigned cases	15.9	46.0	17.0
Nonchargeable noninterviews			
Out of contry	152	6	158
Deceased	13	0	13
Institutionalized	5	0	5
Handicapped	6	0	6
Phone case—no phone	5	0	5
Phone case—unlisted number	0	0	0
Other	20	0	20
Total nonchargeable noninterviews	201	6	207
Percentage of assigned cases	4.1	3.0	4.1
Cases assigned[b]	4860	200	5060
Chargeable cases assigned	4659	194	4853
Response rate (%)			
Overall (total interviewed ÷ cases assigned)	80.0	51.0	78.9
Chargeable[c] (total interviewed ÷ chargeable cases assigned)	83.5	52.6	82.2

[a] These were cases returned to RTI by the Postal Service as undeliverable and RTI's Telephone Tracing Department was unable to determine the sample member's address.

[b] After deletion of late mail returns.

[c] Nonchargeable noninterview cases were excluded in computing the chargeable response rate.

TABLE 14

Results of Premachine Editing of Third Follow-Up Questionnaires

	Mail response		*Personal interview*	
Edit result	*Number*	*Percentage*	*Number*	*Percentage*
Passed	5,970	36.4	3272	82.0
Failed	10,132	61.8	718[a]	18.0
Duplicate questionnaire removed from the edit process	206	1.3		
Other questionnaires removed from the edit process (blanks, ineligible respondents, etc.)	79	0.5		
Total	16,387	100.0	3990	100.0

[a] All questionnaires completed by personal interviews in the field were subjected to a field edit prior to shipment to RTI. Not included in the field edit were cross-year checks that could be made only at RTI. These cross-year edit checks were the reason for over half the 718 fail edit questionnaires found, reducing the actual personal interview edit rate to less than 10%.

mail respondent is $34,061/16,102 = $2.21. Some of the 5075 nonrespondents (those for whom valid addresses were available) were subjected to personal-interview procedures in a final attempt to convert them to respondents, although the additional costs expended are not included in this report. (A total of $6.61 per unit had been spent prior to the personal-interview attempts.)

This presentation shows the expense involved in trying to convert nonrespondents to respondents; indeed, a large portion of the total cost is spent on the "hard-core" nonrespondents who remain nonrespondents.

APPENDIX E. WEIGHT CALCULATIONS

E.1. Introduction

The sample for the NLS study is highly stratified, multistaged, and clustered. As a consequence of the complex design, each observation (response) must be weighted to obtain unbiased sample estimates of population parameters. For all students sampled, the unadjusted weights were calculated as the inverses of the probabilities of being included in the sample. For several sets of nonrespondents, adjusted weights were calculated, using the weighting-class methods described here.

E.2. Unadjusted Student Weights

First, it was necessary to determine which schools and students were "in sample" for the 1972 NLS project. The NLS sample design included 1200 primary sample schools (two per final stratum) and 21,600 students (eighteen per school). The number of schools was increased (up to three or four per stratum) by using backup schools in the base-year and first follow-up surveys and by obtaining responses from additional primary sample schools in the resurvey. Included in the NLS sample were

1153	participating primary sample schools,
21	nonparticipating (no 1972 seniors) primary sample schools,
131	participating backup sample schools,
18	extra base-year backup sample schools,
16	augmentation sample schools,
1339	NLS sample schools.

The 1978 release tape contains data for students representing 1318 schools—all of the 1339 schools in the final NLS sample except the 21 primary sample schools with no 1972 seniors.

There were 23,451 sample members, of whom 16,683 completed a base-year Student Questionnaire, 21,350 completed a First Follow-Up Questionnaire, 20,872 completed a Second Follow-Up Questionnaire, and 20,092 completed a Third Follow-Up Questionnaire. For each of the 23,451 selected students, the unadjusted student weight $W_{\mathrm{u}hij}$ was calculated as

$$W_{\mathrm{u}hij} = \frac{1}{P_{hi}} \frac{N_{hi}}{n_{hi}},$$

where P_{hi} is the sample inclusion probability for school i of stratum h, N_{hi} the number of senior students in school hi, and n_{hi} the number of sampled students in school hi.

The sum of the unadjusted student weights is an estimate of the total number of 1972 high school seniors in the population. If all of the selected students had completed the survey instruments, these weights would be appropriate for the analyses of student data.

E.3. Nonresponse Adjustment Methodology

Handling nonresponse when analyzing survey data is a problem. In general, the mean values of most variables are different for respondents and nonrespondents. If the differences are large or if the survey response rates are low, adjustments are used in an attempt to reduce the bias due to nonresponse.

A weighting-class method was used to adjust the NLS student weights for questionnaire nonresponse but not for item nonresponse within completed questionnaires. Different response rates for students in different weighting classes were reflected in the adjustments. The method involved partitioning the entire sample (respondents and nonrespondents) into weighting classes—homogeneous groups of students with respect to the survey classification variables. For the weighting-class adjustment procedure to be most effective, the classifier variable values used to construct classes must be available for a very large proportion of respondents and nonrespondents alike.

Once the weighting classes have been defined, the weight adjustment procedure involves simply calculating each sample student's adjusted weight as

$$W_{hij(k)} = W_{\mathrm{u}hij} L_{hij} \frac{\sum_{hij} K_{hij} W_{\mathrm{u}hij}}{\sum_{hij} K_{hij} L_{hij} W_{\mathrm{u}hij}},$$

where

$$K_{hij} = \begin{cases} 1 & \text{if student } hij \text{ was assigned to weighting class } k, \\ 0 & \text{otherwise,} \end{cases}$$

$$L_{hij} = \begin{cases} 1 & \text{if student } hij \text{ completed the survey instrument,} \\ 0 & \text{otherwise,} \end{cases}$$

and $W_{hij(k)}$ is the adjusted student weight for student j belonging to the weight class k of stratum h.

Thus the unadjusted weights for all respondent students in a weighting class are simply multiplied by the ratio of the sum of weights for the weighting class for both respondents and nonrespondents to the sum of the respondents' weights for the weighting class. The adjusted weights for all nonrespondents are set equal to zero.

The nonresponse adjustments for the Third Follow-up Survey were analogous to those employed in prior surveys. The same core of unadjusted weights for the entire NLS sample (23,451 sample members) and the previous computational scheme were used. Since no significant classifier data were obtained, the same weighting classes derived in the second follow-up were employed for the third follow-up analyses.

E.3.1. Assembling Classifier Variable Data

The same five classifier variables used in previous follow-ups were used in defining the third follow-up weighting classes. These were

Race: white or nonwhite;
Sex: male or female;
High school curriculum: general, academic, or vocational/technical;
High school grades: B or better or C or below; and
Parents' education: less than high school graduate, high school graduate, some beyond high school, or college graduate. If available, father's education was used; otherwise, mother's education was used.

After several years of survey activity, there were adequate response sources for each of the classifier variables. The following source priorities were used:

Race
(1) Base-Year Student Questionnaire (Q84),
(2) First Follow-Up Questionnaire (Form B, Q95),
(3) Second Follow-Up Questionnaire (Q28);

Sex
(1) Base-Year Student Questionnaire (page 1),
(2) First Follow-Up Questionnaire,
(3) Inferred from name on student rosters,
(4) Second Follow-Up Questionnaire;

High school curriculum
(1) Base-Year Student Questionnaire (Q2),
(2) First Follow-Up Questionnaire (Form B, Q86),
(3) Activity State Questionnaire (Q13),
(4) SRIF (Q7);

High school grades

(1) SRIF (Q1),
(2) Base-Year Student Questionnaire (Q5),
(3) First Follow-Up Questionnaire;

Parents' education

(1) Base-Year Student Questionnaire (Q90A),
(2) First Follow-Up Questionnaire (Q78A),
(3) Activity State Questionnaire (Q14A),
(4) Base-Year Student Questionnaire (Q90B),
(5) First Follow-Up Questionnaire (Q78B),
(6) Activity State Questionnaire (Q14B).

Table 15 shows the number of students for whom a value was ascertained for each of the five classifier variables. The parents' education variable had the highest number of missing values, followed by race, grades, curriculum, and sex. Values of all five classifier variables were determined for over 92% of the 23,451 sample students, and at least three of the classifier variable values were determined for 98.5% of the sample students (Table 16). Because of the missing classifier data, an "unavailable" category had to be included for each of the five classifier variables. Classifier data was not altered by third follow-up responses, although information on the sex variable was obtained for 120 cases for which this variable was previously unavailable. This new information may be used to update the sex composite variable. Altogether, the number of missing classifiers among follow-up respondents is quite small, as seen in the last column of Table 15.

E.3.2. Forming the Weighting Classes

The next step was to cross-classify the 23,451 sample students by the values of the five classifier variables; a total of 540 cells ($3 \times 3 \times 4 \times 3 \times 5$) was

TABLE 15

Availability of Classifier Variable Values by Variable

Classifier variable	*FFU*[a] *value missing*	*SFU*[b] *value missing*	*Value determined*	*Percentage determined*	*Classifier missing among SFU respondents*
Race	2606	1086	22,365	95.4	20
Sex	196	96	23,355	99.6	2
High school curriculum	2376	206	23,245	99.1	22
High school grades	613	613	23,390	99.7	289
Parents' education	1966	1370	22,081	94.2	252

[a] First follow-up.
[b] Second follow-up.

produced, with counts of sample students and responding students for each cell. (Several respondent counts were made since several different adjusted weights were to be calculated; see Section E.3.3.)

An arbitrary rule that each weighting class must contain at least 20 respondents was used to avoid any large weight adjustments. A set of collapsing rules had been predetermined for use in combining "similar" cells that contained fewer than 20 respondents. Cells were combined in the following order until each of the combined cells contained at least 20 respondents:

(1) Add "unavailables" randomly to known category for each variable in proportion to marginal weight totals for each known category.
(2) Change curriculum to two levels:
 (a) Add "general" to "vocational/technical."
(3) Change father's education to two levels:
 (a) Add "less than H.S." to "H.S. graduate."
 (b) Add "some beyond H.S." to "college graduate."
(4) Eliminate grades.
(5) Eliminate curriculum.
(6) Eliminate father's education.
(7) Eliminate sex.
(8) Eliminate race.

It turned out that virtually all of the cells involving one or more "un-

TABLE 16

Availability of Classifier Variable Values by Number of Variables Determined

Number of classifier variables determined	*FFU[a] number of students*	*SFU[b] number of students*	*Percentage of sample students*
5	19,783	21,717	92.7
4	1,210	685	2.9
3	1,281	681	2.9
2	819	189	.8
1	289	138	.5
0	69	41	.2
Total	23,451	23,451	100.0

[a] First follow-up.
[b] Second follow-up.

available" classifier variables had fewer than 20 responding students. Because of this, the combinations specified in step (1) were done for all of those cells. Step (2) was used seven times and step (3) was used two times. The remaining steps, (4)–(8), were never used.

A total of 87 weighting classes were formed using the procedures described in this section. Each weighting class contained 20 or more respondents for each of the data sets described in the next section.

E.3.3. Calculating Adjusted Student Weights

Once the weighting classes had been determined, the adjusted student weights were computed using the expression displayed at the beginning of Section E.3. Since several different data sets could be derived from the NLS base-year, first, second, and third follow-up data base, a total of eight different sets of adjusted weights were computed.

Table 17 lists the data set and variables appropriate to each set of weights resulting from addition of third follow-up data: W13–W20 (weight sets W1–W12 in this table were developed for base-year, first, and second follow-up data and are described by Levinsohn *et al.* (1978).

There was a total of 21,350 respondents to the First Follow-Up Questionnaire, 20,872 respondents to the Second Follow-Up Questionnaire, and 20,092 respondents to the Third Follow-Up Questionnaire. The weight W13 is appropriate for analysis of items in the Third Follow-Up Questionnaire.

Analyses of change (or transition) variables that were derived from all four surveys—base-year, first, second, and third follow-up items—should be carried out by using the W18 weights. For analyses of change variables that are based on second follow-up items in conjunction with third follow-up items, the W16 items should be used. The W20 weights should be used for combined analyses of items across the three follow-up questionnaires. Other new weighting classes are appropriate for analyses involving various follow-up patterns of response, for example, base-year and third follow-up (W14) or first follow-up and third follow-up (W15). Refer to Table 18 for a listing of the 16 response groups that were determined by choice of the questionnaires the students completed.

For each weight set, the adjusted weights for nonrespondents are zero, and the sum of the respondents' adjusted weights equals the sum of the unadjusted weights for the entire sample. The user should choose the set of adjusted weights that is appropriate to the data set and the variables to be analyzed. The weights are adjusted only for questionnaire nonresponse and not for item nonresponse. The same methods could be used to obtain another set of weights, adjusted for both questionnaire nonresponse and item nonresponse, for any questionnaire item or variable.

TABLE 17

Appropriate Weights for Various Data Sets and Variables

Weight set	*Appropriate data set*	*Appropriate variables*
W1	Respondents to base-year Student Questionnaire	Variables defined from base-year Student Questionnaire items
W2	Respondents to the base-year Student Questionnaire and the First Follow-Up Questionnaire (Form B)	Variables defined from base-year Student Questionnaire items 2, 5, 8, 10, 16, 27, 83, 84, 88, and 91–95
W3	Respondents to First Follow-Up Questionnaire	Variables defined from First Follow-Up Questionnaire items
W4	Respondents to the base-year Student Questionnaire and the First Follow-Up Questionnaire	Change variables defined using items from both the base-year Student Questionnaire and the First Follow-Up Questionnaire
W5	Respondents to either (i) the base-year Student Questionnaire and the First Follow-Up Questionnaire (Form A) or (ii) the First Follow-Up Questionnaire (Form B)	Change variables defined using base-year Student Questionnaire items 2, 5, 8, 10, 16, 27, 83, 84, 88, 91–95 and First Follow-Up Questionnaire items
W6	Respondents to either the base-year Student Questionnaire or the First Follow-Up Questionnaire	Variables with value defined from data available for each student in the release file
W7	Raw unadjusted weight for all students, either respondents or nonrespondents	
W8	Respondents to Second Follow-Up Questionnaire	Variables defined from Second Follow-Up Questionnaire items
W9	Respondents to the base-year Student Questionnaire and the Second Follow-Up Questionnaire	Change variables defined from base-year Student Questionnaire and Second Follow-Up Questionnaire
W10	Respondents to the First Follow-Up Questionnaire and the Second Follow-Up Questionnaire	Change variables defined from First Follow-Up Questionnaire and Second Follow-Up Questionnaire
W11	Respondents to all thee questionnaires	Change variables defined using items from all three questionnaires
W12	Respondents to any of the three questionnaires	Variables with values defined from data available for each student in the release file
W13	Respondents to Third Follow-Up Questionnaire ($n = 20{,}092$)	Variables defined from Third Follow-Up Questionnaire items only
W14	Respondents to the base-year Student Questionnaire and the Third Follow-Up Questionnaire ($n = 14{,}839$)	Variables defined from base-year and Second Follow-Up Questionnaire items
W15	Respondents to the First Follow-Up Questionnaire and the Third Follow-Up Questionnaire ($n = 19{,}358$)	Variables defined from First Follow-Up Questionnaire and Third Follow-Up Questionnaire items

TABLE 17 *(Continued)*

Weight set	*Appropriate data set*	*appropriate variables*
W16	Respondents to the Second Follow-Up Questionnaire and Third Follow-Up Questionnaire ($n = 19{,}611$)	Variables defined from Second Follow-Up Questionnaire and Third Follow-Up Questionnaire items
W17	Respondents to base-year or extended base-year item subset (items 2, 5, 8, 10, 16, 27, 83, 84, 88, 91–95) and first, second, and third follow-up questionnaires ($n = 17{,}993$)	Variables defined from questionnaire items from each of the four survey instruments or extended base-year subset and the three follow-up instruments
W18	Respondents to all four survey instruments, base-year through third follow-up ($n = 14{,}112$)	Variables defined from all items located in all four survey instruments
W19	Respondents to any survey instrument, base-year or first follow-up or second follow-up or third follow-up ($n = 22{,}652$)	Variables defined such that they obtain values regardless of which instrument was returned
W20	Respondents to first, second, and third follow-up questionnaires ($n = 19{,}015$)	Variables defined using items from each of the three follow-up instruments

TABLE 18

Response Groups Defined by Base-Year, First Follow-Up, Second Follow-Up, and Third Follow-Up Responses

	Completed questionnaires received for				
Response group	*Base-year*	*First follow-up*	*Second follow-up*	*Third follow-up*	*Number of students*
I	No	No	No	No	799
II	No	No	No	Yes	120
III	No	No	Yes	No	16
IV	No	No	Yes	Yes	118
V	No	Yes	No	No	309
VI	No	Yes	No	Yes	112
VII	No	Yes	Yes	No	391
VIII	No	Yes	Yes	Yes	4,903
IX	Yes	No	No	No	486
X	Yes	No	No	Yes	18
XI	Yes	No	Yes	No	66
XII	Yes	No	Yes	Yes	478
XIII	Yes	Yes	No	No	504
XIV	Yes	Yes	No	Yes	231
XV	Yes	Yes	Yes	No	788
XVI	Yes	Yes	Yes	Yes	14,112
Total					23,451

REFERENCES

Bailar, J. C., and Bailar, B. A. (1978). Comparison of two procedures for imputing missing survey values. *Proceedings of the American Statistical Association*, San Diego.

Cochran, W. G. (1977). *Sampling Techniques*. New York: John Wiley and Sons.

Frankel, M. R. (1971). An empirical investigation of some properties of multivariate statistical estimates from complex samples. Ph.D. dissertation, University of Michigan.

Hartigan, J. A. (1975). *Clustering Algorithms*. New York: John Wiley and Sons.

Hill, C. J. (1978). A report on the application of a systematic method of automatic edit and imputation to the 1976 Canadian Census. *Proceedings of the American Statistical Association*, Section on Survey Research Methods, San Diego.

Levinsohn, J. R., Henderson, L. B., Riccobono, J. A., and Moore, R. P. (1978). National Longitudinal Study: Base Year, First, Second, and Third Follow-up Data File Users Manual. National Center for Education Statistics, DHEW, Washington, D.C.

McCarthy, P. J. (1966). Replication: An approach to the analysis of data from complex surveys. Vital and Health Statistics, Series 2, No. 14, NCHS, DHEW, Washington, D.C.

Plackett, R. L., and Burman, J. P. (1946). The design of optimum multifactorial experiments. *Biometrika* 33: 305–325.

Woodruff, R. S. (1971). A simple method for approximating the variance of a complicated estimate. *Journal of the American Statistical Association* 66: 411–415.

CHAPTER 9

Readership of Ten Major Magazines*

William G. Madow

1. BACKGROUND

When studies are conducted to estimate the number of readers of magazines, the multistage probability design used is in many respects similar to the design used in federal government surveys, such as the Current Population Reports by the Bureau of the Census. However, in readership and in most consumer marketing surveys, an additional stage of selection occurs after the household has been chosen. This stage involves the selection of a particular person to be interviewed. This step is necessary because the information obtained relates to the individual behavior or attitudes of a particular person. As will soon become apparent from the questioning and interviewing procedures described in this chapter, the pertinent information cannot be supplied "by any responsible adult member of the household" or by a surrogate respondent. This requirement makes the task of achieving a high response rate much more difficult than it is in many government surveys.

The audience levels reported in readership studies depend on the general line of questioning employed and on the specific questions used to define and identify readers of particular magazines within the surveyed population. Shaping and wording questionnaires for this purpose is an exercise in the art of communication for which there are no absolute rules and few common standards to practice. Proper execution at the field level by interviewers is essential.

---

* This case study is based on materials from the Advertising Research Foundation (ARF) (1977) and Audits and Surveys (A&S) (1975). A draft was reviewed by Lester R. Frankel of Audits & Surveys.

INCOMPLETE DATA
IN SAMPLE SURVEYS
Volume 1, Part II

ISBN 0-12-363901-8

No current consensus exists on how this kind of research should be done. In planning for this study, Time, Inc., the sponsor, requested a "return to basics." Thus, Audits & Surveys, Inc., used an early research approach in the critically important procedures used to identify people as readers of particular magazines. The line of questioning for reader qualification and the way magazines were shown to respondents were basically those used by Alfred Politz Research in a series of magazine audience studies conducted and widely distributed in the 1950s and 1960s; these Politz procedures differ from those used more recently by other researchers.

The *magazine audience* for this survey is defined as the average issue audience among adults aged 18 and over for each of four monthly and six weekly major publications (see the accompanying tabulation).

Magazine	*Published*
Better Homes and Gardens	Monthly
Business Week	Weekly
Family Circle	Monthly
Newsweek	Weekly
Reader's Digest	Monthly
Sports Illustrated	Weekly
Time	Weekly
TV Guide	Weekly
U.S. News & World Report	Weekly
Women's Day	Monthly

Two basic types of estimates are derived in audience studies. The first type is the "coverage estimate," which is the estimated percentage of the total population or of a segment of the population that reads an average issue of a magazine. The second type of estimate is the "composition estimate," which shows the distribution of the readers of an average issue of a magazine in terms of demographic and other characteristics. The base for this distribution is the estimated number of readers of the average issue who are in the sample.

The questionnaire was relatively short. In addition to demographic questions, the following items were included:

(1) A filter question was used to identify people who are sure they have not looked into any issue of a particular magazine in the last six or seven months, so that those respondents, and only those respondents, could be excused from the next three steps of questioning about reading that publication.

(2) Presentation was made to eligible respondents of whole copies of each of the magazines into which they might have looked, with a sequence of item interest questions to induce through-the-book inspections of the copies.

(3) A question was asked about the respondent's having previously looked into the inspected issue. This question is the one on which reader qualification was based.

(4) Questions were asked to determine where identified issues had been looked into.

The following sections discuss: sampling procedures (Section 3), questioning procedure (Section 4), respondent cooperation program (Section 5), response rates (Section 6), and estimating procedures (Section 7).

2. SAMPLING PROCEDURES

The sample used in this study was derived from the master sample of Audits & Surveys (A&S), which was selected, using materials from the 1970 Census of Population and Housing. The master sample consists of two replicate samples, i.e., two samples selected by the same procedure, each consisting of counties spread throughout the United States, excluding Alaska and Hawaii. The two replicates contain 241 primary sampling units [a primary sampling unit (PSU) is a Standard Metropolitan Statistical Area or a nonmetropolitan county] that were selected by taking into account census geographic division, population size, and metropolitan versus nonmetropolitan identification.

2.1. Selection of the Primary Sampling Units

2.1.1. The Frame of PSUs

According to the 1970 census, there were, in the coterminous United States, approximately 202 million persons. In creating a frame for sampling persons, the first stage was to define a stratified frame of PSUs. All PSUs in the country were first sorted into the nine U.S. Census Divisions and were then sorted by state. In those instances in which a Standard Metropolitan Statistical Area (SMSA) crossed a state line, the SMSA was considered to be in the state of its principal city.

Within each state, the PSUs were then classified as either Standard Statistical Metropolitan Areas (SMSA) or nonmetropolitan counties. In all divisions of the country, SMSAs are defined in terms of county lines. In the case of New England, in which the SMSAs cut across county lines, the New England Metropolitan State Economic Areas, which incorporate entire counties, were used.

Within each state, metropolitan areas were listed geographically in a serpentine manner. The metropolitan areas for all states within a census division were then assembled, preserving the within-state listing, and all divisions were combined. A similar procedure was followed for the counties outside the SMSAs.

Thus in the frame of the PSUs, the first major sort was by the U.S. Census Division, next was the sort by SMSAs and nonmetropolitan counties, and finally, there was the sort by state.

As a result of this listing, each PSU was placed in a specified position in the frame. Now, if each person in the 1970 census is assigned a number from 1 to 202,000,000 and cumulative subtotals are calculated of the 1970 populations of the PSUs in the order in which they appear in the listing of PSUs, then that number specifies the PSU in which the person resides.

2.1.2. *Selection of the PSUs in the Sample*

Two separate and independent samples of 200 points were drawn on a systematic basis. Since each replicate is identified by 200 starting points, the sampling interval was 1,010,000 (i.e., 202,000,000/200 = 1,010,000). The first replicate was drawn by selecting a random number between 1 and 1,010,000 to designate the first point to be selected and every 1,010,000th point was drawn thereafter. The second replicate was drawn by selecting a second random starting point between 1 and 1,010,000 and drawing every 1,010,000th point thereafter. This method of selecting points and including the corresponding PSUs in the sample ensures that PSUs will be represented in proportion to their population size. The total number of PSUs thus selected was 241, since some PSUs contained more than 1 point. Because of the method of stratification employed, as well as the serialization of the PSUs, the location of each point identifies a specific metropolitan area or nonmetropolitan county.

2.1.3. *Selection of the Second-Stage Sample: Minor Civil Divisions and Census Tracts*

The sample of locations used in this study was selected from counties already in Audits & Surveys' master sample of 241 PSUs in such a manner as to make use of a stratification by income.

The income of the household in which a person lives tends to be highly correlated with his readership of various magazines. It would be extremely productive if this variable could be used in the selection of the sample. However, since this measure is not available for each household and since an area sample is employed, the average or median income of the neighborhood is the next best factor. In metropolitan area counties, the median family income is available on a census tract basis. In nonmetropolitan areas, median income is available for counties. Therefore a disproportionate sample of second-stage units within PSU was selected, making use of the information on median incomes of census tracts of SMSAs and median income for nonmetropolitan counties.

For the second stage of sampling, the basic plan was to select a sample of 400 second-stage units. These would consist of census tracts in metropolitan areas and Minor Civil Divisions (MCD) in nonmetropolitan counties. If a proportionate sample was to be selected, then 276 tracts in 117 metropolitan areas and 124 MCDs in 124 nonmetropolitan counties would be drawn.

To determine how the second-stage sampling units would be distributed by income, MCDs in a county were assigned the median income of the county. To obtain the distribution of income by census tract of residence in the metropolitan area counties, three times as many census tracts and MCDs were selected, with probabilities proportionate to population as were needed for the sample. This made possible a two-phase design in the selection of the second-stage units, so as to obtain optimum allocation by oversampling upper-income secondary sampling units.

The actual and proportionate distribution by type of area and median income level of the 400 secondary sampling units (SSUs) is shown in Table 1. The final sample of the 400 SSUs was located in 108 SMSAs and 98 nonmetropolitan counties of the 241 PSUs in the A & S master sample. The reduction in PSUs resulted from the disproportionate selection of secondary sampling units.

2.1.4. *Selection of Third-Stage Units: Blocks and Rural Segments within Secondary Sampling Units*

Within each of the secondary sampling units, three blocks or rural segments were selected either with probabilities proportionate to size or with equal probabilities, depending on whether information on size was available.

Practically all of the census tracts in the sample fell into what the U.S. census refers to as "urbanized areas." This meant that each block in the tract is shown on a census map and the population as of 1970 is indicated. In these cases it was possible to select a sample of blocks with probability proportionate to size.

TABLE 1

Allocation of Sample of Secondary Sampling Units

All areas (medium income)	*Proportionate allocation*	*Sample allocation*
Metropolitan area tracts	400	400
Under $7000	31	15
$7000–12,999	200	199
$13,000 and over	45	88
Nonmetropolitan counties—minor civil divisions		
Under $7000	39	19
$7000 and over	85	79

In this process every block in the tract had a chance of being included in the sample. If a block had no dwelling units or had fewer than 35, it was combined with the next lower numbered block(s), so that each of these block units contained 35 or more dwelling units as of the 1970 census.

For nontracted areas, where no block data were available, the area was divided into smaller areas, or islands, of approximately equal population, and three areas were selected at random.

Thus in the 400 secondary sampling units, 1200 smaller units, defined by block boundaries in urban places or "islands" in rural areas, were selected. From each of these smaller units, a cluster of seven adjacent units was selected.

2.1.5. Selection of Fourth-Stage Dwelling Unit Segments within Third-Stage Units

A segment of seven contiguous dwelling units was selected within each block or block group selected in the tracts and within each of the selected rural segment islands. Each location (block, block group, or rural segment) was divided into definable segments by interviewers, and then one segment was selected at random. A procedure, to be stated later, was used to eliminate any choice on the part of the interviewer as to which segment was to be included in the final sample.

A special detailed map, showing the location and the streets and roadways surrounding it, was prepared in the New York office. A starting point was designated on the map at a corner or intersection within the location. The map showed a route leading from this point through the location. Before doing any interviewing, the interviewer proceeded along this route, creating segments, until the route around and through the location had been completed. The interviewer recorded the first, eighth, fifteenth, etc., dwelling unit on a grouping sheet, so as to divide the area into segments of approximately seven dwelling units each.

When all the segments had been defined and identified on the grouping sheet, a notation previously put on the sheet specified the one particular segment in which to interview. A column on the grouping sheet provided for a series of "X" marks. These "X" marks were put on the sheets in the home office. After forming the segments, the interviewer selected the segment that was on the lowest line with an "X" opposite its starting address. The pattern of "Xs" was such that each segment had an equal probability of selection regardless of the number of segments in the location.

The interviewer began at the dwelling unit that defined the beginning of the segment and contacted for interview all dwelling units within the segment up to, but not including, the starting point of the next segment. By this procedure, the "half-open interval," all households within the boundaries of the segment were to be interviewed, even though they may have been missed in the counting and creating of the segment. The most critical element in this regard was specifying the beginning and ending of the interval.

In specifying this segment, a dwelling unit was counted by a detailed but visual inspection of each residential structure. For this reason, the actual number of dwelling units in each segment can be determined accurately only when personal contact is made and the existence of separate households determined. The total number of actual dwelling units or households generated by all 1200 locations was 8266 instead of 8400.

2.1.6. Selection of Final Sampling Units: Households and Respondents

Each of the sampled segments that were selected as just described contained an expected number of seven dwelling units. The persons living in each occupied dwelling were defined as a household. If a dwelling unit turned out to be unoccupied, a seasonally occupied unit in which no one was living at the time, or not living quarters but a place of business or professional quarters, the unit was ignored (ineligible or out of scope for the survey). No attempt was made to substitute another dwelling unit.

Within each occupied unit or household, one respondent of 18 years of age or older was selected for interview. For each household, the interviewer had a household listing sheet, half of which, on an alternating basis, were designated as "male" and the other half as "female." For a particular household, the interviewer listed all persons of the designated sex and then, through an × ing pattern on the form, selected one specific person to be interviewed. If there were no members of the designated sex living in the household, that household was skipped. No substitutions were made, and only the person designated by this procedure could be the respondent.

2.1.7. Sampling in Time

Although the interviewing was spread over a limited period of time, it was desired to measure as many different issues of the magazines in question as possible. This meant that different issues of the weekly magazines were used each week and that issues of the monthly magazines had to be changed on a monthly basis.

The experimental design called for six weeks of basic interviewing; three weeks for contacting one-half of the respondents followed by one week for clean-up completion of interviews. This was to be followed by another three weeks of contacts with the other half of the respondents plus a one-week clean-up period.

Since magazine issues were changed on a frequent schedule, it was necessary to ensure that each issue was measured by representative samples of the population. In other words, for each of the six weeks, probability samples of the population were required.

The samples of issues in time and the samples of people in space were coordinated in the following manner. It should be recalled that the master sample was a replicated sample and that each replicate yielded 200 secondary sampling units, which were tracts or minor civil divisions. The two replicates contained 400 SSUs, and within each SSU three blocks or rural segments were selected.

Each replicate of 200 SSUs is in itself a probability sample. Since this sample was derived from a PSU sample selected in a systematic manner, a subsample of the 100 obtained by selecting every other SSU will also yield a probability sample of 100 tracts.

The design used to sample magazine issues in time and space can best be described with the following notation:

(1) Let the two replicates be designated as A and B.

(2) Let the 100 odd-numbered SSUs in each replicate be designated as "O" and the even-numbered SSUs as "E."

(3) Within each SSU, three blocks were selected; these are designated as (1), (2), (3).

Thus twelve probability samples were designated: AO_i, AE_i, BO_i, BE_i, $i = 1, 2, 3$, where, for example, BE_2 consists of the blocks or rural segments designated 2 within the even-numbered SSUs of replicate B. Across time, the samples were utilized as shown in Table 2. During each week, two replicated nationwide

TABLE 2

Sample Utilization by Week

	Replicate A		*Replicate B*	
Week	*Designation*	*Number of blocks*	*Designation*	*Number of blocks*
1	AO_1	100	BE_1	100
2	AE_1	100	BO_1	100
3	AO_2	100	BE_2	100
4[a]	—	—	—	—
5	AE_2	100	BO_2	100
6	AO_3	100	BE_3	100
7	AE_3	100	BO_3	100
8[a]	—	—	—	—

[a] Clean-up week.

samples of 200 blocks each were in operation. Thus each issue was observed on the same basis. During the clean-up weeks, new issues of the weekly publications were used.

3. THE QUESTIONING PROCEDURE

The questionnaire used for this study was relatively short. Given the purpose of the study, the questionnaire covered only the following:

(1) magazine readership for ten magazines,
(2) place of reading,
(3) source of copy, and
(4) five demographic characteristics.

Two forms of the questionnaire (Yellow and Green) were used that were identical in all respects, except that the Yellow form contained a forward alphabetic positioning for the ten magazines and the Green form contained a reverse alphabetic positioning for the ten magazines. To the extent that the order in which magazines are presented affects readership claims, the purpose of the two orders is to balance any such effect.

A readership filter question was asked for each magazine separately as follows: The respondent was shown the name of a magazine—reproduced to duplicate exactly the type size and typeface used for the name (logo) on the front cover of the magazine. This use of totally realistic logos is in keeping with the fact that the entire theory of the readership method used is the "Reader Recognition Technique." The order of presentation of the magazine logos was totally controlled by having them printed in a prestapled "Magazine Cards" booklet of which there were two versions to match the rotational variations in the questionnaires. The filter question determined whether or not a respondent stated that he or she had "Surely," "Probably," "Probably Not," or "Surely Had Not" looked into any issue of the magazine in the past "six or seven months," either at home or away from home.

A respondent "passed the filter question" for a given magazine unless the response to this question was "I surely have not looked into any issue of this magazine in the past six or seven months."

The questions and procedures used to determine whether or not a respondent had read the specific issue of a magazine used for each interview were employed for each magazine for which the respondent had "passed the filter question." These procedures involved the following.

(1) The interviewer leafed through the magazine, showing the table of contents plus the beginning pages of all articles in the magazine, telling the respondent to indicate any "item that looks interesting."

(2) The interviewer circled on the questionnaire the number of all items that "looked interesting" to the respondent.

(3) At the top of each questionnaire page associated with these readership procedures and questions, there were specific, preprinted "Comments" for the interviewer to use while leafing through a magazine if the respondent offered that he or she had or had not seen the magazine before. These comments served to allow the interviewer to proceed in showing the respondent all articles in the magazine before the respondent made a final commitment about readership.

(4) After this reader-interest process was completed for a magazine, the readership question was asked: "Now that we've been through the whole issue, are you sure whether or not you happened to look into this particular issue before!"

(5) The response options to the readership question were: I "Surely" looked into this issue before, I'm "Not sure," or I "Did not" look into this issue before. In this study a reader of a magazine was one who said I "Surely looked into this issue before." All others were classified as nonreaders of a magazine.

(6) The question on where respondents may have looked into the specific issue of a magazine included in the interview followed the readership question for each magazine. This was asked before the interview proceeded to the readership question on other magazines or to the next part of the questionnaire, which covered source of copy.

The question on source of copy was asked of all readers of each magazine. It was also asked of all nonreaders. The question for nonreaders was asked in a

TABLE 3

Interviews by Length of Time

Interview length (minutes)	*Percentage of interviews*
10 or less	2
11–15	10
16–20	13
21–25	19
26–30	18
31–35	10
36–40	9
41–45	9
46–50	4
51–55	2
56–60	2
60 or more	2
Total	100

manner that obtained source-of-copy information for the issue in question if the respondent indicated that it "had come into the home" but the respondent had not looked into it. In any instances in which the respondent—reader and nonreader alike—was not sure or did not know about the source of a particular copy, the respondent was asked to obtain this information from other family members.

The final items in the questionnaire dealt with demographic characteristics. Following the order of the questionnaire, they included the following: size of household, age of respondent, education of respondent, employment status and type of work of respondent, and finally, household income. The income question was always asked of the household head if he or she were available, whether or not he or she happened to be the respondent.

The average length of interview for this study was less than 30 minutes. The distribution of length of interview is shown in Table 3, which is based on a sample of the interviews. Included is a case in which an interviewer, who had to stop in the middle of an interview when the respondent had to leave on an emergency, ended up baby-sitting with a young child for over an hour.

4. RESPONDENT COOPERATION PROGRAM

This study employed established and innovative procedures for maximizing respondent cooperation levels. The extensive and intensive program outlined in this section was undertaken to ensure that the results of the study would include the readership habits of those segments of the population from which it is difficult to obtain interviews.

There are a number of factors that have a negative impact on respondent cooperation in survey research. Some people do not want to be interviewed. Some people are regularly not at home during "normal" times of day. Some people do not speak English. Interviewer training, motivation, and behavior represent another set of factors that can influence respondent cooperation rates. This study addressed itself to all of these factors individually. For example, Spanish-speaking interviewers were used in areas where there was a high likelihood of finding households or respondents who speak only Spanish. In addition, one interview was conducted in French and one in Portuguese.

The interviewer training program involved a review of reasons people give for refusing an interview, such as fear, concern that the interview is a disguised sales effort, or a desire to keep their attitudes and behavior private or confidential. Suggested answers to these concerns were discussed and a document was provided to the interviewer entitled "Points to Help Gain Cooperation."

4.1. Incentives Offered

A graduated respondent incentive program was developed for this study. During the first wave of interviews, which covered the first four weeks of the basic interviewing period, incentives were offered only to respondents who initially refused an interview. This offering was made via long-distance telephone calls from New York to all those who refused an interview and whose telephone number could be obtained.

Because the refusal rate for the first wave was somewhat higher than anticipated, incentives were offered during the second wave to all designated respondents at the time of first contact. During both waves, incentives were offered to doormen in apartment buildings in those instances in which their cooperation was required.

The incentives offered in all instances mentioned had a retail value of approximately $9.00 and ranged from such items as a wall clock and a portable radio to a badminton set. The ten incentive options were shown in a four-color, four-page brochure.

In those instances in which an incentive was offered but an interview was not granted, the next contact with the respondent was accompanied by an offer of more attractive incentives, which had a retail value of approximately $16.00, including such items as a digital clock, a steam iron, and an electric drill. These incentives were also shown in a four-color, four-page brochure.

In 22 key areas, those respondents who continued to be not at home after 12 or more personal calls (and 17 or more telephone calls if their telephone number were available), plus those respondents who had refused to be interviewed on prior occasions, received a personally addressed letter and were offered a cash incentive for granting an interview.

4.2. Call and Callback Schedule

In order to maximize the probability of finding respondents at home, the schedule for original calls and callbacks was concentrated during late afternoon and evening hours. However, calls were also made during earlier periods of the day. Although most personal contacts were made during weekdays, weekend callbacks were also made. Telephone callbacks attempting to make appointments for interviews spanned all parts of the day.

A minimum of 12 personal calls was made, if needed, in an attempt to reach a household or respondent, for either an interview or an appointment for an interview. (The phrase "minimum number of calls" means that before contact efforts were terminated, this number of calls was made in an effort to obtain an interview from a dwelling unit recorded either as "Household-Not-At-Home" or "Respondent-Not-At-Home.") The minimum call schedule by wave is shown in Table 4.

Attempts were made to obtain telephone numbers for all households for which a not-at-home condition existed. This included obtaining names and/or telephone numbers from mailboxes, neighbors, or special directories. When a telephone number was obtained, numerous attempts were made to contact the household by telephone in order to make an appointment for an interview. Such efforts were concentrated during the second clean-up week and the follow-up period. During the follow-up period, as necessary, telephone calls were made three times a day from Saturday or Sunday, June 28th or 29th, through Thursday, July 3rd, for a minimum of 15 calls during this period. On each of these days, one call was made in the morning, one in the afternoon, and one in the evening. All contacts during the follow-up period were made by local supervisors and their assistants.

Under some circumstances, contact efforts substantially exceeded this minimum of 12 personal calls plus extensive telephone contacts when telephone numbers were available. These contact efforts occurred in areas where there was a relatively larger number of "Not-At-Homes" and "Refusals" during the basic interviewing period. In these areas, special efforts were made among both "Not-At-Homes" and "Refusals," as described later in this section of the report.

4.3. Refusal Program

The refusal program developed for this study was based on the premises that some people who refuse an interview do so because the interviewer's request for an interview is untimely, some people develop a resistance to being interviewed

TABLE 4

Number of Calls Made by Wave

	Respondents selected for	
Interviewing period	*Wave I*	*Wave II*
Basic		
Week of first call	4	4
First clean-up week	1	—
Second clean-up week	2	2
Follow-up	6	6
Total	13	12

because of their reaction to an individual interviewer, and some people will grant an interview only if a major incentive is offered.

For all households in which the initial response was a refusal by either the household or the designated respondent, interviewers were asked to obtain household telephone numbers and report the number to Audits & Surveys. Long distance telephone calls were made from New York (and identified as such) in an attempt to obtain both an agreement and an appointment for an interview. In these calls and subsequent contacts with "Refusals," stress was placed on the importance of the study, the need to reflect attitudes of all population segments, including those who prefer not to be interviewed, and the fact that no one can be substituted for a scientifically selected respondent. Also, in these calls and in subsequent contacts among this refusal group, incentives to cooperate were offered. Whenever the long distance calls were successful, the information was immediately forwarded to local supervisors, who confirmed or scheduled a time for the interview.

During the follow-up period, all households or designated respondents who were still "Refusals" were personally recontacted in an attempt to obtain an interview. These recontacts were all made by local supervisors or their assistants in order to maximize the likelihood of converting a refusal into an interview.

This basic program of recontacting refusals in order to obtain an interview was supplemented, as was the callback program, by a number of special procedures described next.

4.4. Supplemental Program of Additional Calls on "Not-at-Homes" and "Refusals"

A supplemental program was conducted in locations or areas where the initial cooperation rate was well below average. There were two parts to the supplemental program. One took place during the last week of the basic interviewing period—the week of June 9th—and the other took place during the final period of all interviewing—the end of July and the beginning of August.

During the final week of the basic interviewing period, special efforts were undertaken by local supervisors and their assistants to increase the yield of interviews in low-yield areas. Personal visits and telephone calls were made to all respondents in these areas who were previously not at home or who had previously refused an interview. This special supervisory effort was concentrated in locations in the following cities: Baltimore, Chicago, Detroit, Los Angeles, New York, Philadelphia, and Pittsburgh.

The final effort was designed to obtain interviews from respondents who were chronically not at home and respondents who had refused an interview on prior occasions. This effort occurred in 22 area, which were again areas with relatively large numbers of "hard-if-not-impossible-to-interview" respondents. The program consisted of a personally addressed letter that stressed the importance of the study and the responses of the recipient and that included the offer of a cash incentive. Shortly after the letter was mailed, local supervisors and their assistants contacted these potential respondents in an attempt to obtain an interview. This terminal effort involved some 613 potential respondents and yielded 241 completed interviews.

The levels and types of effort that many interviewers expended to obtain an interview are captured in a report we received about an interview that was conducted in a hospital two days after the respondent had had minor surgery.

5. RESPONSE RATES

The response rate achieved for this study was 82.5%, with 84.6% for females and 80.1% for males. Table 5 details the method of calculation, as well as the disposition of all dwelling units in the 1200 sample locations.

In three locations, the segment of dwelling units designated for interview fell somewhere in apartment buildings that could not be entered by interviewers, despite persistent efforts and offers of appropriate incentives. The yield of households for these three locations was presumed to be 21 eligible respondents. These 21 "potential respondents" were designated as "Household Refusals."

In one of the 400 PSUs selected for this study, there was such severe flooding at the time of the basic interviewing period that the entire area was designated as a disaster area by city, state, and federal authorities. The dwelling units in these three locations had either been destroyed or were uninhabitable throughout the survey period. The yield of households in this area was presumed to be 21 and they were designated as "Vacant."

In nine other locations, flooding contitions also existed at the time designated for interview. However, homes in these areas were not destroyed and it was possible to conduct the interviewing at a later time toward the end of the survey period.

In yet other situations, the elements were adverse, but they did not have an adverse effect on completion rates. For example, an interviewer in Colorado was stranded in a snowstorm and could not get her car back for two days. Another interviewer had the air let out of all four tires on her car. She called a garage and continued interviewing the area—both on that day and on subsequent days.

In many instances interviewers found designated respondents who could neither read nor be interviewed. Examples included respondents who were blind, respondents who were severely ill for an extended period of time, respondents who were senile, and respondents who had no command of English or Spanish. In this study such respondents were classified as "Respondent Not Available." This, of course, had a negative effect on the completion rate.

TABLE 5

Interview and Noninterviews

	Total	*Male*	*Female*
Number of dwelling units in all sample locations	8266	4090	4176
Less			
Vacant	336	181	155
No adult of the designated sex	946	662	284
Respondent away and not expected back until after the field period is over	93	47	46
Base total	6891	3200	3691
Results of the final call made			
No one at home	291	145	146
Household refused[a]	226	146	80
Respondent not available or not at home	104	64	40
Respondent refused	574	278	296
Lost or not ascertained	13	5	8
Interviewed	5683	2562	3121
Response rate[b]	82.5%	80.1%	84.6%

[a] No one in household was willing to answer the questions on which selection of the respondent is based.

[b] Number of respondents interviewed divided by the base total.

TABLE 6

Effort Made to Complete Interviews in Audit Subsample of 20 Locations

			Interview not completed and not tried again						
Call sequence	*Distinct households attempted*	*Interview completed*	*Total*	*Vacant*	*No male/female*[a]	*HH refusal*	*Respondent refusal*	*HH not home*	*Respondent not home*
1	130	33	8	2	6				
2	89	15	6		4	1	1		
3	68	11	5		3		2		
4	52	13	6		1	1	4		
5	33	4	5		1		2	1	1
6	24	6	4				1	3	
7	14	2	3				2		1
8	9	5							
9	4	1							
10	3	2							
11	1	1							
All households	130	93	37	2	2	2	12	4	2

[a] No adult of specified set.

TABLE 7

Sample Recovery by Income

	Adults						*Males*						*Females*					
		Met			*Nonmet*			*Met*			*Nonmet*			*Met*			*Nonmet*	
	Total	*Under 7,000*	*7,000–12,999*	*13,000+*	*Under 7,000*	*7,000+*	*Total*	*Under 7,000*	*7,000–12,999*	*13,000+*	*Under 7,000*	*7,000+*	*Total*	*Under 7,000*	*7,000–12,999*	*13,000+*	*Under 7,000*	*7,000+*
Housing units for which a call record was returned	8265	326	4134	1796	459	1550	4089	156	2053	881	235	764	4176	170	2081	915	224	786
Established																		
as ineligible	*1375*	*99*	*791*	*189*	*82*	*214*	*890*	*63*	*504*	*133*	*53*	*137*	*485*	*36*	*287*	*56*	*29*	*77*
vacant	336	25	186	30	35	60	181	12	96	23	21	29	155	13	90	7	14	31
no male/female	946	69	559	149	40	129	662	49	384	108	29	92	284	20	175	41	11	37
respondent away until after interview period	93	5	46	10	7	25	47	2	24	2	3	16	46	3	22	8	4	9
No interview completed	*1207*	*32*	*636*	*307*	*42*	*190*	*637*	*15*	*340*	*151*	*31*	*100*	*570*	*17*	*296*	*156*	*11*	*90*
no one home	290	17	163	58	14	38	144	9	84	22	9	20	146	8	79	36	5	18
household refused	226	2	120	64	3	37	146	1	75	39	3	28	80	1	45	25	—	9
respondent not home	104	1	48	23	7	25	64	—	29	15	6	14	40	1	19	8	1	11
respondent refused	574	11	294	162	18	89	278	4	149	75	13	37	296	7	145	87	5	52
lost/not ascertained	13	1	11	—	—	1	5	1	3	—	—	1	8	—	8	—	—	—
Interview completed	5683	195	2707	1300	335	1146	2562	78	1209	597	151	527	3121	117	1498	703	184	619
Computation of base																		
1. HH contacted[a]	7546	279	3739	1698	403	1427	3717	133	1849	834	202	699	3829	146	1890	864	201	728
2. without eligible m/f								49	384	108	29	92		20	175	41	11	37
3. with eligible m/f								84	1465	726	173	607		126	1715	823	190	691
4. % with eligible m/f								63.2	79.2	87.0	85.6	86.8		86.3	90.7	95.2	94.5	94.9
5. no one home								9	84	22	9	20		8	79	36	5	18
6. assumed eligible (4 × 5)								6	66	19	8	17		7	72	34	5	17
7. total HH with eligible m/f (3 + 6)	6851	223	3318	1602	376	1332	3171	90	1531	745	181	624	3680	133	1787	857	195	708
8. completion rate (%)	82.9	87.4	81.6	81.1	89.1	86.0	80.8	86.7	79.0	80.1	83.4	84.5	84.8	88.0	83.3	32.0	94.4	87.4

[a] HH contacted = Housing units recorded − HH(vacant + away + no one home).

As many as twelve personal calls were made before terminating efforts to reach and interview the designated respondent. A schedule had been worked out for each interviewing wave, allowing for three callbacks during the week in which the initial call was made and for an additional call during the clean-up weeks. Finally, up to six more calls were allowed during the eight week follow-up period, as shown in Table 6.[1]

The response rates shown in Table 7 were computed as part of the Advertising Research Foundation's audit from detailed tabulations of sample recovery by income stratum within sex. These completion rates are slightly higher than those A & S reported because the audit assumed that a certain proportion of the households not found at home would not have contained an individual of the predesignated sex. Computations are indicated at the bottom of the table.

6. ESTIMATING PROCEDURES (INCLUDING NONRESPONSE ADJUSTMENT)[2]

Audits & Surveys' planned oversampling of higher income areas and undersampling of lower income areas, as well as their selection of one adult from a household regardless of the number eligible, required weighting their tabulated sample to equalize the probabilities of selection of areas and individuals. In addition to these weights, A&S applied weights to compensate broadly for unequal rates of sample completion and to bring some sample distributions more into line with the 1975 population.

Operationally, each individual in the sample received a weight equal to the product of the following:

(1) a "within-household" weight, which weighted each individual by the number of adults of the same sex in his household and thus compensated for the unequal probabilities of individual selection attributable to varying numbers of eligible adults in sample households;

(2) a "secondary-sampling-unit" weight, which combined two weights in one: a weight to correct for the disproportionate allocation of tracts and MCDs to the five income strata and a weight to adjust for unequal completion rates and other factors that caused different proportions of interviews to be obtained in the different strata; and

[1] An audit subsample in twenty locations was selected as part of the evaluation of the study by the Advertising Research Foundation (1977). Table 6 shows the 130 households attempted by number and outcome of calls.

[2] Except for the table number designations and certain styling points, this section is taken verbatim from the Advertising Research Foundation's report, *An Audit of Audits & Survey's Study Readership of Ten Major Magazines*.

(3) a "ratio-estimate" weight, which was determined by the individual's sex, age group, region of the country, and whether he or she lived in a metropolitan or nonmetropolitan area. This weight was intended to project the sample and at the same time balance it by these four characteristics.

6.1. Secondary-Sampling-Unit Weights

In selecting the sample, A&S had divided the secondary sampling units (census tracts and minor civil divisions) into five strata, consisting of three income groups in metropolitan areas and two in nonmetropolitan areas. A&S computed the composite SSU weights by first weighting each individual by his within-household weight and obtaining the distribution of these weighted individuals across the five strata. They then compared this sample distribution with the population distribution across the strata. The population distribution had been inferred from the distribution of the large sample of census tracts and counties from which the A&S SSUs were selected.

The composite SSU weight was then computed by dividing the proportion of total population in a given stratum by the proportion of weighted-sample individuals in the stratum. Every individual in the stratum received this SSU weight. This procedure and the computed weights are shown in Table 8, which is based on a similar table in the A & S technical report.

As a check, ARF constructed the composite SSU weights directly from their two components, as shown in Table 9. For this computation we obtained the "nonresponse" component from the sample completion data supplied by A & S and shown elsewhere in this chapter. Nonresponse among eligible households accounted for only a part of this component, the rest being due to differences in

TABLE 8

Computation of SSU Weights

Stratum (by median 1969 family income in SSU)	*(1) Proportionate distribution of SSUs (%)*	*(2) Distribution of sample adults after application of within-HH weights (%)*	*(3) Composite SSU weight [ratio of (1) to (2)]*
Metropolitan areas			
Under$7,000	7.7	3.3	2.33
$7,000–12,999	50.1	47.2	1.06
$13,000 and over	11.2	23.8	.47
Nonmetropolitan areas			
Under $7,000	9.8	5.8	1.69
$7,000 and over	21.2	19.8	1.07

TABLE 9

Components of SSU Weights

Stratum	*Component due to disproportionate sample allocation*	*Component due to nonresponse*	*Audit composite weight*	*A & S composite weight (Table 8)*
Metropolitan areas				
Under $7,000	2.06	1.09	2.24	2.33
$7,000–12,999	1.01	1.05	1.06	1.06
$13,000 and over	.51	.96	.49	.47
Nonmetropolitan areas				
Under $7,000	2.05	.80	1.66	1.69
$7,000 and over	1.08	.98	1.06	1.07

the proportions of households found and recorded in sample locations and in the proportions of households containing an ineligible adult.

The comparison in Table 9 between the SSU weights computed directly and those obtained by A & S verifies the nonresponse conponent of the A & S weights. The slight difference between the two sets of weights may reflect the fact that the two distributions A & S compared in obtaining their weights were not exactly comparable; that is, the proportionate distribution of SSUs represented a distribution of total persons, whereas the sample distribution was of adults.

The relatively minor adjustment for nonresponse that is reflected in these weights is due at least partly to the level at which the adjustment was made, that is, at the level of the broadly defined income strata. Weighting for differential nonresponse among smaller units, such as SSUs, would have allocated it more specifically to the areas where it occurred but would also have produced greater variation among the weights. The limited extent of nonresponse weighting, along with the limited amount of sample balancing required, as described in the next subsection, contributed to the relatively moderate range of weights used in the study.

6.2. Ratio-Estimate Weights

In projecting the sample, A&S applied different projection weights to different cells of the sample in order to improve its representativeness with respect to four characteristics. These were:

Sex: male, female;
Age: 18–24, 25–34, 35–49, 50–64, 65 and over;
Region: northeast, north central, south, west;
Residence: metropolitan, nonmetropolitan.

TABLE 10

Range from Lowest to Highest Ratio-Estimate Weights in the Five Age Groups (000)

	Males		*Females*	
Region	*Met*	*Nonmet*	*Met*	*Nonmet*
Northeast	19.1–29.3	16.9–39.4	15.3–27.1	17.0–26.7
North central	24.2–30.1	14.6–24.0	18.2–29.6	15.9–27.2
South	17.9–26.5	15.7–32.1	16.0–26.7	18.7–26.2
West	17.5–33.8	14.8–47.4	16.3–25.7	20.1–34.4

A&S divided the sample, after applying the within-household and SSU weights, into the 80 cells formed by the various combinations of these groups. They divided the household population of the coterminous United States into the same 80 cells. The ratio-estimate weight for an individual in any given cell was then determined by dividing the U.S. population in the cell by the corresponding sample estimate.

Audits & Surveys has reported that, because it was unable to obtain sufficiently detailed 1975 distributions by region, the regional distributions reflect 1970 data, but that otherwise the ratio estimates were based on the 1975 population.

Simply to provide the reader with a perspective on order of magnitude, we note that the ratio-estimate weights, because they were used for projection, were large numbers relative to the "within-household" and SSU weights. The combined household and SSU weights averaged 1.17 per respondent and brought the unweighted sample of 5683 to a weighted sample of 6657. This weighted sample was then projected to 141,914,000 adults by means of the ratio-estimating procedure. The average ratio-estimate weight was thus 21,318.

Individual ratio-estimate weights in the 80 cells varied, of course. Two of the cells in which there were small sample bases and relatively large ratio-estimate weights were combined. After this single adjustment, however, the highest ratio-estimate weight, 47,364, was only about 3.2 times the lowest, 14,635. Table 10 shows how these weights varied over the five age groups for each combination of sex, region, and metropolitan/nonmetropolitan residence.

APPENDIX

A.1. Field Personnel

Interviewer training and field interviewing for this study were controlled and operated within the following structure.

Audits & Surveys' seven permanent full-time regional supervisors located in New York, Indiana, California, Massachusetts, Minnesota, Texas, and Georgia all played important roles in training and supervising local supervisors and interviewers. These full-time permanent regional supervisors have been with Audits & Surveys in this capacity for 1.5 to 14 years. All of them have spent considerably longer times in various capacities associated with survey interviewing.

Local control was exercised by 136 local supervisors, all of whom have functioned in this capacity for Audits & Surveys for 1 year or more.

Interviewing was conducted by 373 interviewers, 360 of whom have had 1 year's experience or more and have had their work validated prior to this study. The 13 interviewers with less than 1 year's prior experience received special training by a permanent regional supervisor or a local supervisor.

A.2. Interviewer Training

Every local supervisor and 220 of the interviewers who worked on the study attended 1-day, detailed briefing and training sessions conducted by regional supervisors. The remaining 136 interviewers attended 1-day, detailed briefing and training sessions conducted by local supervisors.

A.3. Interviewer Compensation

Every interviewer on the study was paid on an hourly basis for all work, including briefing and training, practice inteviews, study time, contact time, and interviewing time, plus all travel time. Interviewers were also reimbursed for travel on a mileage basis. This includes one interviewer in Nevada who traveled 850 miles in one week and interviewed two respondents, and one in northern California who traveled 400 miles in one week and obtained one interview. No payments were made to interviewers on a per interview or per contact basis.

A.4. Field Quality Control

The total Field Quality Control program for this study began with a practice interview, a grading form, and an evaluation procedure designed to make sure each interviewer understood and could execute all critical interviewing procedures. It proceeded from validation of the prelisting process to validation of "Refusals" and "Not-at-Homes."

Three aspects of the Field Quality Control program were important. First, every aspect of field work was validated. This included prelisting, household and respondent contact, and interviewing. Second, the levels of validation were very high. For example, 37% of all interviews were validated. Third, every interviewer was clearly aware of the size and intensive character of the field-validation program. All interviewers knew in advance that every aspect of their work would be scrutinized by validation procedures.

CHAPTER **10**

Total Survey Error

Martha J. Banks
Ronald Andersen
Martin R. Frankel

1. THE SURVEY

The 1970 survey conducted for the Center for Health Administration Studies (CHAS) of the University of Chicago by the National Opinion Research Center was the fourth in a series of national surveys collecting information on health services use and expenditures in the United States.[1] In early 1971, 3765 families plus some additional aged individuals (11,610 persons in all) were interviewed in their homes. One or more persons in each family provided information on health services use and related costs during 1970, insurance coverage, attitudes toward providers, and health beliefs.

The sample constituted an area probability sample of the noninstitutionalized population of the United States. It was designed to overrepresent the inner-city poor, the aged, and residents of rural areas in order to permit more detailed analyses of these groups. Weights were developed to correct for the oversampling of these groups and to allow estimates to be made for the total noninstitutionalized U.S. population.

The sample design used consisted of four subsamples, each with its own sampling rate:

(1) The U sample consisted of households chosen in 73 segments in the NORC master sample. These segments were so designated because the 1960

---

[1] The survey was supported by the National Center for Health Services Research under contracts HSM 110-72-392 and HRA 106-74-24.

ISBN 0-12-363901-8

U.S. census data indicated that they contained a high proportion of poor urban families.

(2) The A sample was selected from each of the remaining 730 segments in the NORC national probability sample.

(3) The S sample consisted of families obtained by screening households in the 803 segments mentioned in the two preceding points. Households were screened for poor families and for families containing one or more persons 66 years of age or older.

(4) The R sample was obtained from 193 rural segments in 30 primary sampling units (PSUs) chosen especially for this study. No screening procedure was used for this sample.

The mean cluster size in the U sample was 28.3 assigned households (18.8 interviewed households). These 73 segments had the highest percentage of poor, urban families (based on 1960 census data) of the 803 segments in the NORC master sample. The mean cluster size in the A sample was 2.1 assigned households. The mean number of households interviewed in the 671 segments with interviews was 1.7. About 37% of the S-sample families who were screened were eligible for interview; 340 segments had no households interviewed in this sample. Most of the 463 segments with any S-sample completed cases contained only 1 or 2, though a few contained as many as 5. The mean is 1.7 interviewed households per segment with completed cases. The R sample was chosen independently from all non-Standard Metropolitan Statistical Area (non-SMSA) counties in the continental United States. (Because selection was independent, two of the selected counties are A-sample PSUs as well.) Of the 320 segments selected, only those with rural population were assigned for interview. Of these, interviews were completed in 193 segments. The mean cluster size is 3.1 completed household interviews.

Table 1 presents the final dispositions and response rate (the portion of eligible households with whom an interview was conducted) for each sample. The data were then adjusted to compensate for the varying response rates among the samples. The weights were further revised to equalize (between samples) the weighting of households in the same age–income–residence category. They also adjust for the fact that many of the households originally chosen in the S sample were not interviewed because they fit neither the income nor the age criterion. The results will be referred to as the "intermediate" weights because a further poststratification adjustment was made. Table 2 gives all these initial and intermediate weights and the unweighted number of interviewed households in each category.

Table 3 provides the distribution by location of responding and nonresponding families and the response rates for these categories. Of the 18% not responding, most (four out of five) were refusals. The remainder were never found at home, presented a language problem, were too ill to be interviewed, or were unavailable for the entire field period.

TABLE 1

Final Interviewing Results

Category	Sample A	U	R	S_1[a]	S_2[a]	Total[b]
(1) Number of dwelling units listed in the original sampling frame	1515	2068	810	2887	2451	7280
(2) Dwelling units that were vacant during the interviewing period or had been torn down between the time of listing and the time of interviewing	176	415	126	407	—	1124
(3) Dwelling units where respondents' and interviewers' race did not match; applicable only in some urban segments	5	43	—	9	—	57
(4) Indicates not qualified; Applicable only in the S sample, where families were screened out if they were nonpoor or had no member 66 or over	—	—	—	—	1539	1594[c]
(5) Indicates extra family units, which were added when multiple family dwelling units were discovered at the time of interview or multiple dwelling units within the same structure had originally been listed as single units	42	72	15	68	—	197
(6) Net assignment; categories (1) − (2) − (3) − (4) + (5)	1376	1682	699	2539	912	4702[c]
(7) Number of completed interviews	1119	1376	601	2451	784	3880
(8) Noninterview reports, including "refusals," "breakoffs," "no one home after repeat calls," "language problems," "respondent too ill to be interviewed"	257	306	98	88	128	822[c]
(9) Response rate; categories (7) ÷ (6).	.813	.818	.860	.965	.860	.825[c]

[a] The report on the S sample is divided into two parts: S_1 refers to the screening operation, S_2 to the regular interviewing.

[b] Includes S_1 figures for categories (1)–(3) and S_2 figures elsewhere.

[c] Figure based on the assumption that some of the S sample's noninterview reports are not qualified.

TABLE 2

Initial and Intermediate Weights and Their Corresponding Number of Interviewed Households

Category	*Sample*			
	U	*A*	*R*	*S*
Weight based on selection probability	1.00000	13.86232	7.21427	8.11153
Weight adjusted by differential nonresponse	1.00000	13.96462	6.87833	7.99441
Intermediate weight				
U-type urban low income and/or old	1.00000	—	—	1.00000
U-type urban other	1.12505	—	—	—
A-type urban low income and/or old	—	5.71978	—	5.71978
A-type urban other	—	15.71088	—	—
Rural low income and/or old	—	3.28894	3.28894	3.28894
Rural other	—	5.18478	5.18478	—
Unweighted number of interviewed households				
U-type urban low income and/or old	748	—	—	108
U-type urban other	628	—	—	—
A-type urban low income and/or old	—	275	—	458
A-type urban other	—	544	—	—
Rural low income and/or old	—	132	242	218
Rutal other	—	168	359	—

TABLE 3

Comparisons of Interviewed and Noninterviewed Families

Characteristic of population	*Percentage of those interviewed*	*Percentage of those not interviewed*	*Response rate*
Region			
Northeast	27	31	.804
North Central	27	22	.853
South	30	25	.845
West	16	22	.774
Type of dwelling[a]			
Located on farm	7	3	.917
Nonfarm: single-family house	65	60	.836
Nonfarm: duplex or two-family structure	10	11	.811
Nonfarm: multiunit structure (e.g., apartment)	18	26	.766
Community[b]			
Inside largest city in the primary unit	45	57	.788
In a suburb of the largest city in a primary sampling unit	24	21	.844
In the outskirts (including in all other towns in the primary sampling unit)	22	18	.852
In open country	9	4	.914
Residence			
Urban	71	75	.817
Rural	29	25	.840
Total	100	100	.825

[a] Excludes 186 interviewed and 153 noninterviewed families with "no answer."
[b] Excludes 160 interviewed and 71 noninterviewed families with "no answer."

2. THE VERIFICATION DATA

In addition to data provided by the sample families, a second body of data was collected from family physicians, clinics, hospitals, insuring organizations, and employers about the families' medical care and health insurance for the survey year. This body of data, referred to as the "verification," had two purposes: first, to determine whether or not the reported care was in fact provided during the survey year and, second, to elicit information that was more precise than that which the families were likely to give on diagnoses, costs, kinds of treatment, and sources of payment for services. Approximately 10% of the families in the social survey refused to sign the permission forms needed to obtain information from doctors and hospitals. Other families signed permission forms but failed to provide enough information to enable us to contact their providers. In addition, some providers refused to cooperate with the survey. However, verifying data were obtained for over 90% of the hospital admissions and for two-thirds of the physician visits the sample families reported.

Another difficulty in the verification process was the problem of "false negatives." Although reports of use or expenditures were verified when possible, it was impossible systematically to verify reports of nonuse. Therefore we were more likely to discover persons who reported experiences they did not have (overreporting) than people who failed to report experiences that occurred (underreporting).

The existence of verification data enables us to evaluate nonsampling error. This evaluation is based on the assumption that what doctors and hospitals report is closer to reality than what survey respondents report. Of course, there is also error in the provider reports. However, an operational measure of the "true answer" must exist if we are to assess individual response error.

3. SUMMARY OF THE METHODOLOGICAL STUDY

In Chapter 1 of *Total Survey Error* (Andersen, Kasper, Frankel, and Associates, 1979), a model of survey error is discussed in much more detail. The methodological study was focused on three components of nonsampling bias—nonresponse bias, field bias, and processing bias—and also provided estimates of standard error, a type of variable error. *Nonresponse bias* is the difference between the survey data and data that would have been obtained had observations been obtained from all designated respondents; *field bias* is the difference between the information given by survey respondents and the data collected from the verification sources; and *processing bias* is operationalized as imputation bias, the bias introduced by the use of imputation, compared with verification data. (The ratio of the verification estimate and imputated estimate was used to adjust data for persons who refused to allow verification, the same procedure used for these persons in calculating field bias.)

We have chosen to measure sources of bias and error that may significantly influence the validity and reliability of data and that our survey is particularly suited to evaluate. It is not possible to measure each error component in its entirety, so their sum may not reflect the best available estimate of the "overall" nonsampling bias. These possible discrepancies between the sum of the "main effects" of the three bias components and the overall nonsampling bias would be due in part to the interaction among the "unmeasured" portions of each of these three components and in part to the nonmeasurement of the fourth nonsampling bias component, noncoverage. We attempted to eliminate a portion of the differential noncoverage as well as nonresponse biases by poststratification (ratio-estimation) adjustment on the basis of 16 strata defined by family size, family income, race, and location inside or outside a Standard Metropolitan Statistical Area (SMSA).

This discussion concentrates on nonresponse bias and imputation bias, since these most clearly relate to incomplete data. For information about field bias the reader is referred to *Total Survey Error* (Andersen, Kasper, Frankel, and Associates, 1979, Part 2).

4. STANDARD ERROR

Our only estimates of variable error are those derived from applying standard error formulas, which reflect both the clustered and stratified nature of the sample design and were produced by using a technique described by Keyfitz (1957). An overall design factor of 2.6 was obtained; for nonwhites, the figure was 4.1. These factors were obtained from computations involving many different types of health variables. There was little or no difference in factors for different types of medical variables.

Although standard errors usually are represented as measuring the sampling portion of variable error, the standard error terms, in fact, reflect portions of the nonsampling variable error as well. These nonsampling errors typically include random errors in respondent reporting and a correlated response variance that reflects differences in the performances of individual coders, editors, interviewers, and other survey personnel. Because there is evidence that nonsampling variance may be sizable in some cases, the absence of a direct measure of this error source in our calculations of total error should be noted.

5. NONRESPONSE BIAS

In almost any survey there will be cases designated for interview that were not actually interviewed. Some potential respondents may have refused to be interviewed; a few may initially begin the interview process but break it off before completion; others may not have been at home when repeated interview attempts were made; some may have been away from home during the entire interview period; and a few may not have been interviewed for various other reasons, such as that of a language barrier. Making no adjustment for nonresponse implicitly assumes that nonrespondents do not differ from respondents in any characteristic of interest. The degree to which they *do* differ is proportional to the amount of bias introduced by ignoring nonrespondents.

Any approach to nonresponse adjustment consists of two elements. First, the population must be categorized into subgroups and the response rate for each subgroup must be determined. The categories chosen to form the subgroups should not only be correlated with characteristics of interest in the study (for example, mean number of dental visits) but must also be determined without having to obtain the information from the potential respondents themselves. Dwelling location (for example, in the central city of an SMSA), type (for example, duplex), and state of repair can be observed; such characteristics as race and presence of children can be determined from neighbors; and number of calls made and their results can be obtained as part of the interviewer control and record-keeping process.

The second major element in nonresponse adjustment is that of determining the values to impute to the nonrespondents. It is usually simplest and not unreasonable to assume that respondents and nonrespondents falling into a given category have identical characteristics. That is, nonrespondents living in well-kept housing in central cities have about the same health characteristics as do their responding counterparts; nonresponding and responding persons living in dilapidated central city housing are very similar; and so on. This method (and the rationale for nonresponse adjustment) is summarized in National Center for Health Statistics (1971, p. 22): "The adjustment for nonresponse is intended to minimize the impact of nonresponse on final estimates by imputing to nonrespondents the characteristics of 'similar' respondents, that is, by relating respondents to nonrespondents by ancillary data known for both."

Sometimes, however, we have evidence that respondents and nonrespondents in the same category do *not* have nearly identical characteristics. This evidence may come from external sources or from the current study itself (by choosing a sample of nonrespondents, vigorously attempting interviews with those chosen, and weighting and extrapolating from the results to obtain a final estimate).

Compared with sampling nonrespondents, it is easier and less expensive to use external data, that is, data from a previous study that either used the above approach or had access to the administrative records of both survey respondents and nonrespondents. Care must be taken in adjusting previous data according to current patterns, including nonresponse patterns.

Data from the 1970 CHAS study were adjusted for differential nonresponse by sample. This adjustment ensures that the data have the same relative number of interviewed households by sample as was the case for the assigned households and was made because of the positive correlation between sample designation and most characteristics of interest in the study. The samples grouped geographic areas and persons within them by percentage of elderly, poor, or both, characteristics that influence health and medical care use.

5.1. Adjustments Using External Data

The initial estimates of nonresponse bias in the data were calculated by using external data to estimate how the dependent variable values differed between nonrespondents and respondents. The estimated response rates were based on data that were not only adjusted by sample but also contained a poststratification adjustment. Thus there was an implicit adjustment for differential nonresponse among poststratification categories. We shall illustrate the external imputation method using mean physician visits for those seeing a physician and mean expenditure per hospital admission. Results using these adjustments appear in Andersen, Lion, and Anderson (1976).

5.1.1. Mean Physician Visits

Some evidence concerning the physician utilization experience of families who do not cooperate with health interview surveys is available from a study done of a seventeen-thousand-resident housing estate just outside London (Cartwright, 1959). Eighty-six percent of the people on the estate were registered with six general practitioners, who recorded details of each of their consultations for one calendar year. Interviews were attempted with a sample of three-sixteenths of the families concerning their health and medical care utilization. The nonresponse rate was 14.5%.

The London study showed that nonrespondents were less likely to have seen a physician than were respondents but that the mean number of visits for those seeing the physician was slightly greater for the nonrespondents than for the respondents (Cartwright, 1959, p. 359). The ratio of respondents' visits to nonrespondents' visits was assumed to be the same in the U.S. and London studies. Because the U.S. response rate is known, the mean number of visits for nonrespondents could be calculated. The estimated mean number of visits for nonrespondents in each subgroup was adjusted for the difference in the proportion of people seeing a physician between that subgroup and the total population.

5.1.2. Expenditures per Admission

This bias calculation assumes that per diem costs are the same for respondents and nonrespondents and that differences in expenditures per admission are a function of estimated differences in admission rates and length of stay of nonrespondents. The estimated admission rate for the noninterviewed population was based on a probability selection of 1589 persons with one or more discharges from 21 short-term hospitals as determined from 1958–1959 hospital records (National Center for Health Statistics, 1965). Respondents or proxies for them were interviewed about hospitalizations for the 12-month period prior to the interview. Five percent of the patients were nonrespondents for the household survey, and they were assumed to represent the universe of hospitalizations not reported in health interview surveys (National Center for Health Statistics, 1965, p. 59). The universe of all nonrespondents (both those hospitalized and those not) for the same time period was estimated from the National Health Survey study of hospital discharges from short-stay hospitals for the United States from 1958 to 1960 (National Center for Health Statistics, 1962). The rate for nonrespondents was then adjusted upward for increases in hospital utilization between 1958 and 1970. For subgroup calculations, the ratio of the utilization of nonrespondents to the utilization of respondents was assumed to be the same as that for the total population. Although the 21 hospitals (and thus their patients) are not a probability sample of the hospitalized population, it was the best external data found for estimating nonrespondent's hospital experiences. The average length of stay for nonrespondents was estimated by using data from the study of 21 short-term hospitals (National Center for Health Statistics, 1965). The mean length of stay for nonrespondents in that study was 8.8 days, compared with 7.4 days for respondents. The estimate for nonrespondents for 1970 was then adjusted to take into account changes in length of stay between 1958–1959 and 1970. For each subgroup, the ratio of the utilization rate of the nonrespondents to the respondents was assumed to be the same as that for the total population.

5.2. Adjustments Using Internal Data

With the data available, we chose to try two approaches as alternatives to that described in the previous section. Both assume that respondents and nonrespondents within the same category tend to have the same values. Thus the overall results of the two methods differ only because of different choices of categories. The first alternative creates categories based on geographic location; the second uses information about the reason that no interview was obtained.

The geographic adjustment method used primary sampling unit and sample designation as the category determinants. The use of sample designation effectively grouped cases within PSUs by presence of the poor, the elderly, or both. This method is similar to that used by the Bureau of the Census in many of its surveys (Thompson and Shapiro, 1973, p. 114).

The other method used categories based on the reason that no interview was completed. To adjust for cases in which people refused to be interviewed or in which people broke off in the middle of the interviewing process, we increased the weights of those respondents who were not completely cooperative. "Uncooperative" respondents were those who broke appointments with the interviewer, refused to give permission to verify, and so on. In this study the majority of nonrespondents were refusals. To adjust for other cases, those never found at home or unavailable for the entire field period, we increased the weights of all cases according to the number of calls needed to complete the interviews. We chose to assume that it would have taken 20% more calls to obtain interviews from the never-at-homes. (For example, if there were ten households where no one was ever at home after seven calls, we estimated that it would take an average of 1.4 more calls to obtain interviews. Six would take one more call, and four would take two more.) Kish (1965, p. 537) discusses differentiating between refusals and never-at-homes.

In computing the factors to be used for noninterview adjustment, we used an upper bound in order to control the amount of differential weighting caused by nonresponse adjustment. The United States Bureau of the Census (1978, p. 59) uses an upper bound in its estimation procedures. The upper bound we chose is 2.3995, which is twice the overall average adjustment, that is, twice the number of assigned eligible households divided by the number of interviewed households. When a calculated factor exceeded this bound, we reduced it to the largest acceptable figure (2.3995) and increased the factor of a similar category appropriately.

Table 4 compares the percentages of interviewed households that received various levels of noninterview adjustment weights according to each method. Most of the sample households were given weights between 1.02 and 2.00 by the geographic method; most received a weight outside this range (1.02 or less or more than 2.00) from the adjustment, based on the reason that no interview was obtained.

In Tables 5–9 we attempt to assess the impact of each of the three noninterview adjustment methods. We present the results that we obtained by their

TABLE 4

Percentage Distribution of Sample Households by Noninterview Adjustment Factor

Noninterview adjustment factor	*Geographic (%)*	*Reason no interview obtained (%)*
1.00	15.8	67.3
Over 1.00 through 1.02	.0	12.7
Over 1.02 through 1.15	24.8	3.4
Over 1.15 through 1.30	34.1	1.0
Over 1.30 through 2.00	25.3	2.9
Over 2.00 through 2.40	.1	12.8
Total	100.0	100.0

TABLE 5

Effect of Noninterview Adjustment on Distribution of Sample Persons

	Percent of Weighted Sample Persons			
	Unadjusted Difference	*Estimates Adjusted for Nonresponse*		
Characteristic		*Adjusted Using External Data*	*Adjusted Using Geographic Information*	*Adjusted Using Reason No Interview Obtained*
Demographic				
Age of oldest family member				
Less than 65 years	86.2%	86.5%	86.2%	86.3%
65 years or more	13.8	13.4	13.8	13.7
Family income				
Nonpoor	77.0	77.3	77.1	77.8
Poor	23.0	22.8	22.9	22.2
Race				
White	87.9	87.9	88.0	88.3
Nonwhite	12.1	12.1	12.0	11.7
Residence				
Rural nonfarm	24.5	23.7	24.4	24.4
Rural farm	6.8	6.2	6.8	6.6
SMSA central city	29.8	31.2	29.4	29.7
SMSA other urban	26.9	27.0	27.2	26.5
Urban nonSMSA	12.1	12.0	12.2	12.8
Perceived and evaluated health				
Perception of health				
Excellent	37.8	—[a]	37.9	38.1
Good	43.0	—	42.7	42.9
Fair	12.4	—	12.4	12.1
Poor	3.9	—	3.9	3.9
Worry about health				
A great deal	8.4	—	8.4	8.1
Some	18.2	—	18.3	18.3
Hardly any	20.6	—	20.6	20.2
None	50.1	—	50.1	50.6
Severity over all diagnoses				
Elective only	23.9	—	23.9	24.6
Elective and mandatory	20.3	—	20.3	20.2
Mandatory only	20.2	—	20.2	19.7
Number of diagnoses				
1	28.2	—	28.2	28.5
2	17.6	—	17.7	18.2
3	9.2	—	9.1	9.0
4 or more	9.0	—	9.0	8.4

(*Continues*)

TABLE 5 *(cont.)*

	Percent of Weighted Sample Persons			
	Unadjusted Difference	*Estimates Adjusted for Nonresponse*		
Characteristic		*Adjusted Using External Data*	*Adjusted Using Geographic Information*	*Adjusted Using Reason No Interview Obtained*
Level of use				
Physician visits per person				
0	32.4	—	32.4	32.1
1	18.6	—	18.5	19.0
2 to 4	24.1	—	24.1	24.1
5 or more	24.9	—	25.0	24.8
5 to 13	18.7	—	18.7	18.9
14 or more	6.2	—	6.3	5.9
Hospital admissions per person				
0	88.7	—	88.7	88.9
1	9.0	—	9.1	8.7
2 or more	2.3	—	2.3	2.4
Total	100.0%	100.0%	100.0%	100.0%

[a] Data necessary to provide these estimates were unavailable.

use without intending to imply what the expected results would be if these methods were applied to a number of similar data sets.

Table 5 provides information about the effect of noninterview adjustment on the distribution of sample persons. The column labeled "Adjusted Using External Data" was obtained by multiplying the unadjusted social survey estimates by the overall family response rate and dividing by the family response rate for the characteristic. These rates appear in Table 3. Because they are family response rates, this column in Table 5 assumes implicitly that the mean number of persons per household is the same for interviewed and noninterviewed households of the same type. The estimated percentages of persons with adjustment, using geographic information and information about the reason why no interview was obtained, are the result of applying noninterview adjustment factors to all persons in a given category of households.

All differences shown in the table are very small; however, the population estimates that were adjusted by using geographic data are more similar to the unadjusted estimates than are those that were adjusted by using the reason for noninterview. The latter increased the proportion of the sample who were above poverty, were white, lived in small towns, never worried about their health, had one or two diagnoses, saw a physician once, and were not admitted to a hospital during the year. It reduced the suburban population, those in fair health, those

TABLE 6

Effect of Nonresponse Adjustment on Estimates of Mean Number of Physician Visits for Persons Seeing a Physician

Characteristic	*Physician Visits per Person with One or More Visits*			
	Unadjusted Social Survey Estimate	*Estimates Adjusted for Nonresponse*		
		Adjusted Using External Data	*Adjusted Using Geographic Information*	*Adjusted Using Reason No Interview Obtained*
Demographic				
Age of oldest family member				
Less than 65 years	5.4	5.5	5.4	5.3
65 years or more	7.8	7.9	7.8	7.7
Family income				
Nonpoor	5.5	5.6	5.6	5.5
Poor	6.5	6.6	6.6	6.5
Race				
White	5.7	5.7	5.7	5.6
Nonwhite	6.0	6.1	5.9	5.9
Residence				
Rural nonfarm	5.4	5.4	5.4	5.3
Rural farm	5.6	5.6	5.6	5.7
SMSA central city	6.0	6.1	6.1	6.0
SMSA other urban	5.6	5.6	5.6	5.4
Urban nonSMSA	6.1	6.2	6.2	6.1
Perceived and evaluated health				
Perception of health				
Excellent	3.6	—[a]	3.6	3.5
Good	5.3	—	5.3	5.2
Fair	9.1	—	9.2	8.9
Poor	14.2	—	14.2	14.5
Worry about health				
A great deal	11.7	—	11.8	11.5
Some	7.4	—	7.5	7.3
Hardly any	4.9	—	5.0	4.9
None	3.6	—	3.6	3.6
Severity over all diagnoses				
Elective only	3.2	—	3.2	3.2
Elective and mandatory	8.3	—	8.4	8.1
Mandatory only	7.2	—	7.2	7.3
Number of diagnoses				
1	3.8	—	3.8	3.8
2	5.6	—	5.7	5.8
3	8.0	—	8.0	7.8
4 or more	11.6	—	11.7	11.6

(*Continues*)

TABLE 6 (*cont.*)

Characteristic	*Physician Visits per Person with One or More Visits*			
	Unadjusted Social Survey Estimate	*Estimates Adjusted for Nonresponse*		
		Adjusted Using External Data	*Adjusted Using Geographic Information*	*Adjusted Using Reason No Interview Obtained*
Level of use				
Mean number of visits				
1	1.0	—	1.0	1.0
2 to 4	2.8	—	2.8	2.8
5 or more	12.1	—	12.2	12.1
5 to 13	8.0	—	8.0	8.1
14 or more	24.8	—	24.7	24.9
Total	5.7	5.8	5.8	5.7

[a] Data necessary to provide these estimates were unavailable.

who worried a great deal about their health, those with three or four diagnoses, those who did not see a physician, and those who had exactly one hospital admission during the year. The estimates that were adjusted by using geographic information decreased the central city figures and increased the suburban figures.

Tables 6 and 7 show the effect of nonresponse adjustment on two different estimates—the mean number of physician visits for persons seeing a physician and the mean hospital expenditures per person. Both tables show that the adjustment based on geographic information has less effect on the means than does the adjustment based on the reason that no interview was obtained. The nonwhite mean hospital expense per admission is clearly the most volatile. The estimate varies by as much as $100, depending on the nonresponse adjustment method. In general, the difference between the unadjusted estimates and the estimates adjusted by using external data is greater than the difference between the unadjusted figures and those adjusted by using geographic information or the difference between the original figures and those adjusted based on the reason for nonresponse. This is expected because the external data did not assume that respondents and nonrespondents in the same adjustment category have identical health experiences, but the other adjustments used this assumption. The noninterview adjustment seems to have had more effect on hospital expenditures than on physician visits, at least for totals and among the demographic characteristics.

The effect of nonresponse adjustment on differences of mean physician visits and of mean hospital cost among population groups appears in Tables 8 and 9. Following each difference is the ratio of difference to the estimated standard

TABLE 7

Effect of Nonresponse Adjustment on Estimates of Mean Hospital Expenditures per Admission

Characteristic	*Expenditures per Admission*			
	Unadjusted Social Survey Estimate	*Estimates Adjusted for Nonresponse*		
		Adjusted Using External Data	*Adjusted Using Geographic Information*	*Adjusted Using Reason No Interview Obtained*
Demographic				
Age of oldest family member				
Less than 65 years	$ 640	$662	$ 642	$ 624
65 years or more	863	886	877	857
Family income				
Nonpoor	674	697	679	643
Poor	715	737	717	749
Race				
White	684	709	691	659
Nonwhite	688	711	669	770
Residence				
Rural nonfarm	557	573	565	534
Rural farm	575	586	562	574
SMSA central city	727	757	745	735
SMSA other urban	910	941	904	904
Urban nonSMSA	444	458	446	431
Perceived and evaluated health				
Perception of health				
Excellent	488	—[a]	495	475
Good	609	—	619	614
Fair	698	—	712	682
Poor	868	—	877	862
Worry about health				
A great deal	1,036	—	1,049	978
Some	560	—	567	544
Hardly any	442	—	440	506
None	441	—	437	449
Severity over all diagnoses				
Elective only	499	—	501	482
Elective and mandatory	550	—	554	539
Mandatory only	830	—	833	807
Number of diagnoses				
1	626	—	643	617
2	573	—	578	620
3	563	—	559	544
4 or more	883	—	886	839

(*Continues*)

TABLE 7 *(cont.)*

Characteristic	Expenditures per Admission			
	Unadjusted Social Survey Estimate	Estimates Adjusted for Nonresponse		
		Adjusted Using External Data	Adjusted Using Geographic Information	Adjusted Using Reason No Interview Obtained
Level of use				
Expenditures per admission				
$0 to 300	184	—	183	182
301 to 600	426	—	426	423
601 or more	1,556	—	1,574	1,480
601 to 1200	843	—	841	845
1201 or more	2,801	—	2,812	2,704
Total	$ 685	$707	$ 689	$ 670

[a] Data necessary to provide these estimates were unavailable.

error of the difference. Table 8 shows little or no noninterview effect on differences in mean visits. All differences, except those by race and age, are highly significant statistically.

Because there were many fewer instances of hospital admissions in the sample than of persons seeing a physician, the statistical significance of the differences appearing in Table 9 would be expected to be less than in Table 8, even if the population represented by the sample differed equally in these two health characteristics. So either for this reason or because there really is less difference in hospital expenditures between groups than there is difference in physician visits or partially for both reasons, the significance ratios tend to be smaller in Table 9 than in Table 8.

Among the demographic variables, the differences seem to be larger in Table 9 for the estimates adjusted based on noninterview reason, especially differences based on family income and on race. In the health categories, the differences based on geographic nonresponse adjustment generally are more like the unadjusted differences than either group is like the differences involving reason of noninterview. All three are relatively quite similar, however.

5.3. Summary and Recommendations

None of the preceding discussion attempts to suggest which type of adjustment produces the most accurate estimates or even whether the time spent doing any type of adjustment for nonresponse is time well spent. In fact, in most data collection, there is no practical way to find out what would be the responses of all nonrespondents, so there is no way to determine the improvement in the estimates caused by noninterview adjustment. We can only measure the change it makes in the unadjusted estimates. In our examples, few of the estimates that were adjusted for nonresponse varied markedly from the unadjusted estimates. However, a well-thought-out plan for noninterview adjustment is usually worth making because doing so is fairly simple. Further, the benefits of noninterview adjustment are increasing because the response rates of most surveys have been declining for at least a decade (Daniel, 1975, p. 292). The larger the sample size, the more impact on the total error a noninterview adjustment will have because the standard error component of the total error will be smaller.

A fairly firm plan for noninterview adjustment should be devised before the study interviews are conducted, so the desired information can be collected both for those who were interviewed and for those who either refused or were never found at home. The type of information collected depends on the correlations of important variables in the study to variables that could be given values for the nonrespondents as well as for the respondents (type of dwelling, for example). The larger the correlation, the more gain noninterview adjustment produces. Categories can be chosen by using data from previous surveys to estimate correlations.

As previously stated, noninterview adjustment also requires the values to impute to the nonrespondents in each category. Unless there is firm evidence to the contrary from a fairly recent and similar survey or a pattern is found in the data according to ease of obtaining the interview, it would be best to assume that respondents and nonrespondents falling into the same category are otherwise identical. That is, such an adjustment assumes that respondents and nonrespondents differ *overall* and that this difference can, in large part, be explained by the differing distribution of respondents and nonrespondents between the chosen adjustment categories.

In the majority of cases, it is preferable to use internal data for nonintevіew adjustment rather than external data. Definitional differences between the present data and the external data make it difficult to form noninterview categories. (An example of this can be seen in the fact that we are able to present data adjusted for nonresponse using external data only for demographic categories.) Procedural differences usually lead to such problems as differing response rates, which might result in unlike types of persons being nonrespondents in the two surveys. Thus it is unlikely that the cost and time spent locating and adapting external data could be justified.

It is difficult to choose between the two adjustment methods that use internal data, given the limited information available on the effect of each. We feel that using either is somewhat preferable to performing no noninterview adjustment at all, but either set of categories could be used. When it is possible to do so, the effective nonresponse rate might be reduced by subsampling initial

TABLE 8

Effect of Nonresponse Adjustment on Differences between Groups for Mean Number of Physician Visits for Persons Seeing a Physician[a]

Characteristic	Physician Visits per Person with One or More Visits			
	Unadjusted Difference	Difference Adjusted for Nonresponse		
		Adjusted Using External Data	Adjusted Using Geographic Information	Adjusted Using Reason No Interview Obtained
Demographic				
Age of oldest family member				
Less than 65 versus				
65 or more	− 2.4(6.6)[a]	−2.5(6.6)	− 2.4(6.2)	− 2.4(5.9)
Family income				
Nonpoor versus poor	− 1.0(3.7)	−1.0(3.8)	− 1.1(3.7)	− 1.0(3.3)
Race				
White versus nonwhite	− .3(.6)	− .3(.7)	− .2(.4)	− .3(.5)
Residence				
SMSA other urban versus:				
Rural nonfarm	.2(.4)	.1(.4)	.2(.6)	.1(.3)
Rural farm	.0(.0)	.0(.0)	.0(.1)	− .3(.4)
SMSA central city	− .5(1.1)	− .5(1.2)	− .5(1.1)	− .6(1.3)
Urban nonSMSA	− .6(1.1)	− .6(1.0)	− .6(1.2)	− .7(1.0)
Perceived and evaluated health				
Perception of health				
Excellent versus:				
Good	− 1.7(6.7)	—[b]	− 1.8(6.8)	− 1.7(2.3)
Fair	− 5.5(10.8)	—	− 5.6(11.0)	− 5.4(11.2)
Poor	−10.6(9.3)	—	−10.6(8.7)	−11.0(6.8)

Worry about health				
None versus:				
Hardly any	– 1.3(4.4)	—	– 1.4(4.5)	– 1.3(3.7)
Some	– 3.8(10.2)	—	– 3.8(10.1)	– 3.6(9.8)
A great deal	– 8.1(11.3)	—	– 8.2(10.0)	– 7.9(9.9)
Severity over all diagnoses				
Elective only versus:				
Elective and mandatory	– 5.1(15.9)	—	– 5.2(15.7)	– 5.0(15.1)
Mandatory only	– 4.0(11.6)	—	– 4.1(11.6)	– 4.1(10.8)
Number of diagnoses				
1 versus:				
2	– 1.8(5.2)	—	– 1.9(5.2)	– 2.0(4.5)
3	– 4.1(9.6)	—	– 4.2(10.0)	– 4.0(10.6)
4 or more	– 7.8(20.6)	—	– 7.9(20.8)	– 7.9(19.2)
Level of use				
Physician visits per person				
1 versus:				
2 to 4	– 1.8(83.1)	—	– 1.8(93.8)	– 1.8(71.5)
5 or more	–11.1(39.5)	—	–11.2(39.4)	–11.1(33.1)
5 to 13	– 7.0(72.8)	—	– 7.0(72.9)	– 7.1(60.8)
14 or more	–23.8(29.3)	—	–23.7(27.7)	–23.9(25.8)

[a]Numbers in parentheses are the ratios of the differences to the estimated standard errors of the differences.
[b] Data necessary to provide these estimates were unavailable.

TABLE 9

Effect of Nonresponse Adjustment on Differences between Groups for Mean Hospital Expenditures per Admission[a]

Characteristic	Expenditures per Admission			
	Unadjusted Difference	Difference Adjusted for Nonresponse		
		Adjusted Using External Data	Adjusted Using Geographic Information	Adjusted Using Reason No Interview Obtained
Demographic				
Age of oldest family member				
Less than 65 versus 65 or more	$− 222(3.0)[a]	$− 225(3.0)	$− 235(3.0)	$− 233(3.3)
Family income				
Nonpoor versus poor	− 41(.6)	− 40(.6)	− 37(.5)	− 106(1.3)
Race				
White versus nonwhite	− 3(.0)	− 2(.0)	22(.3)	− 111(.7)
Residence				
SMSA other urban versus:				
Rural nonfarm	353(2.2)	368(2.2)	339(2.1)	371(2.4)
Rural farm	335(2.1)	355(2.2)	342(2.1)	330(2.1)
SMSA central city	183(1.2)	184(1.2)	160(1.0)	169(1.1)
Urban nonSMSA	467(2.9)	483(3.0)	458(2.9)	473(3.1)
Perceived and evaluated health				
Perception of health				
Excellent versus:				
Good	− 122(1.1)	—[b]	− 124(1.0)	− 139(1.5)
Fair	− 210(2.1)	—	− 217(1.9)	− 207(2.3)
Poor	− 380(3.6)	—	− 382(3.2)	− 387(3.8)

Worry about health				
None versus:				
Hardly any	– 1(.0)	—	– 3(.0)	– 58(.7)
Some	– 119(1.6)	—	– 129(1.6)	– 95(.7)
A great deal	– 595(6.6)	—	– 612(6.8)	– 530(5.8)
Severity over all diagnoses				
Elective only versus:				
Elective and mandatory	– 51(.6)	—	– 54(.6)	– 57(.7)
Mandatory only	– 331(3.5)	—	– 333(3.6)	– 325(3.7)
Number of diagnoses				
1 versus:				
2	53(.5)	—	64(.6)	– 3(.0)
3	63(.7)	—	83(.8)	72(.9)
4 or more	– 257(2.2)	—	– 243(2.0)	– 222(2.2)
Level of use				
Expenditures per admission				
$0 to 300 versus:				
301 to 600	– 242(36.2)	—	– 243(35.9)	– 241(33.5)
601 or more	– 1,371(12.2)	—	– 1,391(12.6)	– 1,298(11.6)
601 to 1200	– 659(35.3)	—	– 658(34.2)	– 662(32.6)
1201 or more	– 2,616(11.1)	—	– 2,539(11.8)	– 2,522(9.9)

[a] Numbers in parentheses are the ratios of the differences to the estimated standard errors of the differences.
[b] Data necessary to provide these estimates were unavailable.

nonrespondents and by looking for a pattern in responses by the ease in obtaining an interview.

Because there usually is no way to determine the potential responses of nonrespondents, it is extremely difficult to decide on the best nonresponse adjustment procedure. However, a few studies have been conducted that *do* give data estimates for both respondents and nonrespondents. (For instance, administrative records might be available for all persons designated for interview.) In these instances, it can be determined whether or not there are significant differences between survey respondents and nonrespondents and what choice of adjustment categories would be best. However, not only are such studies infrequent, but it is also rarely clear to what degree the results are generalizable to other surveys.

The amount of available information about how nonrespondents differ from respondents needs to be increased appreciably. Researchers who conduct interviews with persons for whom administrative records are available should compare entries on nonrespondents' records with those of respondents. The data should also be examined by ease of getting the interview and by reason for noninterview so that the results can be generalized to other surveys that may have different response rates, different patterns of reasons for noninterview, or both. The results should be made available to other researchers.

Noninterview adjustment is a relatively inexpensive method of reducing nonresponse bias somewhat, but it certainly is no substitute for obtaining actual responses from as many designated individuals as possible.

6. IMPUTATION BIAS

Complete noncooperation or nonavailability of designated respondents is most likely the largest source of the bias component we can designate as nonsampling bias due to nonobservation. It is not, however, the only source. Even if a respondent is willing to be interviewed, that respondent may be unable or unwilling to provide information for *every* survey question. The fact that most analytical survey software packages provide for the use of "missing-data" codes attests to the prevalence of item-specific nonresponse.

Three basic methods traditionally have been used to deal with this problem:

(1) Restrict variable-specific calculations to those cases for which the specific variable(s) have been provided by the respondent.

(2) Do not accept any data cases that contain missing data. In other words, restrict variable-specific calculations to those cases for which *all* variables have been provided by the respondent.

(3) Use a procedure to "impute" missing-data values on a case-specific or subgroup-specific basis. In the health care and expenditures study just described, a form of this method was used to deal with item-specific nonresponse for certain important variables.

Although it may not be immediately obvious, all three methods involve a form of "data adjustment." In fact, a strong case can be made that the third method, which initially appears to involve the most adjustment, is in fact the most conservative of the available methods. If we do not use the third method, we are in fact imputing the mean (median, and so on) value among all cases for which a response exists to all cases for which responses are missing. Even when the relationship between the variable in question and other survey variables used in imputation is weak, the implicit imputation induced by the elimination of case-specific nonresponse from the analyses (the first method) may be inappropriate. (This statement refers to univariate analyses. The case against elimination of such cases is not nearly as strong when we consider estimates of relationships.)

This is illustrated by the following example. Consider a population composed as shown in the accompanying tabulation. The first method would result

	Item-specific respondents		*Item-specific nonrespondents*	
Group	*Number*	*Mean value*	*Number*	*Mean value*
1	3	100	2	95
2	4	110	1	115

in an estimated overall mean of $[(3 \times 100) + (4 \times 110)] \div 7 = 105.7$; the true mean is $[(3 \times 100) + (4 \times 110) + (2 \times 95) + (1 \times 115)] \div 10 = 104.5$. A simple form of the third method would yield a mean of $\{[(3 + 2) \times 100] + [(4 + 1) \times 110]\} \div 10 = 105.0$. Thus the third method produces less imputation bias in this case than does the first. However, the true mean of 104.5 is not normally known, so a case could be made that differences between variable-specific responders and nonresponders were such that the estimate using the first method (105.7) might be closer to reality than the mean obtained by group-specific imputation (105.0). Such an argument would, however, require substantial outside evidence.

6.1. Effect of Imputation on Hospital Expenditures

We chose to examine the effect of imputation on the variable "hospital expenditures" because this variable required a great deal of imputation. As a result we felt that differences among alternative forms of imputation would be relatively pronounced. Of all hospital admissions, 32.0% (weighted) required imputation. The dollar figures imputed for these admissions represented 35.1% of all hospital admission dollars.

In Table 10, we show the number of cases for which imputation of hospital expenditures was necessary. These numbers are provided for the total sample and various population subgroups. This table also shows the percentage of all admissions (weighted and unweighted) requiring imputation, as well as the percentage of total expenditure dollars these groups represent.

6.1.1. *Method Used in the Original Imputation Procedure*

The original CHAS imputation method involved use of external data sources. We obtained two tapes from the American Hospital Association (AHA) one gave hospital characteristics and the other contained cost information. A per diem inpatient charge for each hospital was generated from these tapes. If a hospital did not appear on the AHA tapes, data from the American Hospital Association (1970) were used to determine a per diem according to the type of facility (federal, long-term care, and so on).

To impute the cost of a hospital admission when such information was not reported, we multiplied the per diem by the number of days the respondent reported being in the hospital. When the length of stay was not reported, NCHS data on average length of stay by diagnosis was used to impute days and the per diem rate was applied. Imputation was rare for hospital days but frequent for hospital expenditures.

6.1.2. *Alternative Procedures*

The original CHAS imputation method used "external" data in the imputation procedure, that is, data external to the social survey and subsequent validation. There exists another set of data imputation techniques based on the use of "internal" survey data. (Much of the basic development of internal imputation techniques took place at the U.S. Bureau of the Census.) In many situations, the use of internal survey data is necessary (if imputation is to be done at all) because external data are not available, are of dubious quality, or involve definitions that differ widely from the internal data. It is only when reasonably good, compatible outside data are available that the choice of imputation data source is relevant. In this case it may *still* be better to use internal data. The term "good" is relative; the good outside data may be based on averages. To the extent that persons with item nonresponse are not "average," the use of internal imputation methods may be preferable.

TABLE 10

Imputed Social Survey Expenditures for Hospital Admissions[a]

Characteristic	*Percent of Admissions with Imputed Expenditures*	*Unweighted Admissions with Imputed Expenditures*		*Percent of Expenditures Imputed*
		Percent	*N*	
Demographic				
Age of oldest family member[a]				
Less than 65 years	28%	39%	493	32%
65 years or more	46	49	241	45
Family income				
Nonpoor	25	31	297	27
Poor	51	55	437	56
Race				
White	30	35	441	32
Nonwhite	52	57	293	66
Residence				
Rural nonfarm	29	29	129	33
Rural farm	23	23	39	32
SMSA central city	37	53	406	32
SMSA other urban	28	39	90	43
Urban nonSMSA	42	44	70	39
Perceived and evaluated health				
Perception of health				
Excellent	25	31	84	24
Good	24	34	220	24
Fair	42	47	215	47
Poor	44	51	137	56
Worry about health				
A great deal	36	45	277	35
Some	28	40	200	29
Hardly any	30	40	94	46
None	44	36	127	33
Severity over all diagnoses				
Elective only	42	45	43	55
Elective and mandatory	27	36	240	28
Mandatory only	36	45	435	38
Number of diagnoses				
1	34	44	203	34
2	30	40	169	44
3	28	41	141	40
4 or more	34	41	206	30

(*Continues*)

TABLE 10 (*cont.*)

Characteristic	*Percent of Admissions with Imputed Expenditures*	*Unweighted Admissions with Imputed Expenditures*		*Percent of Expenditures Imputed*
		Percent	*N*	
Level of use				
Expenditures per admission				
$0 to 300	32	38	249	32
301 to 600	31	42	231	32
601 or more	33	46	254	34
Total	32%	42%	734	35%

[a] Note: All hospital expenditures (weighted) = $5,492,054. Total unweighted social survey admissions = 1769. Total weighted social survey admissions = 8021.

We examined the effect on mean hospital expenditures for three related subclass-specific imputation techniques that use internal data. One of the procedures used the cell-mean approach; the other two were based on alternative forms of the so-called "hot-deck" method. The same set of classification cells was used throughout, so differences in results are due solely to the differing methods.

6.1.3. *Choosing Imputation Categories*

Imputation categories should be chosen in a manner that maximizes the homogeneity of cases within the category with respect to the variable to be imputed and maximizes the heterogeneity of cases in different categories.

In choosing imputation categories for use with estimates of total hospital expenditures, we began with about a dozen variables that we felt could be used for prediction. Looking at the stays with expenditures reported by the respondents, length of hospital stay was by far the best predictor of stay cost. We therefore decided that one of our options would be to use length of stay as the sole determinant of imputation category. As an alternative, we considered length of stay combined with residence (SMSA, non-SMSA) and another variable, which for pregnancy-related stays was a three-category variable telling reason for stay (minor, moderate, serious) and for all other stays was number of operations (none, one, two or more). Data were collapsed such that each of the two possible sets of imputation categories contained 26 cells. In the former set of categories, this simply meant a collapsing by length of stay. In the latter set, it meant a more complex collapsing; the data were categorized according to fairly complicated combinations of criteria. Initial categories were combined based on trends in the mean expenditures of cases that had expenditure information reported.

We used the latter method of category designation in preference to that involving only length of stay. The categories created using other variables as well not only produced larger correlation statistics but also increased the number of reported stays relative to the number of imputed stays in the categories in which each single imputation would be expected to have the greatest impact—the categories with the largest expenditures.

6.1.4. *Using Imputation Categories*

Given the 26 cells, the actual imputation is itself quite straightforward. For the cell-mean method, we first computed the mean expenditures per admission for each cell. (These calculations were, of course, restricted to those admissions for which expenditures were reported.) The resulting cell means were then used as the values assigned to all admissions in the cell for which expenditure data were not available.

The hot-deck technique was carried out by first ordering admissions on the basis of the interview's geographic location. Two alternative orderings were used: (1) in PSU order, that is, east to west among SMSAs followed by east to west among non-SMSAs, and (2) in reverse PSU order. For each admission for which expenditure data were not available, a backward search was in effect carried out until the first non-missing-data admission falling into the same classification cell was encountered. The expenditure value associated with this non-missing-data admission was used as the imputation value. (In actual practice, the entire ordered data set is passed through an algorithm that saves in storage the most current non-missing-data value associated with each cell. When a missing-data case is encountered, reference to the appropriate cell storage provides the required value.)

6.1.5. *Comparison of Results by Imputation Method*

Table 11 shows the overall estimates of expenditures per admission that result from the different forms of item imputation. These results are shown for the entire sample and for the standard analytical groups used throughout this study. Table 12 shows the impact of these methods on the standard subgroup comparisons used in this report. Given the large proportion of the sample for which imputation of hospital expenditures was necessary, the differential effect of the three methods is remarkably small.

For the 74.3% (weighted) (65.0%, unweighted) of the imputed hospital admissions for which verification expenditure estimates were available, the mean expenditure estimates were as shown in the accompanying tabulation.

Thus for those cases in which outside verification was available, the use of external imputation (in other words, the procedure actually used) produces overall mean expenditure estimates that are closer to the results obtained by verification than are any of the internal imputation methods.

			Internal imputation		
				Hot deck	
Type of estimate	*Verification*	*Original (external) imputation*	*Cell mean*	*PSU order*	*Reverse PSU order*
Mean expenditures estimate	$677	$694	$778	$746	$727

The general tendency for the hot-deck method to produce results that are closer to the external imputation results than does the cell-mean method is moderately surprising. We had expected the opposite result.

The general reason often given for preference of the hot-deck method over a cell-mean approach involves the assumption that the hot-deck method will better preserve "relationships" between imputed and nonimputed variables. It is often conjectured that the cell-mean approach will either attenuate or artificially strengthen these relationships. The apparently superior performance of the hot-deck method with respect to the imputation itself is a welcomed result.

We are still somewhat puzzled by the relatively large differences between results obtained by the two different PSU orderings used in the hot-deck procedure. PSU, ordered low to high, represents general east-to-west order. The general tendency for this ordering to produce higher imputed expenditures than does the west-to-east ordering (reverse PSU ordering) probably reflects a geographic relationship with respect to hospital charges.

On strictly empirical grounds, the reverse PSU ordering seems to be the preferable choice for these data. We expect, however, that some type of joint method would actually be preferable. Such an approach might impute a data value equal to the value of the closest within-category respondent, whether preceding or following the respondent for which imputation is required.

6.1.6. Recommendations

Our investigation of alternative imputation approaches is limited. Cost and time considerations might make the choice among imputation procedures more difficult than the preceding suggests because locating, adapting, and using external data are much more laborious than is using internal data for imputation. Additionally, using these same methods of internal imputation on

TABLE 11

Effect of Item Imputation on Estimates of Hospital Expenditures per Admission

Characteristic	*Expenditures per Admission*			
	Using External Item Estimation	*Using Internal Item Imputation*		
		Cell Mean Method	*Hot Deck Method*	
			PSU Numbers Ordered Low to High	*PSU Numbers Ordered High to Low*
Demographic				
Age of oldest family member				
Less than 65 years	$ 640	$ 654	$ 652	$ 638
65 years or more	863	973	932	931
Family income				
Nonpoor	674	708	697	684
Poor	715	745	736	730
Race				
White	684	721	715	700
Nonwhite	688	687	629	658
Residence				
Rural nonfarm	557	582	567	566
Rural farm	575	634	611	642
SMSA central city	727	761	754	747
SMSA other urban	910	933	930	873
Urban nonSMSA	444	495	480	516
Perceived and evaluated health				
Perception of health				
Excellent	488	496	502	483
Good	609	623	614	600
Fair	698	785	759	762
Poor	868	929	861	910
Worry about health				
A great deal	1,036	1,097	1,076	1,072
Some	560	602	593	579
Hardly any	442	433	419	413
None	441	442	445	437
Severity over all diagnoses				
Elective only	499	484	521	462
Elective and mandatory	550	575	558	555
Mandatory only	830	875	866	853
Number of diagnoses				
1	626	692	703	641
2	573	551	561	527
3	563	606	573	619
4 or more	883	924	896	908

(*Continues*)

TABLE 11 (*cont.*)

Characteristic	*Expenditures per Admission*			
	Using External Item Estimation	*Using Internal Item Imputation*		
		Cell Mean Method	*Hot Deck Method*	
			PSU Numbers Ordered Low to High	*PSU Numbers Ordered High to Low*
Level of use				
Expenditures per admission				
$0 to 300	184	220	225	209
301 to 600	426	447	443	431
601 or more	1,556	1,599	1,562	1,561
601 to 1200	843	874	856	867
1201 or more	2,801	2,865	2,796	2,773
Total	$ 685	$ 717	$ 707	$ 696

other variables might produce a very different assessment of the relative merit of each. We urge further investigation.

6.2. Imputation of Other Variables

Table 13 shows the amount of imputation needed for various key variables. The preceding discussion could have been extended to these data as well had resources for doing so been available.

7. POSTSTRATIFICATION ADJUSTMENT

Another way in which to decrease bias in a sample is to use poststratification. Compared with the original population, a sample chosen from it will exhibit chance differences in nearly all possible variables. Usually there are a number of characteristics that are correlated with the dependent variables of interest in the study and for which more reliable estimates exist. Thus the study data generally can be improved by applying a set of factors that adjusts the sample distribution according to the more reliable data, although in some instances the improvement may be slight. These adjustment factors are called "poststratification," or "ratio-estimation," factors.

Data used for calculating such factors should, of course, be based on the very same population represented in the sample. Each of the characteristics chosen to form the categories should be correlated fairly highly with statistics of interest but not so highly correlated with each other that some are superfluous.

An upper and lower bound may be used in calculating poststratification adjustment factors. Although doing so lessens the sample's correspondence to the more reliable information in regard to these characteristics, it controls the amount of differential weighting this set of factors causes. We used an upper bound of 2 and a lower bound of $\frac{1}{2}$.

7.1. Original Approach

Estimates appearing in all previous tables were adjusted according to the family characteristics of race, residence, size, and income. Table 14 contains the 16 categories used and the factors[1] computed by comparing CHAS data with those of the 37,000-household Current Population Survey (CPS) sample for a similar time period. The CPS figures can be found in United States Bureau of the Census (1972, Table 16).

7.2. Suggested Alternative Approach

Most of the CHAS data have been examined on the individual, rather than the family, level. Thus it would appear to be preferable to adjust the data on the individual level. In order to choose categories, the mean numbers of disability days, physician visits, and hospital days were examined for persons, based on their age, sex, race, educational level, and relationship to household head. Age and race (and sex during women's childbearing years) seemed the most relevant variables, based on their ability to form categories of persons with different mean levels of health and medical care use.

Table 15 indicates the 17 suggested categories based on the individual characteristics of age, sex, and race. The category boundaries were chosen (1) according to changes in mean disability days, physician visits, and hospital days and (2) so that each category involved about the same number of unweighted sample persons, involved at least 100 unweighted sample persons, and had roughly the same importance in the final statistics.

The poststratification adjustment factors given in Table 15 were calculated by comparing CHAS data for persons (weighted by the intermediate weights of Table 2) with the CPS figures appearing in United States Bureau of the Census (1971, Tables 2 and 3).

[1] Nonresponse adjustment should be performed before poststratification adjustment. Thus the poststratification factors used with the data presented previously differ from those given in Table 13.

TABLE 12

Effect of Item Imputation on Estimates of Differences of Hospital Expenditures per Admission

Characteristic	Expenditures per Admission			
	Using External Item Estimation	Using Internal Item Imputation		
		Cell Mean Method	Hot Deck Method	
			PSU Numbers Ordered Low to High	PSU Numbers Ordered High to Low
Demographic				
Age of oldest family member				
Less than 65 versus 65 or more	$– 222(3.0)	$– 319(3.8)	$– 281(3.4)	$– 294(3.7)
Family income				
Nonpoor versus poor	– 41(.6)	– 37(.5)	– 39(.6)	– 46(.7)
Race				
White versus nonwhite	– 3(.0)	34(.5)	86(1.5)	42(.7)
Residence				
SMSA other urban versus:				
Rural nonfarm	353(2.2)	350(2.2)	362(2.2)	306(1.9)
Rural farm	335(2.1)	298(1.7)	318(1.9)	230(1.3)
SMSA central city	183(1.2)	171(1.1)	176(1.2)	126(.8)
Urban nonSMSA	467(2.9)	437(2.9)	450(2.9)	357(2.2)
Perceived and evaluated health				
Perception of health				
Excellent versus:				
Good	– 122(1.1)	– 127(1.2)	– 112(1.1)	– 118(1.1)
Fair	– 210(2.1)	– 289(2.6)	– 257(2.5)	– 279(2.4)
Poor	– 380(3.6)	– 433(3.6)	– 360(3.2)	– 428(3.5)

Worry about health				
None versus:				
Hardly any	– 1(.0)	9(.2)	26(.6)	24(.4)
Some	– 119(1.6)	– 159(2.2)	– 148(2.2)	– 142(1.8)
A great deal	– 595(6.6)	– 655(7.2)	– 631(6.9)	– 416(4.5)
Severity over all diagnoses				
Elective only versus:				
Elective and mandatory	– 51(.6)	– 91(1.3)	– 38(.5)	– 93(1.2)
Mandatory only	– 331(3.5)	– 391(4.6)	– 345(3.7)	– 391(4.3)
Number of diagnoses				
1 versus:				
2	53(.5)	141(1.5)	141(1.5)	115(1.2)
3	63(.7)	86(.9)	130(1.3)	22(.2)
4 or more	– 257(2.2)	– 232(1.9)	– 193(1.6)	– 267(2.2)
Level of use				
Expenditures per admission				
$0 to 300 versus:				
301 to 600	– 242(36.2)	– 227(19.6)	– 219(13.4)	– 222(18.3)
601 or more	– 1,371(12.2)	– 1,379(12.8)	– 1,337(12.7)	– 1,352(11.9)
601 to 1,200	– 659(35.3)	– 654(23.1)	– 631(20.5)	– 658(25.2)
1,201 or more	– 2,616(11.1)	– 2,646(12.2)	– 2,571(10.5)	– 2,565(12.1)

TABLE 13

Extent to which Selected Utilization and Independent Variables Were Imputed in Social Security Data

Variable	*Type of imputation done*	*Percent of unweighted observations for which an imputation was made*
Dentist visits, mean number	If the dentist was seen in 1970 but number of visits was not stated, visits were imputed	1.1
Family income	All families who did not answer this question had income imputed for them. Earned family income for at least one family member was imputed for 402 families. Other family income was imputed for 268 families	3.4 earned income; 2.3 other income. Since some families had both earned and other income imputed, the percentage of families with any portion of their income imputed lies between 3.4 and 5.7
Physician visits, some with:		
Hospitalized illness	These are physician visits outside of the hospital in conjunction with an illness for which the patient was hospitalized	8.6 have visits imputed
Major illness	These are physician visits outside of the hospital for a chronic or expensive illness	1.3 have visits imputed
Pregnancies terminating in 1970	These are prenatal care visits and include the delivery and in-hospital visits. They include, in addition to live births, still births, miscarriages, and abortions, both legal and illegal.	2.9 have visits imputed
Minor illnesses, routine checkups, shots, tests, and routine visits to an opthalmologist for eye refraction	In order to be counted as a doctor visit, the test must have been administered in a doctor's office.	2.2 have visits imputed
Hospital days per admission	If number of days was unknown but expenditure was given, the per diem for that hospital was divided into expenditure to obtain total days. If neither days nor expenditure was known, number of days was based on diagnosis combined with age. If diagnosis was not known, number of days was based on age, sex, and race.	0.9 of admissions have days imputed

TABLE 14

Original Poststratification Adjustment Categories

Characteristic				*Post-Stratification Adjustment Factor*	*Percent of Sample Households*	
Race	*Residence*	*Family Size*	*Family Income*		*Unweighted, Unadjusted*	*Weighted, Adjusted*
White	SMSA	1	Under $3,000	1.622	6.0%	5.9%
White	SMSA	1	$3,000 and over	1.191	5.6	8.0
White	SMSA	2+	Under $3,000	1.136	3.3	2.5
White	SMSA	2+	$3,000-14,999	1.160	18.6	29.8
White	SMSA	2+	$15,000 and over	1.181	5.6	12.3
White	NonSMSA	1	All categories	.967	5.9	5.7
White	NonSMSA	2+	Under $3,000	1.080	3.3	2.7
White	NonSMSA	2+	$3,000-14,999	.750	21.9	18.4
White	NonSMSA	2+	$15,000 and over	1.147	2.4	3.8
Nonwhite	SMSA	1	Under $3,000	1.200	3.3	1.2
Nonwhite	SMSA	1	$3,000 and over	.714	2.2	1.0
Nonwhite	SMSA	2+	Under $3,000	.714	4.3	1.0
Nonwhite	SMSA	2+	$3,000-14,999	.600	13.6	4.1
Nonwhite	SMSA	2+	$15,000 and over	1.167	.8	.7
Nonwhite	NonSMSA	1	All categories	1.000	.6	.6
Nonwhite	NonSMSA	2+	All categories	.917	2.6	2.2
Total				1.000	100.0%	100.0%

TABLE 15

Alternative Poststratification Adjustment Categories

Characteristic			*Post-Stratification Adjustment Factor*	*Percent of Sample Persons*	
Race	*Sex*	*Age*		*Unweighted, Unadjusted*	*Weighted, Adjusted*
White	Both	Less than 6	1.0344	6.5%	8.7%
White	Both	6 to 11	.9699	7.9	10.2
White	Both	12 to 17	.9563	8.3	10.2
White	Male	18 to 29	1.1487	5.0	7.8
White	Female	18 to 29	1.1417	5.5	8.3
White	Male	30 to 44	1.0336	5.1	7.3
White	Female	30 to 44	1.0036	5.4	7.5
White	Both	45 to 54	1.1228	6.7	10.3
White	Both	55 to 64	1.0330	6.6	8.3
White	Both	65 to 74	1.0245	6.5	5.5
White	Both	75 or more	1.1384	4.2	3.5
Nonwhite	Both	Less than 9	.7906	7.3	2.8
Nonwhite	Both	9 to 17	.6803	8.4	2.6
Nonwhite	Male	18 to 44	.9781	3.8	2.0
Nonwhite	Female	18 to 44	.7465	6.0	2.3
Nonwhite	Both	45 to 64	.7459	4.7	2.0
Nonwhite	Both	65 or more	.7410	2.3	.8
Total			1.0000	100.0%	100.0%

TABLE 16

Effect of Poststratification Adjustment on Distribution of Sample Persons

Characteristic	*Percent of Weighted Sample Persons*		
	Without Post-Stratification Adjustment	*With Post-Stratification Adjustment*	
		Original Categories	*Alternative Categories*
Demographic			
Age of oldest family member			
Less than 65 years	86.8%	86.2%	86.5%
65 years or more	13.3	13.8	13.5
Family income			
Nonpoor	75.7	77.0	77.1
Poor	24.3	23.0	22.9
Race			
White	83.8	87.9	87.5
Nonwhite	16.2	12.1	12.5
Residence			
Rural nonfarm	25.8	24.5	26.5
Rural farm	7.8	6.8	7.9
SMSA central city	29.1	29.8	27.7
SMSA other urban	23.3	26.9	23.8
Urban nonSMSA	13.9	12.1	14.2
Perceived and evaluated health			
Perception of health			
Excellent	37.0	37.8	37.4
Good	43.3	43.0	43.1
Fair	12.7	12.4	12.7
Poor	3.8	3.9	3.8
Worry about health			
A great deal	8.3	8.4	8.3
Some	18.2	18.2	18.4
Hardly any	20.7	20.6	20.8
None	50.1	50.1	49.8
Severity over all diagnoses			
Elective only	23.4	23.9	23.4
Elective and mandatory	19.8	20.3	20.3
Mandatory only	20.1	20.2	20.3
Number of diagnoses			
1	27.9	28.2	28.0
2	17.5	17.6	17.6
3	8.9	9.2	9.2
4 or more	8.6	9.0	8.9

(*Continues*)

TABLE 16 (*cont.*)

Characteristic	*Percent of Weighted Sample Persons*		
	Without Post-Stratification Adjustment	*With Post-Stratification Adjustment*	
		Original Categories	*Alternative Categories*
Level of use			
Physician visits per person			
0	33.1	32.4	32.6
1	18.5	18.6	18.4
2 to 4	23.9	24.1	24.1
5 or more	24.5	24.9	24.9
5 to 13	18.4	18.7	18.7
14 or more	6.1	6.2	6.2
Hospital admissions per person			
0	88.6	88.7	88.3
1	9.1	9.0	9.3
2 or more	2.3	2.3	2.4
Total	100.0%	100.0%	100.0%

7.3. Comparisons of Results

Tables 16–18 present the effect of the use of poststratification adjustment on our data. We do not intend to suggest that these specific results would occur if such adjustment were used with any or all similar data.

Table 16 shows the effect of poststratification adjustment on the weighted distribution of sample persons. The most pronounced effect is on the estimate of racial composition: Both adjustments reduce the estimated percentage of the sample that is nonwhite. The original poststratification adjustment reduced the proportion living in rural areas but the alternative increased it slightly. Among the perceived and evaluated health categories, the original poststratification produced slightly more people in the most favorable groups than did the alternative method. The use of the original poststratification categories also resulted in an estimate that fewer people saw a physician more than once or were admitted to a hospital. Only in the case of racial composition do we know which adjustment gave better results, because this is the only case presented in which a poststratification category was used as an analytic category.

In Table 17 the effect of poststratification on both estimates of physician visits and estimates of expenditure per hospital admission was examined. Poststratification seems to have little effect on physician visits. The hospital expenditure figures are more affected by the adjustments, at least by the original adjustment. The original adjustment tended to increase the mean expenditure in most categories. The main effect of the alternative poststratification adjustment was to increase the mean for nonwhites. (The original adjustment appeared to decrease this same mean.)

Table 18 considers poststratification according to its effect on the reliability of differences between groups. This table does not indicate that there was any great change in the data as a result of using either of the two sets of poststratification adjustment factors or any measurable consistent increase in reliability (that is, declines in standard errors of differences).

7.4. Recommendations

In order to determine definitively the effect of alternative poststratification adjustments, we would need the results of a complete census, using the sample survey questionnaire. Numerous samples could be formed from the census data; poststratification could be performed on each; and the results could be compared with the census data and with the unadjusted sample data. We would decide whether or not to use poststratification (and, if so, which set of categories to use in the adjustment) by examining the distribution of the differences between the census value and the various estimates.

It is difficult to make this same decision with the limited information actually available. We have had to examine the effect of poststratification by comparing adjusted and unadjusted estimates from a single sample. Also, we were able to look at only two different types of estimates—those for mean number of physician visits and mean total hospital expense per admission.

Our attempt at measuring the impact of poststratification was further confounded by the fact that noninterview adjustment would usually be performed first, but in this study we had considered each separately. Had we computed estimates using combinations of noninterview adjustments and poststratification adjustments, some combination of the two might have interacted in such a way that such adjustment would have had a larger effect than either individual adjustment would have suggested.

This information about the effect of poststratification, limited though it is, is a useful first step in developing a more thorough investigation of the expected effect of poststratification adjustment of different types of data.

Although it is difficult to assess the effect of poststratification on the data and even more difficult to predict what its use would mean to other surveys, we suggest that it be done. Poststratification is an extremely inexpensive procedure and should result in at least some small improvement in the data, so long as there is more homogeneity of persons within than between categories. For this particular study, this means that persons in the same age–sex–race group or the same race–residence–family size–income group should have more similar medical experiences than do people in general. The choice of categories depends on the nature of the survey because the categories should be delineated by characteristics correlated with important estimates in the study.

TABLE 17

Effect of Poststratification Adjustment on Estimates of Physician Visits for Persons Seeing a Physician and on Estimates of Hospital Expenditures per Admission

Characteristic	*Physician Visits per Person*			*Expenditures per Admission*		
	Without Post-Stratification Adjustment	*With Post-Stratification Adjustment*		*Without Post-Stratification Adjustment*	*With Post-Stratification Adjustment*	
		Original Categories	*Alternative Categories*		*Original Categories*	*Alternative Categories*
Demographic						
Age of oldest family member						
Less than 65 years	5.4	5.4	5.4	$ 616	$ 640	$ 616
65 years or more	7.7	7.8	7.7	830	863	831
Family income						
Nonpoor	5.5	5.5	5.6	649	674	647
Poor	6.3	6.5	6.3	685	715	691
Race						
White	5.7	5.7	5.7	652	684	650
Nonwhite	5.9	6.0	5.9	702	688	730
Residence						
Rural nonfarm	5.4	5.4	5.4	544	557	540
Rural farm	5.5	5.6	5.5	555	575	566
SMSA central city	6.0	6.0	6.0	706	727	711
SMSA other urban	5.5	5.6	5.5	920	910	915
Urban nonSMSA	6.1	6.1	6.2	447	444	447
Perceived and evaluated health						
Perception of health						
Excellent	3.5	3.6	3.6	467	488	472
Good	5.2	5.3	5.3	600	609	596
Fair	8.9	9.1	9.0	670	698	682
Poor	14.1	14.2	14.2	816	868	809

Worry about health						
A great deal	11.6	11.7	11.6	973	1,036	980
Some	7.3	7.4	7.3	539	560	537
Hardly any	4.9	4.9	5.0	487	442	481
None	3.6	3.6	3.7	427	441	427
Severity over all diagnoses						
Elective only	3.1	3.2	3.2	472	499	473
Elective and mandatory	8.3	8.3	8.3	536	550	538
Mandatory only	7.1	7.2	7.1	789	830	792
Number of diagnoses						
1	3.8	3.8	3.8	605	626	600
2	5.6	5.6	5.7	573	573	574
3	8.1	8.0	8.0	543	563	540
4 or more	11.6	11.6	11.5	836	883	839
Level of use						
Physician visits per person						
1	1.0	1.0	1.0	—	—	—
2 to 4	2.7	2.8	2.7	—	—	—
5 or more	12.1	12.1	12.1	—	—	—
5 to 13	8.0	8.0	8.0	—	—	—
14 or more	24.8	24.8	24.7	—	—	—
Expenditures per admission						
$0 to 300	—	—	—	184	184	185
301 to 600	—	—	—	427	426	427
601 or more	—	—	—	1,525	1,556	1,525
601 to 1200	—	—	—	841	843	842
1201 or more	—	—	—	2,769	2,801	2,771
Total	5.7	5.7	5.7	$ 658	$ 685	$ 658

TABLE 18

Effect of Poststratification Adjustment on Differences between Groups for Mean Number of Physician Visits for Persons Seeing a Physician and on Differences between Groups for Hospital Expenditures per Admission[a]

Characteristic	Physician Visits per Person			Expenditures per Admission		
	Without Post-Stratification Adjustment	With Post-Stratification Adjustment		Without Post-Stratification Adjustment	With Post-Stratification Adjustment	
		Original Categories	Alternative Categories		Original Categories	Alternative Categories
Demographic						
Age of oldest family member						
Less than 65 versus 65 or more	− 2.4(6.4)[a]	− 2.4(6.6)	− 2.3(6.2)	$− 214(3.2)	$− 222(3.0)	$− 215(3.1)
Family income						
Nonpoor versus poor	− .8(3.1)	− 1.0(3.7)	− .7(2.9)	− 36(.5)	− 41(.6)	− 43(.6)
Race						
White versus nonwhite	− .2(.5)	− .3(.6)	− .2(.4)	− 50(.5)	− 3(.0)	− 80(.7)
Residence						
SMSA other urban versus:						
Rural nonfarm	.2(.5)	.2(.4)	.2(.4)	377(2.3)	353(2.2)	375(2.3)
Rural farm	.0(.0)	.0(.0)	.1(.1)	365(2.2)	335(2.1)	349(2.1)
SMSA central city	− .4(1.1)	− .5(1.1)	− .5(1.2)	214(1.3)	183(1.2)	205(1.3)
Urban nonSMSA	− .6(1.2)	− .6(1.1)	− .7(1.2)	474(2.9)	467(2.9)	467(2.9)
Perceived and evaluated health						
Perception of health						
Excellent versus:						
Good	− 1.7(6.9)	− 1.7(6.7)	− 1.7(7.0)	− 132(1.4)	− 122(1.1)	− 123(1.2)
Fair	− 5.4(11.7)	− 5.5(10.8)	− 5.4(11.8)	− 203(2.3)	− 210(2.1)	− 209(2.2)
Poor	−10.6(9.7)	−10.6(9.3)	−10.6(9.2)	− 348(3.7)	− 380(3.6)	− 337(3.5)

Worry about health						
None versus:						
Hardly any	−1.3(4.0)	−1.3(4.4)	−1.3(3.9)	−60(.6)	−1(.0)	−54(.6)
Some	−3.7(9.9)	−3.8(10.2)	−3.6(9.5)	−112(1.7)	−119(1.6)	−110(1.7)
A great deal	−8.0(12.1)	−8.1(11.3)	−8.0(11.8)	−547(7.1)	−595(6.6)	−663(6.9)
Severity over all diagnoses						
Elective only versus:						
Elective and mandatory	−5.2(17.0)	−5.1(15.9)	−5.2(16.6)	−65(.9)	−51(.6)	−65(.9)
Mandatory only	−4.0(12.3)	−4.0(11.6)	−3.9(12.1)	−318(3.8)	−331(3.5)	−319(3.9)
Number of diagnoses						
1 versus:						
2	−1.8(5.2)	−1.8(5.2)	−1.9(5.0)	32(.3)	53(.5)	26(.3)
3	−4.2(10.0)	−4.1(9.6)	−4.2(9.8)	63(.8)	63(.7)	60(.8)
4 or more	−7.8(19.5)	−7.8(20.6)	−7.7(19.8)	−230(2.3)	−257(2.2)	−239(2.3)
Level of use						
Physician visits per person						
1 versus:						
2 to 4	−1.7(8.7)	−1.8(83.1)	−1.7(86.2)	—	—	—
5 or more	−11.1(40.2)	−11.1(39.5)	−11.1(40.1)	—	—	—
5 to 13	−7.0(78.3)	−7.0(72.8)	−7.0(77.9)	—	—	—
14 or more	−23.8(30.9)	−23.8(29.3)	−23.7(29.8)	—	—	—
Expenditures per admission						
$0 to 300 versus:						
301 to 600	—	—	—	−242(38.1)	−242(36.2)	−242(38.2)
601 or more	—	—	—	−1,341(12.2)	−1,371(12.2)	−1,340(12.0)
601 to 1200	—	—	—	−657(36.2)	−659(35.3)	−657(37.1)
1201 or more	—	—	—	$−2,585(10.4)	$−2,616(11.1)	$−2,586(10.1)

[a] Numbers in parentheses are the ratios of the differences to the estimated standard errors of the differences.

REFERENCES

American Hospital Association (1970). *Guide to the Health Care Field.* Chicago: American Hospital Association.

Andersen, R., Kasper, J., Frankel, M. R., and Associates (1979). *Total Survey Error: Applications to Improve Health Surveys.* San Francisco: Jossey-Bass.

Andersen, R., Lion, J., and Anderson, O. W. (1976). *Two Decades of Health Service: Social Survey Trends in Use and Expenditure.* Cambridge, Mass: Ballinger.

Cartwright, A. (1959). The families and individuals who did not cooperate on a sample survey. *Milbank Memorial Fund Quarterly* 37, 347–368.

Daniel, W. W. (1975). Nonresponse in sociological surveys: A review of some methods for handling the problem. *Sociological Methods and Research* 3, 291–307.

Keyfitz, N. (1957). Estimates of sampling variance where two units are selected from each stratum. *Journal of the American Statistical Association* 52, 503–510.

Kish, L. (1965). *Survey Sampling.* New York: Wiley.

National Center for Health Statistics (1962). Series B, No. 32.

National Center for Health Statistics (1965). Series 2, No. 6.

National Center for Health Statistics (1971). Series 2, No. 43.

Thompson, M. M., and Shapiro, G. (1973). The current population survey: An overview. *Annals of Economic and Social Measurement* 2, 105–129.

U.S. Bureau of the Census (1978). *The Current Population Survey—Design and Methodology.* Technical Paper No. 40. Washington, D.C.: U.S. Government Printing Office.

U.S. Bureau of the Census (1971). Series P-20, No. 225.

U.S. Bureau of the Census (1972). Series P-60, No. 80.

CHAPTER 11

An Investigation of Nonresponse Imputation Procedures for the Health and Nutrition Examination Survey*

David W. Chapman

1. INTRODUCTION

1.1. Brief Description of the Health and Nutrition Examination Survey[1]

The National Center for Health Statistics (NCHS) has the responsibility for conducting a continuing National Health Survey to help determine the level of health of the people of the United States. One program developed by NCHS in this regard is the Health Examination Survey (HES). The HES is a national probability sample of the noninstitutional population of the United States. The survey respondents are given health and dental examinations, tests, and measurements. Three HES cycles were conducted between 1959 and 1970, each cycle lasting roughly three years. The age groups covered by the three HES cycles were 18–79, 1–5, and 6–11.

* This chapter is a condensed version of the final project report that was prepared on August 16, 1974, for the Division of Health Examination Statistics of the National Center for Health Statistics as the final task in a project performed by Westat, Inc., of Rockville, Maryland, under Contract HSM-110-73-371. This study was conducted during 1973 and 1974 with the assistance of Stephen F. Kaufman, under the general direction of Morris H. Hansen, both of Westat, Inc. There was close cooperation and communication between the NCHS and the Westat project staffs during the design and conduct of the study. In particular, the cooperation and advice of Mr. Garrie Losee and Mr. Wesley Schaible of NCHS are gratefully acknowledged.

[1] This description is extracted from NCHS (1973).

INCOMPLETE DATA
IN SAMPLE SURVEYS
Volume 1, Part II

ISBN 0-12-363901-8

A separate survey, the National Nutrition Surveillance Survey (NNSS), was also developed to investigate national health characteristics. The purpose of this survey is to estimate, on a continuing basis, the nutritional status of all those 1 year of age and over in the noninstitutional population of the United States.

Chiefly because of operational considerations and of the compatibility and similarity of the HES and the NNSS, the HES and the NNSS were combined into one survey, called the Health and Nutrition Examination Survey (HANES). It is intended that HANES be a continuing national probability survey with approximately two-year cycles, the first cycle beginning in 1971. The sample to be selected for each of HANES is a multistage probability sample of about 30,000 persons ages 1–74. Each two-year cycle is to be divided randomly into two annual rounds in such a way that each round constitutes a probability sample from the entire population. There is to be oversampling of (1) persons in poverty-level areas, (2) preschool children, (3) women of child bearing age, and (4) the elderly. The data-collection and health-examination procedures for the sample cases for HANES included the following:

(1) *Census interview:* obtains household composition and demographic information and to determine eligibility for HANES.

(2) *HANES interview:* sets up an examination appointment and to obtain medical history and other data.

(3) *General nutrition exam:* consists of a general physical examination; dermatological, ophthalmological, and dental examinations; body measurements; biochemical assessments; and dietary intake measurements.

(4) *Detailed medical exam:* is given to a subsample of about 6000 persons ages 25–74 in Cycle I. (Data from the detailed medical exams were not used in this study.)

Cycle I of HANES involves 65 primary sampling units (PSUs), 15 of which were large enough to be selected with certainty. For the first round of the cycle, 35 PSUs were selected, 10 with certainty. The Round I sample size turned out to be 14,147 persons. Of these, 10,126 (72%) received the nutrition examination and 12,296 (87%) completed (or had completed for them) a medical-history questionnaire. In addition, some of the demographic data (e.g., age, race, and sex) were obtained for all 14,147 sampled persons. Other demographic characteristics (e.g., family income and education of the head of the household) were obtained for most but not all of the sampled persons.

1.2. The Current Treatment of Nonresponse for HANES and Possible Alternatives

When there is nonresponse in a survey, there will be a nonresponse bias in the resulting survey estimates. This bias cannot be eliminated by imputing for nonrespondents. However, an attempt is usually made to impute for nonrespondents in such a way that the associated nonresponse bias is reduced without increasing the variance of a survey estimate excessively. In this study the following two basic approaches to making imputations for nonresponding units have been considered:

(1) Assume that within the entire population (or within subgroups of the population) nonrespondents are a random sample of the population (or population subgroup).

(2) Develop relationships for various survey measurements among the respondents that can be used to extrapolate measurements to the nonrespondents.

The Division of Health Examination Statistics (DHES) has developed a nonresponse imputation procedure to use for HANES. The DHES procedure, which is based on the first of the two approaches just listed, involves adjusting the basic sampling weights (i.e., the inverses of the selection probabilities) of the respondents, to "account for" the nonrespondents. The weight adjustments of the respondents, or examined persons, are made separately within several classes defined by cross-classifications of a number of characteristics. These adjustments, which are described in detail in Section 4.1.1, are made in three separate stages.

The purpose of this project was to consider imputation procedures using both approaches just given and to develop those that appear to be promising. The selected procedures along with the current DHES procedure were evaluated and compared, principally by estimating the nonresponse biases of the procedures. Five alternative imputation methods were developed that were based on the first approach given. These procedures, which are basically similar to, but perhaps less complicated than, the current DHES procedure, involve alternative choices of nonresponse weight-adjustment classes. These choices were based, to a large extent, on (1) some computer calculations produced by the AID programmed procedure and (2) some "explained" variance and relcovariance calculations. Both the AID computations and the variance and relcovariance calculations are described in Section 2.

The use of an iterative adjustment procedure, called "raking," was also considered as a variant of the first approach. However, as indicated later, no raking procedures were evaluated with the other alternatives because it appeared that little could be added by an iterative procedure.

Using the second approach to nonresponse imputation, two basic procedures were considered. The first was to identify, among respondents, any relationship between selected examination characteristics and the number of household calls the person required to make and keep an examination appointment, referred to as "degree of persuasion." Any relationship discovered between survey characteristics and degree of persuasion could be used to impute characteristics to unexamined persons. However, as demonstrated later, no meaningful relationships were found for the characteristics investigated, and consequently this type of procedure was not developed further.

The other procedure considered, which was based on the second approach, was to use multiple regression in some way to develop a relationship to project survey measures for unexamined persons from those of examined persons. As discussed later, it was decided that the use of regression in this way was not feasible in this situation.

Therefore, the comparisons made were among six procedures developed from the first approach listed at the beginning of this section, one procedure being the current DHES procedure. Comparisons of approximate nonresponse bias were made from calculations involving both the examination characteristics and the medical-history information. The basic result was that all six procedures compared, which included a simple single-stage adjustment to 20 age–race–sex census counts, provided similar bias-reduction levels. The estimated relative biases were meaningful for estimated means for medical-history characteristics (many falling between .5 and 2.0%) but were quite small for estimated means for examination characteristics. The details of these comparisons, including estimated biases, are given in Section 4.

2. IDENTIFICATION OF IMPORTANT VARIABLES TO USE IN DEFINING WEIGHTING CLASSES

2.1. The Use of Weighting Classes for Nonresponse Adjustments

With the use of weighting classes to make nonresponse adjustments, an attempt is made to partition the entire sample into classes, based on available characteristics, which group together respondents and nonrespondents having similar survey measurements. Within each class, the survey characteristics of the respondents are imputed to the nonrespondents by adjusting the basic sampling weights of the respondents. Specifically, if N_c denotes the sum of the sampling weights of all sampled units contained in weighting class c and if N'_c represents the sum of the sampling weights of all survey respondents contained in class c, the weight-adjustment factor u_c is computed as

$$u_c = N_c/N'_c. \tag{1}$$

The sampling weight of each respondent in class c is multiplied by u_c. This "weights up" the respondents in class c to the total sample in class c. This process may be applied once or may be used in two or more successive stages of weight adjustments, utilizing different variables to define weighting classes at each stage.

The use of the adjusted respondent weights in computing an estimated mean, total, or percentage for a characteristic is equivalent to assigning the (weighted) mean value of the characteristic among the respondents in each class to the nonrespondents in the same class and then making estimates from all sampled units, using the basic sampling weights. Assuming that respondents and nonrespondents are more alike in the classes than they are over the entire sample, this procedure will be better than that which makes only one overall nonresponse weight adjustment. In addition to grouping together respondents and nonrespondents that have similar survey characteristics, it is also important to form weighting classes for which the response rates vary. Otherwise the nonresponse adjustment given in Eq. (1), which is simply the inverse of the class weighted response rate, would be constant from one class to the next. This would be equivalent to making one overall nonresponse adjustment.

Because of the initial census household interview, there is a substantial amount of demographic information available for the HANES nonrespondents to use in defining weighting classes. The methods used to help identify those demographic variables to use in defining weighting classes are described in Sections 2.2 and 2.3.

2.2. The Use of the AID Programmed Procedure to Define Weighting-Class Variables

2.2.1. *Summary of the AID Procedure*[2]

Basically, the AID programmed procedure (or tree analysis) produces a continuing sequence of splits of subgroups of a file into two parts. One dependent variable and a set (or pool) of independent variables are specified for use in the program. At each stage, the subgroup that is split is the one that, for the dependent variable, has the largest variance among its members. Once the particular subgroup to be split is chosen, the particular split made is determined by checking all possible splits into two subgroups that can be made, based on the values of each variable in the pool of independent variables. The split chosen is the one that maximizes the sum of squares between the means of the two parts. The sequence of splitting subgroups into smaller and smaller subgroups ends when each of the final subgroups cannot be subdivided further because of one of the following three reasons:

(1) The subgroup has fewer observations than a specified minimum number (25 for this study).

(2) The sum of squares of the observations within the subgroup is less than a specified minimum percentage of the sum of squares of the observations in the total file (.5% for this study).

(3) The sum of squares between means for all possible splits of the subgroup into two parts does not exceed a specified minimum percentage of the total sum of squares for the subgroup (2.0% for this study).

[2] A more detailed description of this technique is given by Morgan and Sonquist (1963).

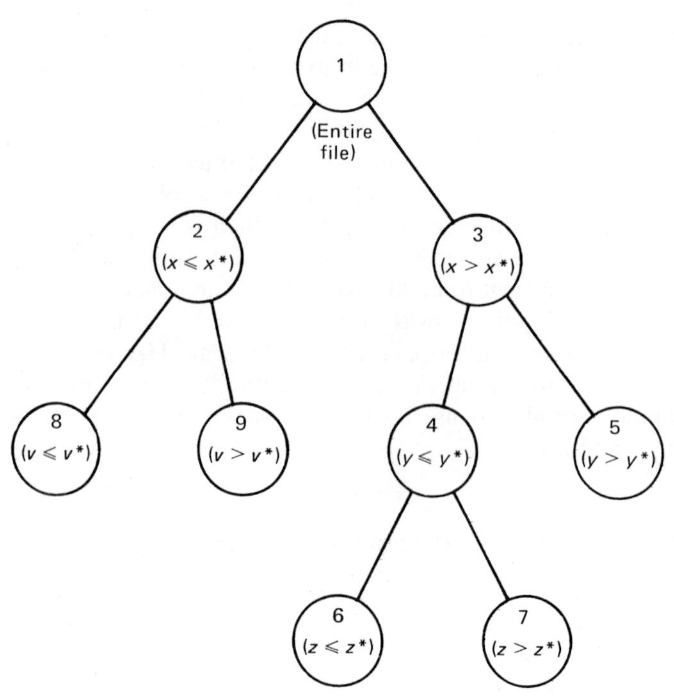

Fig. 1. *An example of the splitting of a file into five subgroups by the AID procedure.*

As an example, suppose that the AID procedure split a file into five final groups, as illustrated in Fig. 1. In this case, the entire file (subgroup 1) was split into subgroup 2 and subgroup 3, based on the independent variable x. Subgroup 3, which has a larger within sum of squares than subgroup 2, was then split into subgroups 4 and 5, based on the independent variable y. Then subgroup 4 was split into subgroups 6 and 7 based on variable z. Finally, subgroup 2 was split into subgroups 8 and 9, based on variable v. The splitting variables x, y, z, and v are not necessarily different. Based on the criteria for further splitting of subgroups, no other subgroups were eligible for subdivision, leaving subgroups 5, 6, 7, 8, and 9 as final.

The AID printout includes the following information:

(1) the name and corresponding values of the independent variable used to define the two subgroups created at each split,

(2) the number of observations in each of the subgroups,

(3) the sums of the weights associated with each of the subgroups,

(4) the sum of squares within each of the subgroups, and

(5) the sum of squares between subgroup means for the two subgroups created at each split, along with a listing of the between-subgroup sum of squares for all the splits tested.

2.2.2. *The Application of AID to the HANES Data*

As indicated in Section 2.1, it is important to define weighting classes for which response rates vary substantially in order to avoid having approximately the same nonresponse adjustment in all weighting classes. The AID procedure was applied to help identify the demographic variables that are most useful in defining weighting classes having substantially different response rates. The dependent variable for the AID procedure was the 0–1 response variable (i.e., whether or not a sampled person was examined). There were 12 demographic variables that made up the pool of independent variables. The names of these variables and the number of categories used for each are given in Table 1. The definitions of the categories for each of the 12 variables are given in Appendix A.

The particular version of the AID procedure used for this project would not accept data files that had more than 8000 records. Therefore, the HANES file of 14,147 records was subdivided into the following four sets of subgroups, with the tree analysis being applied separately to each subgroup:

(1) the odd- and even-numbered records (that is, alternate records in the file were assigned to two groups),

(2) the records contained in each of the four census regions: Northeast, North Central, South, and West,

(3) the records associated with each of five age groups: 1–5, 6–19, 20–44, 45–64, and 65–74; and

(4) the records contained in each of the 35 stands listed in Appendix A.[3]

The AID analyses computed for the first three of the four sets of subgroups listed are summarized in Table 2. Six of the eleven tree diagrams corresponding to the AID calculations summarized in Table 2 are given in Appendix B. The values of the independent variable, the subgroup sizes, and the subgroup means (i.e., response rates) associated with each split for each tree are also given in Appendix B.

For the set of subgroups defined by the 35 stands, the AID results varied substantially from those given in Table 2 for the other three sets. Overall, there was a considerably higher percentage of explained variance for the AID final subgroups within the stands than there was for the other sets of subgroups. Across the 35 stands, the percentages of explained variance ranged from 11.1 to 57.2%, and averaged about 37%. The number of splits and the number of resulting final groups were also considerably higher for the AID analyses within stands.

Our hypothesis regarding the AID results is that the higher percentages of explained variance for the stand analyses appear to be due, to a large extent,

[3] Stands, or stand locations, are counties, cities, or towns in which the examination centers are located. Sample areas from which examinees are drawn for the stands consist of the PSUs, which may include several counties.

TABLE 1

The 12 Demographic Variables Available for the AID Programmed Procedure

Category number	*Demographic variable*	*Number of categories*
1	Stand location	35
2	Age	5
3	Race	3
4	Sex	2
5	Family income	13
6	Education of household head	19
7	Degree of urbanization	8
8	SMSA code	3
9	Region	4
10	Household size	17
11	Number of sampled persons in household	6
12	Poverty index	4

TABLE 2

Summary of AID Results

Subgroups	*Number of final subgroups*	*Variables used to split subgroups*[a]	*Percentage of variation "explained"*
Odds–evens[b]			
Odds	6	2, 5, 9, 8, 6	6.0
Evens	7	5, 10, 2, 7, 6, 7	7.3
Age classes			
1–5	13	1, 1, 3, 10, 12, 10, 1, 7, 8, 1, 9, 5	15.5
6–19	9	1, 6, 10, 1, 5, 10, 7, 1	10.0
20–44	7	1, 10, 1, 1, 6, 5	8.2
45–64	6	1, 6, 1, 5, 5	9.4
65–74	8	1, 5, 1, 1, 6, 6, 1	11.5
Regions			
Northeast	11	1, 10, 5, 10, 5, 6, 6, 2, 7, 7	13.0
North central	5	2, 5, 1, 1	6.8
South	7	6, 1, 2, 10, 5, 1	6.7
West	11	5, 1, 10, 8, 2, 6, 1, 8, 1, 5	14.0

[a] The independent variables are identified in Table 1 and are given here in the sequence they were used to split subgroups.

[b] Variable 1 (Stand Location) was not included in the pool of independent variables for this AID analysis.

to the smaller total group sizes. The average number of records (i.e., sampled persons) per stand is 404 (14,147 ÷ 35) as compared with averages of 3537 per region and 2829 per age group. With the smaller total group size, the random variation in the sample alone can result in a considerably higher sum of squares between subgroup means, relative to the total sum of squares, than it would with a larger group size. For the four sets of subgroups examined, the percentage of variance explained appears to be inversely related to the subgroup size.

2.2.3. Conclusions Based on the AID Results

The use of AID to define specific weighting classes for nonresponse adjustments does not seem to be feasible. First, the relatively low proportions of explained variance in Table 2 indicate that it is difficult to split the larger subgroups into finer subgroups having substantially different response rates. This implies that if the AID final groups were used as nonresponse weight-adjustment classes, the impact on estimated means and proportions would probably not be very substantial.

Second, many of the subgroup splits that occur are selected over alternative splits by a very narrow margin. The random variation in the sample could easily determine the split selected in such cases. A particular split selected leads to a sequence of splits and resulting final subgroups that, in general, are quite different from those that would follow from an alternative split. This influence of sampling variation on the tree analysis is demonstrated by comparing the analyses for the odd- and even-numbered records (given in Appendix B and summarized in Table 2). These two sets of records differ only by random variation in the selection of households within segments and in the selection of eligible persons within selected households. It is therefore reasonable to expect the tree analyses for the two groups to be similar. However, the very first split is different for the two groups. Also, the two sets of final groups formed are quite different.

A third difficulty with the use of AID to define weighting classes is that a split will sometimes yield a subgroup containing very few records. For the stand subsets, many splits yielded subsets with less than 10 records. Even if the specified minimum size group allowed to be split is set higher than 25 records, small subsets are still possible. Small weighting-class groups are undesirable because of the possibility of erratic weight adjustments that cause an increase in the variance of survey estimates.

Another substantial problem in applying the AID procedure would be the difficulty in defining classes that would vary both with respect to response rates and with the mean values of the examination characteristics. Since the AID procedure allows only one dependent variable, the tree analysis would have to be carried out separately for each dependent variable of interest. It would be extremely difficult to combine two or more tree analyses in some way to obtain some joint (or overall) tree analysis to use in defining weighting classes.

Finally, whenever the number of splits is very large, the definitions of the specific classes become rather complex and could be rather awkward to work with in practice. The use of cross-classifications of two or more variables provides a more convenient set of weighting classes to work with than does the AID procedure.

Perhaps the most useful information from the AID results is obtained by noting which independent variables are used most often to define the "optimal" splits in the tree analysis. An inspection of Table 2 reveals that the following five demographic variables are used most often: (1) stand location (variable number 1), family income (number 5), age (number 2), education of head of household (number 6), and household size (number 10). These five variables, and perhaps degree of urbanization (number 7), appear to be those most important in defining subclasses of the sample having varying response rates. This conclusion will be substantiated later, using other criteria.

2.3. Using Variances and Covariances to Select Variables to Use in Defining Weighting Classes

2.3.1. *Proportions of "Explained" Variance in the 0–1 Response Variable and in Selected Examination Characteristics*

In this section a description is given of a method used to identify variables that appear to be most effective in defining subclasses of the sample that vary both with respect to response rates and with the means for selected examination characteristics. This method associates with each set of subclasses of the sample a proportion of the variance in the 0–1 response variable, or in a selected examination characteristic, that is "explained" or "accounted for" by the set of subclasses (or subclass means). The sets of subclasses used are those defined by the categories of the 11 demographic variables listed in Table 3. This set of variables is the same one used for the AID pool of independent variables, except that the variable "number of sampled persons in household" has been dropped, because it is dependent on the sample design. Also, some of the categories of several of the variables used were combined for these variance calculations. In Appendix A the categories and collapsing pattern used for each of the 11 variables are given.

Ten examination characteristics were chosen as dependent variables to be included in the variance calculations. These 10 variables, which are given in Table 4, were selected jointly by Westat and DHES project personnel. An attempt was made to select items that would be "representative" of each of the four major classes of examination characteristics indicated in Table 4.

TABLE 3

The 11 Demographic Variables Included in the Variance Analysis

Demographic variable	*Number of categories*[a]
Stand location	35
Age	5
Race	3
Sex	2
Family income	9(13)
Education of household head	7(19)
Degree of urbanization	6(8)
SMSA code	3
Region	4
Household size	9(17)
Poverty index	3(4)

[a] The number of categories used in the AID analysis is given in parentheses, if different from the number used for the variance analysis.

The proportion P_r of total variance in the 0–1 response variable "explained" by a particular set of classes was computed as

$$P_r = s_{r[c]}^2 / s_R^2, \tag{2}$$

where $s_{r[c]}^2$ is the weighted sum of squares between the weighted response rates associated with a particular set of classes and s_R^2 the total weighted sum of squares for the 0–1 response variable for the entire sample.

Similarly the proportion P_X of variance in each of the 10 dependent variables explained by a particular set of classes was computed as

$$P_X = s_{\bar{x}[c]}^2 / s_X^2, \tag{3}$$

where $s_{\bar{x}[c]}^2$ is the weighted sum of squares between the weighted means of characteristic X for a particular set of classes and s_X^2 the total weighted sum of squares for characteristic X for the entire sample.

The percentage of total variance in the 0–1 response variable and in each of the 10 examination characteristics explained by the categories of each of the 11 demographic variables is given in Table 5. The explained proportions for the 0–1 response variable were based on all sample units, whereas, the explained proportions for the 10 examination characteristics were based on sample respondents. The first row of Table 5 indicates that the variable "stand location" provides the highest percentage of explained variance, 4.9, of the 0–1 response variable. The variables "age," "family income," "household size," "education of household head," and "poverty index" explain about 2% or slightly more. The variables "degree of urbanization" and "region" explain 1.6% of the variance. The other three demographic variables explain a substantially smaller

TABLE 4

The Set of 10 Selected Examination Characteristics

Examination characteristic	*Values of the characteristic*
Body measurements	
Triceps skinfold	0–65 mm
Body weight	15–400 lb
General medical examination	
Systolic blood pressure[a]	65–300 mm Hg
Head, eyes, ears, nose, and throat findings	Findings (1), no findings (2)
Chest evaluation	Findings (1), no findings (2)
Examiner's subjective impression of nutritional status	Normal (1), abnormal (2), not obtained (8)
Biochemistry measurements	
Hemoglobin	0–20 g %
Cholesterol	10–600 mg per 100 ml
Dietary measurements	
Iron	0–100 mg
Calcium	0–7000 mg

[a] Not obtained for examined persons 5 years of age or less.

proportion of the variance. With the exception of the relatively high proportion explained by the variable "poverty index," these results are consistent with those from the AID analysis.

An inspection of the other ten rows of Table 5 reveals that the variable "age" is apparently the most important variable for defining weighting classes. The five age categories explain the largest percentage of variance for seven of the ten examination characteristics. The variable "sex" explains the most variance for two of the three remaining characteristics. The variables "stand location" and "household size" are consistently higher with respect to explained variance than most of the others. Of the remaining seven independent variables, "income" and "education" appear, on average, to explain the most variance. These approximate rankings are also reasonably consistent with the findings from the AID calculations.

2.3.2. *Approximate Relcovariances between Class Response Rates and Class Means*

The "explained variance" calculations that were made helped to identify variables that can be used to define subclasses of the sample that vary with respect to (1) response rates and (2) the means of selected examination characteristics. By looking at the impact on survey estimates of the use of nonresponse weight-adjustment classes, a criterion was developed that incorporates the simultaneous effect of these two criteria.

TABLE 5

Percentage of "Explained" Variance for One-Way Classifications of the 11 Demographic Variables

Dependent variables	Independent variables										
	Stand No.	Age	Race	Sex	Income	Urbanization	SMSA	Region	Education	HH size	Poverty index
Response (0–1)	4.9[a]	2.2	.23	.075	2.4	1.6	.84	1.6	1.9	2.3	2.2
Triceps skinfold	1.0	12.0	.60	23.7[a]	.11	.30	.11	.05	.15	2.7	.38
Weight	.80	56.5[a]	.17	2.2	.56	.02	.10	.05	.29	9.2	1.4
Blood pressure	5.7	33.4[a]	.04	.49	2.3	.83	1.7	1.1	2.2	12.2	.04
Head, eyes, etc., findings	10.1[a]	3.4	.03	.003	.10	.97	.45	2.7	.23	1.7	.03
Chest evaluation	2.1	5.3[a]	.04	.43	.30	.16	.18	.13	.39	1.9	.006
Nutritional status	5.1	6.3[a]	.04	.71	.86	.51	.43	1.3	1.3	1.2	.01
Hemoglobin	6.3	19.0[a]	4.3	14.5	1.9	.95	.82	.69	1.1	4.9	3.1
Cholesterol	1.8	31.2[a]	.03	.08	.59	.15	.01	.15	.39	11.4	.70
Iron	2.3	4.6	.53	9.4[a]	1.1	.44	.06	.71	.68	.51	.85
Calcium	4.1	6.6[a]	1.9	5.1	1.5	.26	.55	1.7	2.5	1.7	.51

[a] The maximum value in the row.

Recall the fundamental assumption underlying the use of weight-adjustment classes to adjust for nonresponse: Within each weighting class the survey characteristics—examination measurements in this case—of the respondents and nonrespondents are similar, or at least more alike than they are across all classes. Under this assumption, it seems reasonable to attempt to define weighting classes that yield survey estimates that vary from those computed from one overall adjustment. If this assumption is true, presumably, although not necessarily, the more the estimates obtained from weighting classes vary from those involving only a single nonresponse adjustment, the smaller will be the nonresponse bias associated with the weighting-class procedure. Therefore, the criterion of choosing weighting classes that maximize the difference between weighting-class estimates and the estimates corresponding to a single nonresponse adjustment was developed. The quantity chosen to measure this difference was the relative difference Δ in estimated totals:

$$\Delta = (x' - \tilde{x})/\tilde{x}, \tag{4}$$

where x' is an estimated total based on nonresponse weighting-class adjustments and $\tilde{x}$ an estimated total based on a single nonresponse adjustment.

In most instances the value of the single nonresponse adjustment will be approximately equal to the simple average of the weighting-class nonresponse adjustments. If this assumption is made, it can be shown that Δ can be written as

$$\Delta = \text{Relcov}(u_c, x'_c), \tag{5}$$

where u_c is the weight adjustment for class c, x'_c the weighted sum of values of X among respondents in class c, and Relcov(u_c, x'_c) the covariance between the u_c and x'_c values computed over the adjustment cells and divided by the product of the mean of the u_c values and the mean of the x'_c values.

Under the assumption that respondents and nonrespondents are more alike within weighting classes than they are across the entire sample the approximate relcovariance, $(x' - \tilde{x})/\tilde{x}$, between u_c and x'_c, gives an indication of the reduction in relative bias when the weighting-class adjustments are used instead of a single adjustment. Therefore, the higher (in absolute value) the relcovariance is between the u_c and x'_c values, the lower the bias of the weighting-class-based estimate is. Consequently, based on this assumption, it is not enough to select weighting classes that vary both with respect to response rates and examination measurements; in addition, the weight adjustments and weighted cell totals must be correlated. (If the cells are about the same size, in terms of sums of weights, this is equivalent to the requirement that the weight adjustments be correlated to the cell means.)

The relative difference Δ between x' and $\tilde{x}$ was computed for each of the 10 examination characteristics and for each set of classes defined by the categories of each of the 11 demographic variables. These values are given as percentages in Table 6. It is apparent from inspection of Table 6 that except for the variable "age" none of the variables defines a set of weighting classes that provides

TABLE 6

Approximate Relcovariances between u_c and x'_c in Percentages $[100(x' - \tilde{x})/\tilde{x}]$ of One-Way Classifications of the 11 Demographic Variables

	Independent variables										
Dependent variables	*Stand No.*	*Age*	*Race*	*Sex*	*Income*	*Urbanization*	*SMSA*	*Region*	*Education*	*HH size*	*Poverty index*
Triceps skinfold	−.04	1.6	.11	1.2	.01	−.05	−.10	−.02	.08	.80	.08
Weight	−.06	2.5	.05	−.04	.13	−.07	−.07	−.02	.05	1.2	.22
Blood pressure	−.11	[a]	−.002	−.004	.02	−.09	−.14	−.04	.02	.19	.005
Head, eyes, etc., findings	−.18	−.39	.003	−.001	−.06	.10	.09	−.15	−.01	−.28	−.05
Chest evaluation	.03	−.18	−.004	−.004	−.008	.03	.03	−.002	.02	−.14	−.007
Nutritional status	−.09	−.67	−.02	.01	−.02	−.11	−.11	.02	.02	.30	−.01
Hemoglobin	−.04	.37	.07	−.02	.06	−.04	−.05	.07	−.03	.22	.09
Cholesterol	.13	1.2	.01	.004	.12	.05	.008	.08	.09	.75	.14
Iron	.26	.53	.11	−.09	.09	.18	.06	.21	−.11	.30	.23
Calcium	.07	−1.5	.28	−.09	.005	.01	.06	.23	−.19	−.76	.11

[a] This number was not computed properly because there were unexpectedly about 15 persons aged 1–5 who had blood pressure measurements. It was intended that this age group be ignored for blood pressure. Instead, these few measurements were weighted up to all persons aged 1–5.

important relative differences between the weighting-class estimates x' and the single adjustment estimates $\tilde{x}$.

2.3.3. *Approximate Relcovariances for Two-, Three-, and Four-Way Classifications*

In the previous section, only one-way classifications of the 11 demographic variables were included in the relative difference calculations. To obtain more information regarding the selection of these variables to form weighting classes, relative differences in weighting-class and single-adjustment estimates were also calculated for selected two-, three-, and four-way classifications.

To reduce the number of cells that would be involved in the two-, three-, and four-way classifications, some of the categories used for one-way classifications of the following four variables were collapsed for use in two-, three-, and four-way classifications:

(1) Stand location: from 35 locations to five stand groups. The grouping was done by listing the stands in order by weighted response rates. Four breaks in the list were made to form five stand groups having approximately equal numbers of examined persons.

(2) Family income: from nine categories to four categories:
- (a) less than $7000,
- (b) $7000–14,999,
- (c) $15,000 or more,
- (d) not obtained.

(3) Degree of urbanization: from six categories to four categories:
- (a) urbanized areas with one million or more population,
- (b) other urbanized areas,
- (c) urban areas outside of urbanized areas,
- (d) rural areas.

(4) Household size: from nine categories to three categories.
- (a) 1–2 persons,
- (b) 3–4 persons,
- (c) 5 or more persons.

Before these four new sets of categories were adopted for further analysis, the approximate relcovariances between x'_c and u_c [i.e., $(x' - \tilde{x})/\tilde{x}$] for the new sets of categories were computed and compared with those for the prior sets. These sets of approximate relcovariances are given as percentages in Table 7.

An inspection of the relative differences given in Table 7 indicates that there are no important decreases in these differences (approximate relcovariances). In fact, for those relative differences equal to .1% or more (in absolute value), the new and prior relative differences are essentially the same, except for one figure in the "stand-group" column. Therefore, it was decided to use the revised

TABLE 7

Approximate relcovariances between u_c and x'_c in Percentages $[100x' - \bar{x})/\bar{x}]$ for One-Way Classifications Defined by Stand Group, Revised Income, Urbanization Code, and Household Size[a]

	Independent variables			
Dependent variables	*Stand group*	*Income*	*Urbanization*	*Household size*
Triceps skinfold	−.04(−.04)	−.005(.01)	−.13(−.05)	.73(.80)
Weight	−.04(−.06)	.12(.13)	−.009(−.04)	1.1(1.2)
Blood pressure	−.09(−.11)	.004(.02)	−.09(−.09)	.17(.19)
Head, eyes, etc., findings	−.20(−.18)	−.044(−.06)	.03(.10)	−.25(−.28)
Chest evaluation	.008(.03)	−.002(−.008)	.013(.03)	−.13(−.14)
Nutritional status	−.052(−.09)	−.03(−.02)	−.12(−.11)	.27(.30)
Hemoglobin	−.04(−.04)	.05(.06)	−.02(−.04)	.20(.22)
Cholesterol	.14(.13)	.10(.12)	.05(.03)	.71(.75)
Iron	.13(.26)	.15(.19)	.06(.18)	.27(.30)
Calcium	−.06(.07)	.06(.005)	.05(.01)	−.70(−.76)

[a] Prior numbers are in parentheses.

sets of categories for these four variables in the calculations for the multiway classifications.

Because of the apparent value of the use of "age" in defining weighting classes, one set of two-way classifications used to define weighting classes was "age" crossed with each of the other ten demographic variables. The resulting approximate relcovariances are given in Table 8. Also, since "stand location" explained a relatively high proportion of variances of the 0–1 response variable, relcovariance calculations were calculated for "stand group" crossed with the other ten demographic variables. The results are given in Table 9.

The approximate relcovariances (i.e., relative differences) in Table 8 indicate that crossing "age" with the other demographic variables provides only small increases, if any, in the absolute values of the relcovariances associated with the age classifications alone.

On the other hand, when the variable "stand group" is crossed with the other demographic variables, the approximate relcovariances (Table 9) are increased substantially over those corresponding to "stand group" alone. Also, the approximate relcovariances corresponding to the cross-classification of "stand group" by "age" are higher (in absolute value) than those corresponding to all other two-way classifications in Table 9 for each of the 10 dependent variables.

Based on these two-way classification calculations, the results from the one-way classifications, and the AID calculations, three- and four-way classifications were selected for application of the relcovariance calculations. These multiway classifications were identified at a joint meeting with Westat and DHES personnel. Only one three-way classification was included in the analysis:

TABLE 8

Approximate Relcovariance between u_c and x'_c in Percentages $[100(x - \tilde{x})/\tilde{x}]$, with Weighting Classes Defined by "Age" Crossed with Other Demographic Variables

	Independent variables: age by each variable										
Dependent variables	*Stand group*	*Race*	*Sex*	*Income*	*Urbanization*	*SMSA*	*Region*	*Education*	*HH size*	*Poverty index*	*Age alone*
Triceps skinfold	1.6	1.7	1.9	1.5	1.5	1.5	1.6	1.6	1.5	1.5	1.6
Weight	2.4	2.5	2.5	2.4	2.4	2.4	2.4	2.4	2.5	2.4	2.5
Blood pressure	−8.7	−8.6	−6.6	−8.7	−8.7	−8.6	−8.6	−8.6	−8.7	−8.7	[a]
Head, eyes, etc., findings	−.58	−.38	−.39	−.42	−.36	−.30	−.55	−.33	−.42	−.41	−.39
Chest evaluation	−.18	−.18	−.15	−.17	−.16	−.15	−.18	−.15	−.18	−.18	−.18
Nutritional status	.57	.64	.72	.59	.52	.53	−.65	.63	.59	.60	.67
Hemoglobin	.33	.42	.37	.40	.35	.33	.43	.33	.40	.41	.37
Cholesterol	1.3	1.2	1.3	1.3	1.3	1.3	1.3	1.2	1.2	1.3	1.2
Iron	.61	.61	.58	.65	.55	.59	.69	.47	.69	.67	.53
Calcium	−1.7	1.3	−1.5	−1.4	−1.6	−1.5	−1.5	−1.6	−1.4	−1.4	−1.5

[a] This value was not computed properly; see the footnote to Table 6.

TABLE 9

Approximate Relcovariances between u_c and x'_c in Percentages $[100(x' - \tilde{x})/\tilde{x}]$, with Weighting Classes Defined by "Stand Group" Crossed with Other Demographic Variables

	Independent variables: stand group by each variable										
Dependent variables	*Age*	*Race*	*Sex*	*Income*	*Urbanization*	*SMSA*	*Region*	*Education*	*HH size*	*Poverty index*	*Stand group alone*
Triceps skinfold	1.6	.11	.09	−0.8	−.07	−.06	−.01	.07	.67	−.02	−.04
Weight	2.4	.02	−.08	.05	−.08	−.06	−.03	.05	.99	.14	−.04
Blood pressure	−8.7	−.11	−.09	−.08	−.15	−.13	−.09	−.03	.09	−.09	−.09
Head, eyes, etc., findings	−.57	−.19	−.21	−.25	−.24	−.20	−.18	−.19	−.47	−.23	−.20
Chest evaluation	−.18	.006	.008	.007	.02	.02	.02	.01	−.12	.007	.008
Nutritional status	.57	−.08	−.03	−.05	−.06	−.06	−.07	−.01	.21	−.05	−.05
Hemoglobin	.33	.06	−.07	−.001	−.04	−.06	−.02	−.09	.15	.03	−.04
Cholesterol	1.3	.16	.17	.22	.12	.12	.15	.22	.82	.25	.14
Iron	.61	.34	.04	.21	.18	.14	.21	.03	.33	.32	.13
Calcium	−1.7	.39	−.23	−.09	.10		.03	−.31	−.84	.008	.06

TABLE 10

Approximate Relcovariances between u_c and x'_c in Percentage $[100(x' - \bar{x})/\bar{x}]$ for Multiway Classifications

	Independent variables						
Dependent variables	*Age × stand Group × sex*	*Age × stand Group × sex × income*	*Age × stand Group × sex × HH size*	*Age × stand Group × sex × urbanization*	*Age × stand Group × sex × region*	*Age* *Alone*	*Age* *× Sex*
Triceps skinfold	1.7	1.7	1.6	1.8	1.8	1.6	1.9
Weight	2.4	2.3	2.4	2.3	2.4	2.5	2.5
Blood pressure	−8.6	−8.7	−8.6	−8.7	−8.6	[a]	−8.6
Head, eyes, etc., findings	−.60	−.65	−.68	−.70	−.58	−.39	−.39
Chest evaluation	−.15	−.15	−.16	−.15	−.14	−.18	−.15
Nutritional status	.61	.57	.56	.58	.58	.67	.72
Hemoglobin	.33	.35	.35	.33	.35	.37	.37
Cholesterol	1.43	1.5	1.4	1.4	1.4	1.2	1.3
Iron	.65	.63	.72	.81	.75	.53	.58
Calcium	−1.7	−1.6	−1.5	−1.2	−1.6	−1.5	−1.5

[a] This value was not obtained; see the footnote to Table 6.

"age" by "stand group" by "sex." The four-way classifications for which approximate covariances (i.e., relative differences) were calculated are those formed by crossing each of the four variables "family income," household size," "degree of urbanization," and "region" with the three-way classification "age" by "stand group" by "sex." The calculations for the three- and four-way classifications are given in Table 10.

An inspection of Table 10 reveals that the approximate relcovariances are fairly constant from one set of weighting classes to the next, including the weighting classes determined by (1) the age categories alone and (2) the cross-classification of age and sex. Furthermore, for some of the dependent variables, the age and the age-by-sex classifications provide larger relcovariances in absolute value than for any of the three- or four-way classifications investigated. However, these differences are small and are likely due to random variation in the sample. Even so, it seems that for the ten examination measurements included in the calculations, most if not all the relative differences between a survey estimate using nonresponse weighting classes and one using only a single adjustment can be accounted for by using age by sex to define weighting classes or, perhaps, by using age alone. However, there are other reasons, to be discussed later, why it would be desirable to use additional variables to define weighting classes.

3. INVESTIGATION OF THE USE OF DEGREE OF PERSUASION OR MULTIPLE REGRESSION IN NONRESPONSE IMPUTATION

3.1. Introduction

In Section 2, the basic approach of using weight-adjustment classes to make nonresponse adjustments was developed. In this section, another basic approach to the problem of imputation for nonresponse will be investigated. This approach involves the search for some relationship (1) among survey respondents for predicting one or more survey characteristics or (2) among all sample members for predicting the 0–1 response variable. Any relationship developed is used to impute for the characteristics of the nonrespondents.

Two types of procedures based on this approach were considered and subsequently rejected. One procedure involved the use of a measure of the difficulty required to obtain cooperation as an independent variable to project survey characteristics to unexamined persons. The other procedure involved the use of a multiple regression on the 0–1 response variable to determine nonresponse weight adjustments for unexamined persons.

3.2. An Investigation of the Use of Degree of Persuasion

3.2.1. The Two Imputation Procedures Considered

Included in the basic HANES data tape is an indication, for each examined person, of the number of household calls required for the person to make and keep the examination appointment. The required number of calls is referred to as "degree of persuasion." If whether or not a sampled person participates in the survey is correlated to the measurements taken, then it seems plausible that the degree of persuasion required to obtain participation would also be correlated to the measurements taken. If so, then any relationship that exists between the degree of persuasion and the average of a survey measurement might make it possible to project a mean value of the survey measurement to the nonrespondents.

The other procedure is based on the assumption that, relative to the survey characteristics, nonrespondents are more like those respondents requiring a high degree of persuasion than like those requiring lower levels of persuasion. The nonrespondents and those respondents requiring high degrees of persuasion would be grouped into weight-adjustment classes like those developed in Section 2. The sampling weights of only the respondents that are "late cooperators" would be adjusted, class by class, to account for the nonrespondents.

3.2.2. The Means of Examination Characteristics by Degree of Persuasion

To determine the possible usefulness of these procedures, the means for each of the 10 selected examination characteristics were computed for respondents requiring various degrees of persuasion. These means were examined for possible trends. The examination characteristics used for the calculations of means by degree of persuasion are the 10 used in the weighting-class investigations and are listed in Table 4. The weighted means for each degree of persuasion level and for each of the 10 examination characteristics are given in Table 11, along with the number of respondents for each degree of persuasion. Some type of trend or pattern should be present for either of the two imputation procedures based on degree of persuasion to be useful.

In viewing Table 11, it should be noted that there are only 12 respondents who required five calls and only 93 who required four calls. Therefore, the last column in Table 11 and, to some extent, the next to last column are rather unreliable and erratic. There do not appear to be any simple trends in Table 11 relating degree of persuasion and the corresponding means of the examination characteristics. There are possible upward trends in triceps skinfold and weight as degree of persuasion increases. However, it seems likely that this can be accounted for by the relationship between degree of persuasion and age since younger respondents require, on average, a lower degree of persuasion than older respondents and have lower weights and smaller triceps skinfold measurements than do older respondents.

TABLE 11

Examination Means by Degree of Persuasion

Examination characteristic	*Degree of persuasion*				
	1	2	3	4	5
Triceps skinfold	15.409	14.848	15.269	15.817	16.208
Body weight	125.134	129.372	127.207	129.037	134.783
Blood pressure	122.582	123.408	122.284	116.724	134.813
Head, eyes, etc., findings	1.755	1.757	1.807	1.754	1.682
Chest evaluation	1.945	1.945	1.959	1.971	2.000
Nutritional status	1.138	1.139	1.099	1.165	1.068
Hemoglobin	14.300	14.483	14.373	14.297	14.344
Cholesterol	204.054	205.381	197.843	189.024	231.452
Iron	11.808	12.269	12.186	11.399	11.436
Calcium	867.853	892.429	889.631	668.437	793.053
Number of examined persons	7569	1980	471	93	12

3.2.3. The Decision Not to Use Degree of Persuasion for Imputation

The apparent lack of any meaningful trends between characteristic means and degree of persuasion made the determination of a predictive relationship rather difficult. Even if there would have appeared to be some trends, there are other difficulties with this procedure. Regarding the first use of degree of persuasion that was considered, projecting examination measurements from apparent trends, a separate imputation would have to be made for each examination characteristic. This would require that a separate projection model be developed for each survey characteristic to apply to each nonrespondent. Furthermore, it is not clear how one would extrapolate from a table (or graph) of means for different levels of degree of persuasion to a mean for nonrespondents. It seems questionable to assume that nonrespondents would be represented by the next (i.e., the sixth) degree of persuasion. In fact, there does not seem to be any specific degree-of-persuasion level that would logically represent the unexamined group.

The other procedure considered, accounting for nonresponse by adjusting the sampling weights of only the "late cooperators," also seemed to be inappropriate. If there are no meaningful trends, then it does not seem as though, relative to survey characteristics, the unexamined persons are more like the examined persons who required higher (rather than lower) degrees of persuasion. If in fact the nonrespondents do not tend to be more like the "late" cooperators, then the procedure that adjusts only the sampling weights of the late cooperators is inadvisable; it will increase the variances of survey estimates due to the differential weighting introduced with little, if any, associated reduction in nonresponse bias. Consequently, it was decided at a joint meeting between Westat and DHES project personnel to not pursue further the use of degree of persuasion in nonresponse imputation.

3.3. Consideration of the Use of Multiple Regression

The basic idea considered here in the application of multiple regression to nonresponse imputation is to take the 0–1 response variable as the dependent variable and to take some or all of the demographic variables as the independent variables. Those demographic variables that are not quantitative (e.g., region and sex) would require the use of 0–1 ("dummy") independent variables. The resulting regression equation would be of the form

$$y = a_0 + a_1x_1 + a_2x_2 + a_3x_3 + \cdots + a_nx_n, \tag{6}$$

where y is the 0–1 response variable, x_i the ith independent variable, and a_i the regression coefficient for the ith independent variable. Some cross-product terms might also be introduced in the regression equation to allow for some interactions.

The value of the regression equation for each sampled person would be interpreted as an estimate of his or her probability of response. Therefore, the nonresponse weight adjustment for each examined person would be the inverse of the value of the regression equation.

The main advantage of this procedure is that it allows the use of a large number of independent variables simultaneously to make weight adjustments. There is no concern about weighting classes becoming too small, resulting in erratic nonresponse adjustments. This is because the regression equation provides a type of "smoothing" of the weighting-class adjustments.

The primary disadvantage of the use of multiple regression is the linearity assumption. To see the effect of this assumption, assume, for simplicity, that the only independent variable is "age class" in the regression equation. In Fig. 2, hypothetical means of the y values (i.e., the weighted response rates) are plotted for the age classes. The resulting regression would approximate the

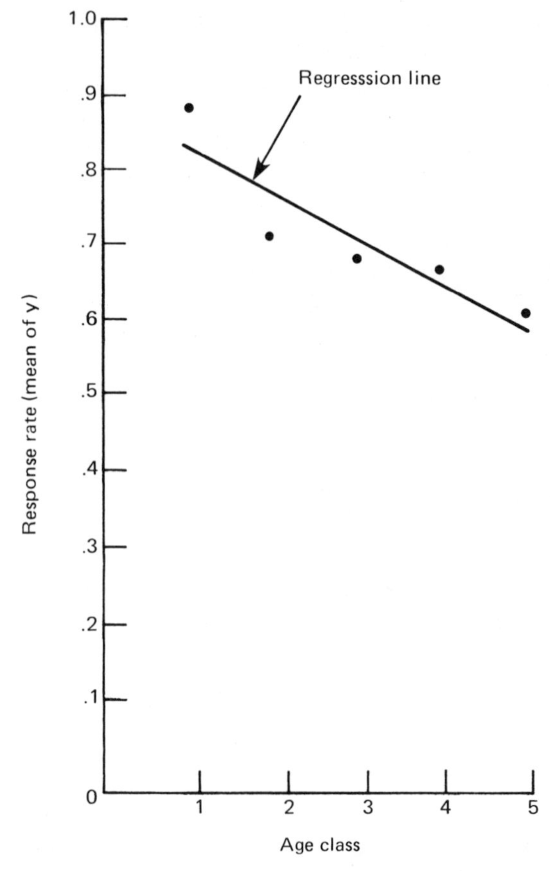

Fig. 2. *Hypothetical regression of the 0–1 response variable on "age," where the response rates are not the actual survey rates for the five age groups.*

line in Fig. 2. For age classes for which the point representing the response rate does not fall on the line, the nonresponse weight adjustment for that class (i.e., the value of the regression equation) would differ from the inverse of the response rate. Therefore, the total of the adjusted weights for the respondents in such a class would not equal the total of the sampling weights of all sampled persons in the class. This appears to be undesirable, especially in cases in which deviations from the regression plane are substantial.

It therefore seems that this multiple-regression approach is not as good as the weighting-class approach, at least in cases in which the number of respondents per class is not too small. Since it was shown in Section 2 that only a few variables appear to be useful in defining weighting classes and that the absolute values of relcovariances are not increased by additional cross-classification, it is unlikely that many independent variables will be used in a single stage of adjustment. Consequently, the number of respondents in each weighting class is not likely to be small.

Therefore, it was decided that the multiple-regression approach would not be useful in this situation. It should be noted, however, that the application of regression methods to make imputations for item nonresponse could be more fruitful.

4. DEVELOPMENT AND COMPARISON OF ALTERNATIVE IMPUTATION PROCEDURES

4.1. The Imputation Procedures Selected for Comparison

4.1.1. The Current DHES Weight-Adjustment Procedure

The procedure currently used by DHES to make adjustments to the basic weights of persons examined in Round I of HANES is a three-stage procedure. In the first stage, sampled persons within each stand are grouped into 20 weighting classes defined by (1) age (five categories: 1–5, 6–19, 20–44, 45–64, 65–74 years), (2) sex (two categories), and (3) family income (two categories: \$3000 or less and greater than \$3000). Within each of the first-stage classes, the weight adjustment u_c is computed in the usual manner as

$$u_c = N_c/N'_c, \tag{7}$$

where N_c and N'_c are the sums of basic sampling weights for all sampled persons and examined persons. If N_c/N'_c exceeds 3.0, u_c is set equal to 3.0. Also, there are no weight adjustments for classes that contain no examined persons.

The second-stage adjustments are carried out within eight classes defined by the four census regions and two SMSA codes (i.e., in SMSA or not in SMSA).

The weight adjustment v_k for the kth class at the second stage is

$$v_k = N_k / N'_k, \tag{8}$$

where N_k is the sum of basic sampling weights of all sampled persons in the kth second-stage weighting class[4] and N'_k the sum of the weights, adjusted at the first stage, of all examined persons in the kth second-stage weighting class. The eight second-stage adjustments are all very close to 1.0.

The third stage of this procedure is a "poststratification"-type weight-up to independent census estimates. Twenty classes for this adjustment are defined by the three-way classification of age (five categories), race (two categories), and sex (two categories). The five age categories are the same as those used for the first-stage adjustment. The two race categories are white and nonwhite. The resulting weight adjustment z_g for the gth third-stage class is computed as

$$z_g = N^*_g / N'_g, \tag{9}$$

where N^*_g is the March 1, 1972, census estimate for the gth third-stage cell and N'_g the sum of weights, adjusted at both the first and second stages, of all examined persons in the gth third-stage cell.

The final weight for each examined person is computed by multiplying his or her basic sampling weight by the three appropriate adjustment factors: u_c, v_k, and z_g. (The stages of the DHES adjustment procedure are summarized in Table 12 in the next section.)

4.1.2. *Five Alternative Imputation Procedures*

The AID calculations discussed in Section 2.2 and the variance and approximate relcovariance analyses described in Section 2.3 led to the conclusion the variables "age," "stand location," and "sex" are probably important variables to use in defining weighting classes. Also, the variables "family income," "household size," "degree of urbanization," and "region" seem to be of some use in defining weighting classes, especially in terms of proportions of explained variance and the AID results.

Some consideration was given to the use of an iterative weight adjustment (or "raking") procedure. With this type of procedure, weight adjustments are made in stages, using different variables or sets of variables at each stage and iterating the stages of adjustment two or more times. This allows the use of many variables without creating the problem of small cells through too many cross-classifications. However, since only a few variables were identified as being useful in defining classes, there seemed to be little chance for important gains from using raking relative to an unrepeated two-stage adjustment procedure.

[4] The definition given here of the numerator of the ratio N_k is a minor simplification. The precise definition of N_k is given in Schaible (1973).

It was decided that the final-stage adjustments to census counts carried out in the DHES procedure were very worthwhile and would be included in all the alternative procedures developed. This decision was based primarily on the apparent importance of the variables "age" and "sex" in defining weighting classes. The first procedure selected for the analysis consists only of a single-stage adjustment to the 20 age–race–sex census counts. It was felt that it would be worthwhile to compare this simple procedure with the more complex ones chosen.

The other four are all two-stage procedures, with the second stage of each of these procedures being the weight-up to the 20 age–race–sex census counts. In selecting variables to use in defining weighting classes for the first-stage adjustment, it was decided to eliminate from consideration the variables used for the second stage (i.e., "age," "race," and "sex"). There were only a few of the remaining demographic variables that indicated in the earlier analyses some possible usefulness in defining weighting classes. It was therefore decided that since "stand location" showed substantially higher percentages of explained variance than did "stand group," "stand location" would be used to define weighting classes at the first stage of each of these procedures. "Stand location" has the additional advantage of accounting for any useful effect that could be attributed to the variables "region," "SMSA code," and, to some extent, "degree of persuasion."

One of the four two-stage procedures chosen uses only stand location to define the weighting classes for the first stage. The other three procedures use stand by family income, stand by education, and stand by household size to define weighting classes at the first stage. These four two-stage procedures, along with the DHES and the single-stage procedures, are summarized in Table 12. Also, the categories used for household size, family income, and eduction of household head are given in Table 13.

The three categories used for household size are the same ones used for the relcovariance calculations for multiway classifications. However, the family income and education classes were modified. For family income, an income class was added to those used for multiway classifications (listed in Section 2.3.3). Regarding education, the number of classes was reduced from the seven used in multiway classifications (listed in Appendix A) to four. Whenever family income was unreported, it was imputed from the education level indicated and from whether or not the sampled person resided in a "poverty" segment. Likewise, whenever the education of the household head was missing, it was imputed from the income reported and the "poverty" classification of the segment. The imputation schemes used are given in Appendix C.

The first-stage weight adjustment u_c for each of these four two-stage procedures was computed in exactly the same way as were the first-stage adjustments for the DHES procedure, including the upper bound of 3.0. The second-stage adjustments z_g to the 20 age–race–sex cells were also computed in the same way as for the DHES procedure.

TABLE 12

The Six Alternative Imputation Procedures

Imputation procedure	Variables and corresponding number of classes used at each stage		
	Stage 1	*Stage 2*	*Stage 3*
1 (DHES)	Stand (35) × age (5) × sex (2) × income (2)	Region (4) × SMSA–non-SMSA (2)	Age (5) × race (2) × sex (2)
2	Stand (35) × income (4)	Age (5) × race (2) × sex (2)	
3	Stand (35) × education (4)	Age (5) × race (2) × sex (2)	
4	Stand (35) × household size (3)	Age (5) × race (2) × sex (2)	
5	Stand (35)	Age (5) × race (2) × sex (2)	
6	Age (5) × race (2) × sex (2)		

4.2. Comparisons of the Six Nonresponse Imputation Procedures

4.2.1. Summary of the Three Comparisons

Based on the AID results and the variance and approximate relcovariance calculations, five alternative nonresponse imputation procedures were developed. To test the effectiveness of these five procedures and the DHES procedure and to compare the performance of all six, three difference comparisons were carried out. The first comparison involved the calculation of survey means using the six alternative procedures. The six means were calculated for each of the 10 selected examination characteristics listed in Table 4. These calculations indicated how much variation there is among survey esti-

TABLE 13

Household Size, Income, and Education Classes

Class code	Household size	Income	Education
1	1–2	Less than $7,000	Eighth grade or less
2	3–4	$7,000–9,999	Some high school
3	5+	$10,000–14,999	High school graduate
4	—	$15,000 and over	One or more years of college

mates computed by these alternative procedures. Of course, for the differences that are observed it is not possible to know which estimates are better.

The second comparison made use of the medical-history information provided by over half of the unexamined persons. For each of 15 medical-history characteristics, the survey mean was computed, using each of the six imputation procedures. In these calculations, the set of all sampled persons who supplied medical-history data was treated as the entire sample and the set of examined persons was treated as the respondents. This provided a valid indication as to the levels of nonresponse bias for the six procedures with regard to estimates for the medical-history respondent population.

The third comparison used the 10 selected examination characteristics listed in Table 4. In this comparison the set of all examined persons was treated as the entire sample. The set of respondents was taken to be those examined persons who required only one call to make and keep an examination appointment. For each of the 10 examination characteristics, appropriate imputation estimates were made based on each of the six procedures and were compared with the estimate based on the entire "pseudo"-sample. The basic rationale for this type of comparison was that if only one call were used in HANES, those responding to it would be the set of examined persons. Those who cooperated on subsequent calls would then have been part of the nonrespondents. This type of comparison provides a useful measure of the nonresponse biases associated with the six procedures. The details and results of these three comparisons will be given in the next three sections.

4.2.2. *Comparisons of Survey Estimates for the Alternative Procedures*

The first comparison of the six procedures involved the 10 selected examination measurements. For each of these 10 measurements, weighted survey estimates of means, using the entire Round 1 file, were computed for each of the six imputation procedures. These means were calculated as simple weighted averages. Specifically, the estimated mean $\bar{x}_a$ for examination measurement X, based on the ath nonresponse weight-adjustment procedure was computed as

$$\bar{x}_a = \sum_{j=1}^{n'} W_{aj}^{(\mathrm{f})} x_j \Big/ \sum_{j=1}^{n'} W_{aj}^{(\mathrm{f})}, \tag{10}$$

where $W_{aj}^{(\mathrm{f})}$ is the final weight for the jth examined person using alternative imputation procedure a, x_j the value of the examination measurement X for the jth examined person, and n' the total number of examined persons.

Also computed for each of the 10 measurements was the mean $\bar{\bar{x}}$, using unadjusted weights (i.e., basic sampling weights). Each unadjusted mean was computed as specified in Eq. (10), with the basic sampling weights substituted for the final weights.

For each of the 10 examination characteristics, the relative difference between the mean based on each of the six alternative procedures and the unadjusted mean was computed. This relative difference d_a for examination measurement X was calculated as

$$d_a = (\bar{x}_a - \bar{\bar{x}})/\bar{\bar{x}}. \tag{11}$$

The relative difference in percentage between each of the 60 estimated means and the corresponding unadjusted mean is given in Table 14, which includes 10 unadjusted means also.

An inspection of Table 14 indicates that there is very little difference in the estimates resulting from the six alternative imputation methods for any of the 10 examination characteristics. However, there are some noticeable differences (e.g., 2–3%) between the weighting-class estimates and the unadjusted estimates. It is not clear whether these differences arise mostly because of the nonresponse weight adjustments or because of the poststratification weight adjustment. We presume that they are larger than those that could result from the reduction in the sampling error in a poststratification weight adjustment with no nonresponse. Therefore, it appears that these differences represent to some extent the effect of the nonresponse adjustments.

4.2.3. Evaluation of the Six Imputation Procedures Using the Medical-History Characteristics

This comparison was based on the HANES data collected for the set of 15 medical-history characteristics given in Table 15. There were separate medical-history questionnaires for persons in the age classes 1–5 years, 6–11 years, and 12–74 years. The 15 medical-history items are grouped in Table 15 by these three age classes.

As indicated earlier, medical-history characteristics were provided by over half (i.e., 55%) of the 4021 unexamined persons. The comparisons of the alternative procedures were made by treating the medical-history respondents (about 87% of the sampled persons) as if they comprised the entire sample. The set of examined persons was still taken as the set of respondents. The weights of the examined persons were then adjusted up to this pseudosample of medical-history respondents by using each of the six alternative procedures. Since about 72% of the sample persons were examined, the sampled persons constitute about 83% (72 ÷ 87) of this pseudosample.

For each medical-history characteristic, the weighted mean was computed for each of the six alternative adjustment procedures and compared with the "true," or "standard," mean for the characteristic derived from all medical-history respondents. The standard mean was computed as a weighted mean, the weights being the basic sampling weights *adjusted to the 20 age–race–sex census counts*. Also computed for each medical-history characteristic was the

TABLE 14

Relative Differences, in Percentages, between the Estimated Means Computed from the Alternative Procedures and the Unadjusted Means

Examination characteristics	Alternative procedures[a]						Unadjusted mean
	DHES (*1*)	*Stand–income* (*2*)	*Stand–education* (*3*)	*Stand–HH size* (*4*)	*Stand* (*5*)	*Race–sex–age* (*6*)	
Triceps skinfold	1.8	2.0	1.8	1.8	1.9	2.0	15.29
Body weight	3.2	3.3	3.3	3.2	3.3	3.4	126.2
Blood pressure	.81	.79	.77	.76	.74	.90	122.7
Head, eyes, etc., findings	−.63	−.57	−.63	−.63	−.57	−.40	1.76
Chest evaluation	−.10	−.10	−.10	−.10	−.10	−.10	1.95
Nutritional status	.53	.62	.62	.53	.53	.70	1.14
Hemoglobin	.56	.56	.56	.56	.56	.63	14.31
Cholesterol	1.5	1.6	1.6	1.5	1.5	1.4	203.4
Iron	1.5	1.5	1.4	1.5	1.5	1.3	11.92
Calcium	−1.2	−1.1	−1.1	−1.1	−1.1	−1.2	872.4

[a] The six procedures are identified by their first-stage weight adjustment. Also, the number given in parentheses corresponds to the number of the procedure given in Table 12.

TABLE 15

The 15 Medical-History Characteristics Used in the Comparison Analysis

Medical history characteristic	*Value of the characteristic*	
Ages 1–5		
Has the sampled person had a		
(1) DPT immunization?	1. Yes	2. No
(2) Polio immunization?	8.[a] Not obtained	
(3) Smallpox immunization?	9.[a] Don't know	
(4) Has the sampled person ever had a bad accident?	1. Yes	2. No
	9.[a] Don't know	
(5) Has the sampled person ever stayed in a hospital overnight?	1. Yes	2. No
(6) Number of colds in last 6 months?	0–7	
Ages 6–11		
(7) Has the sampled person ever had a bad accident?	1. Yes	2. No
	9.[a] Don't know	
(8) Has the sampled person ever stayed in a hospital overnight?	1. Yes	2. No
(8) Number of colds in last 6 months?	0–8	
Ages 12–74		
Has a doctor ever told you that you had		
(10) Arthritis?	1. Yes	
(11) A heart murmur?	2. No	
(12) High blood pressure?	8.[a] Not obtained	
(13) Have you ever had anemia?	1. Yes	2. No
	9.[a] Don't know	
(14) Have you used medication for nerves in the past 6 months?	1. Regularly	2. Occasionally
	3. No	8.[a] Not obtained
(15) Are you on a special diet?	1. Yes	2. No

[a]Examined persons with this code were not included in the calculation of estimates for the particular medical-history characteristic.

"unadjusted" mean, which was the weighted mean calculated from the unadjusted basic sampling weights.

For each medical-history characteristic, the six means for the alternative weighting procedures, the standard mean, and the unadjusted mean are given in Table 16. Also, using the standard mean as the base, estimated relative biases, in percentages, were computed for all estimates and are given in Table 17. The estimated relative bias for each mean was calculated by dividing the difference between each mean and the standard mean by the standard mean and converting the quotient to a percentage.

Inspection of Table 16 reveals, as was the case for the previous comparison, that the differences in the estimates using the alternative weighting procedures are, in general, quite small. The approximate biases given in Table 17 are also

TABLE 16

Estimated Means Computed by Each of Six Imputation Procedures (Plus Unadjusted and Standard Mean) for Each of 15 Selected Medical-History Characteristics

		Alternative procedures[a]							
Age	*Examination characteristics*	*DHES* (*1*)	*Stand–income* (*2*)	*Stand–education* (*3*)	*Stand–HH size* (*4*)	*Stand* (*5*)	*Race–sex–age* (*6*)	*Standard mean*	*Unadjusted mean*
1–5	DPT	1.050	1.051	1.051	1.050	1.051	1.051	1.057	1.052
	Polio	1.076	1.077	1.076	1.076	1.076	1.077	1.083	1.078
	Smallpox	1.321	1.322	1.324	1.322	1.321	1.327	1.336	1.326
	Bad acc.	1.858	1.857	1.857	1.856	1.857	1.857	1.861	1.850
	Hosp. ovnt.	1.532	1.535	1.533	1.535	1.536	1.535	1.527	1.534
	No. colds	.974	.975	.976	.971	.978	.978	.959	.980
6–11	Bad acc.	1.798	1.799	1.800	1.800	1.800	1.799	1.793	1.798
	Hosp. ovnt.	1.532	1.535	1.533	1.535	1.536	1.535	1.527	1.534
	No. colds	.974	.975	.976	.971	.978	.978	.959	.980
12–74	Arthritis	1.838	1.838	1.837	1.837	1.838	1.838	1.845	1.844
	Murmur	1.933	1.934	1.933	1.934	1.933	1.933	1.937	1.934
	High bld. pr.	1.858	1.857	1.858	1.857	1.857	1.856	1.853	1.861
	Anemia	1.817	1.819	1.819	1.819	1.819	1.819	1.830	1.821
	Nerves	2.845	2.846	2.845	2.845	2.846	2.845	2.844	2.849
	Spec. diet	1.892	1.892	1.892	1.891	1.892	1.891	1.894	1.894

[a] The six procedures are identified by their first-stage weight adjustment. Also, the number given in parentheses corresponds to the number of the procedure given in Table 12.

TABLE 17

Estimates of Relative Bias in Percentages for Each of the Six Alternative Procedures (and the Unadjusted Procedure) Used to Estimate Medical-History Characteristics

		Alternative procedures[a]						
Age	*Examination characteristics*	*DHES* (*1*)	*Stand–income* (*2*)	*Stand–education* (*3*)	*Stand–HH size* (*4*)	*Stand* (*5*)	*Race–sex–age* (*6*)	*Unadjusted procedure*
1–5	DPT	−.66	−.57	−.57	−.66	−.57	−.57	−.47
	Polio	−.65	−.55	−.65	−.65	−.65	−.55	−.46
	Smallpox	−1.1	−1.0	−.90	−1.0	−1.1	−.67	−.75
	Bad acc.	−.16	−.21	−.21	−.27	−.21	−.21	−.11
	Hosp. ovnt.	.42	.36	.24	.12	.36	.24	.42
	No. colds	1.1	1.1	1.1	.91	.84	1.3	1.8
6–11	Bad acc.	.28	.33	.39	.39	.39	.33	.28
	Hosp. ovnt.	.33	.52	.39	.52	.59	.52	.46
	No. colds	1.6	1.7	1.8	1.3	2.0	2.0	2.2
12–74	Arthritis	−.38	−.38	−.43	−.43	−.38	−.38	−.05
	Murmur	−.21	−.15	−.21	−.15	−.21	−.21	−.15
	High bld. pr.	.27	.22	.27	.22	.22	.16	.43
	Anemia	−.71	−.60	−.60	−.60	−.60	−.60	−.49
	Nerves	.035	.070	.035	.035	.070	.35	.18
	Spec. diet	−.11	−.11	−.11	−.16	−.11	−.16	.0

[a] The six procedures are identified by their first-stage weight adjustment. Also, the number given in parentheses corresponds to the number of the procedure given in Table 12.

rather small, but not negligible. These levels of biases are about the same from one procedure to the next, including those associated with the unadjusted means. Therefore, from these calculations, it appears that there is very little difference in the performance of the alternative weighting procedures and that the gains are slight, if any, over the unadjusted estimating procedure.

4.2.4. Comparison of the Alternative Weighting Procedures Using Data from Early Cooperators

This method of comparing alternative procedures is similar to the previous one in that a subset of the sampled persons is treated as the entire sample, with a subset of this pseudosample being treated as the set of respondents. In this case, the set of 10,126 examined persons is treated as the entire sample, with the 7569 examined persons who required only one call to make and keep an examination appointment being treated as the set of respondents. This pseudogroup of respondents constitutes about 75% of the pseudosample (i.e., 7569 ÷ 10,126). Appropriate weight adjustments were calculated for the pseudosample members using each of the six alternative procedures.

TABLE 18

Estimated Examination Means Computed by Each of Six Imputation Procedures (Plus Unadjusted and Standard Means) Based on Examined Persons Requiring Only One Call

Examination characteristics	Alternative procedures[a]						Standard mean	Unadjusted mean
	DHES (1)	*Stand–income* (2)	*Stand–education* (3)	*Stand–HH size* (4)	*Stand* (5)	*Race–age–sex* (6)		
Triceps	15.61	15.60	15.59	15.61	15.60	15.60	15.59	15.41
Body weight	130.6	130.5	130.5	130.5	130.5	130.6	130.5	125.1
Blood pressure	123.8	123.6	123.6	123.6	123.6	123.6	123.8	122.6
Head, eyes, etc., findings	1.750	1.748	1.746	1.747	1.748	1.747	1.750	1.755
Chest evaluation	1.944	1.945	1.945	1.945	1.945	1.944	1.944	1.945
Nutritional status	1.146	1.144	1.144	1.145	1.145	1.146	1.145	1.148
Hemoglobin	14.39	14.38	14.38	14.38	14.39	14.39	14.40	14.27
Cholesterol	206.3	206.3	206.5	206.4	206.3	206.4	206.3	203.5
Iron	12.06	12.05	12.06	12.05	12.05	12.06	12.07	11.81
Calcium	859.4	860.1	858.6	858.8	859.3	862.8	862.4	867.9

[a] The six procedures are identified by their first-stage weight adjustment. Also, the number given in parentheses corresponds to the number of the procedure given in Table 12.

For each of the 10 selected examination characteristics, the sample mean was computed from the pseudorespondents, using the weights derived from each of the six alternative nonresponse weight-adjustment procedures. In addition to these, a "true" (or "standard") mean was calculated as a weighted mean computed across the pseudosample (i.e., the set of examined persons), using basic sampling weights adjusted to the census counts. (The standard mean in this case is equivalent to the single-stage survey mean based on the age–race–sex census counts.) Also computed was an "unadjusted" mean for each of the 10 examination characteristics, calculated as a simple weighted average among examined persons in the pseudosample. The weights used were the initial sampling weights.

Table 18 contains eight estimated means for each of the 10 examination characteristics: six means based on the procedures being compared, an unadjusted mean, and a standard mean. The estimates of relative nonresponse bias associated with each of the six weighting procedures and with the unadjusted estimate are given as percentages in Table 19. As before, measures of bias based on this evaluation procedure were obtained by computing the difference between an estimated mean and the standard mean and dividing this difference by the standard mean.

An inspection of Tables 18 and 19 reveals that, as in the previous comparison, the means and estimated biases do not vary much among the six procedures. In this case, however, the relative biases of the procedures appear to be negligible, except perhaps for those of the last measurement, "calcium," for which the estimated biases are as high as one-third or one-half of 1% in

TABLE 19

Estimates of Relative Bias, in Percentage, for Each of the Six Alternative Imputation procedures (and the Unadjusted Procedure) Used to Estimate Examination Measurements

	Alternative procedures[a]						
Examination characteristics	*DHES* (*1*)	*Stand–income* (*2*)	*Stand–education* (*3*)	*Stand–HH size* (*4*)	*Stand* (*5*)	*Race–sex–age* (*6*)	*Unadjusted procedure*
Triceps	.13	.06	.0	.13	.06	.06	−1.2
Body weight	.11	.04	−.01	.04	.01	.06	−4.1
Blood pressure	−.01	−.13	−.15	−.12	−.15	−.12	−.97
Head, eyes, etc., findings	.0	−.11	−.23	−.17	−.11	−.17	.29
Chest examination	.0	.05	.05	.05	.05	.0	.05
Nutritional status	.09	−.09	−.09	.0	.0	.09	.26
Hemoglobin	−.07	−.14	−.14	−.14	−.07	−.07	−.90
Cholesterol	.04	−.005	.11	.07	.03	.07	−1.4
Iron	−.08	−.17	−.08	−.17	−.17	−.08	−2.2
Calcium	−.34	−.27	−.43	−.41	−.35	.05	.64

[a] The six procedures are identified by their first-stage weight adjustment. Also, the number given in parentheses corresponds to the number of the procedure given in Table 12.

some instances. Also, each unadjusted mean differs from the corresponding standard mean by a greater absolute amount than does the mean corresponding to any of the weighting procedures. This indicates that the weighting procedures do provide important decreases in biases over those associated with the unadjusted estimates.

The single-stage procedure (i.e., the simple weight-up to age–race–sex census counts) compares quite favorably with the other procedures. In fact, this procedure provided the estimate having the smallest estimated bias (by a narrow margin) for more of the characteristics than did any of the other procedures. However, the estimates derived from all of the procedures are so close that the observed differences could be attributed to sampling error.

5. SUMMARY AND RECOMMENDATIONS

In the first phase of this project, criteria were used to try to determine the demographic variables that appeared to be most useful in defining weighting classes to use for nonresponse weight adjustments. Also, two other approaches to nonresponse imputation, the uses of degree of persuasion and regression, were considered but not pursued further in developing imputation procedures.

In the second phase of the study, five specific nonresponse weight-adjustment procedures were proposed for comparison, along with the current DHES procedure. The selection of the weighting-class variables for these procedures was based on the AID, explained variance, and approximate relcovariance calculations that were carried out in the first phase. These procedures were then evaluated and compared, based primarily on the results of two methods of estimating the relative nonresponse biases of the procedures. The first method was based on the imputation of the medical-history characteristics of about half of the unexamined persons, using the medical-history characteristics of all examined persons. The second method was based on the imputation of selected examination characteristics of those examined persons requiring more than one call to make and keep an examination appointment, using the examination data of those requiring only one call. As part of the analyses, there were some comparisons of estimated means computed from the alternative weighting procedures with those computed from unadjusted basic weights. The procedure that uses unadjusted weights to compute means is equivalent to the procedure that makes a single overall weight adjustment.

Based on the evaluations carried out, we have drawn the following general conclusions:

(1) All of the proposed procedures show substantially lower estimated biases for the examination variables than does the procedure that makes a single overall weight adjustment (i.e., the unadjusted-weights procedure).

(2) The procedure that involves a single stage of weight adjustments, those being to census estimates of age–race–sex counts, is approximately as good as the other procedures, on the basis of the evaluative results in this study. In fact, all six procedures yielded estimates that did not differ substantially.

(3) For the case in which medical-history characteristics were being estimated and used for evaluation, there are some nontrivial estimated nonresponse biases remaining after the nonresponse adjustment procedures have been applied. Some of these estimated biases are as large as .5–2.0%. Such biases would presumably be substantially greater than the corresponding sampling errors for these estimates. We believe that because of the similarity of results for the alternative weight-adjustment procedures, this bias cannot be reduced noticeably by using alternative adjustment variables; the bias can perhaps be reduced only by increasing the survey response rate.

Because of the similarity of performance among the alternative procedures, it might be tempting to recommend the easiest procedure to use: the single-stage weight-up to the 20 age–race–sex census counts. However, we believe that it would be useful to include a geographic variable to define weighting classes. Consequently, we recommend that stand be included in an adjustment procedure since (1) nonresponse rates vary substantially by stand, (2) its joint use with other variables is not difficult, (3) its use has an intuitive appeal, and (4) its use may be important in reducing nonresponse bias for some estimates.

In general, it seems reasonable to assume that the use of additional adjustment variables might be helpful in making estimates for some examination measurements not studied here and, in any event, should not introduce biases. It is possible, however, to add to the variances of estimates by using additional variables that create weighting classes containing too few respondents. But with the use of weight classes such that the minimum weighting class size is 25 respondents, this variance increase is trivial and can be neglected. All of the procedures we have proposed appear to meet this minimum requirement. On the other hand, the current DHES procedure does provide very small classes in some cases. We therefore recommend that one of the three two-stage procedures, using a two-way classification at the first stage, be adopted. Of these three procedures, we have a slight preference for the stand–income or possibly the stand–education first-stage adjustments over the stand–household-size alternative.

Minor modifications in the classes defined in any of the procedures might be helpful. For example, it might be useful to add some age categories, since "age" appeared to be the most useful adjustment variable.

APPENDIX A. THE 12 AID INDEPENDENT VARIABLES

Table 20 provides a listing of the categories for each of the 12 demographic variables used in AID analysis, including the number of sampled persons and weighted response rates for each category. (Braces indicate the combining of categories for the variance and covariance calculations.)

TABLE 20

Categories and Associated Response Roles for the 12 AID Independent Variables

Categories of demographic variables	*Number of sampled persons*	*Weighted response rate (%)*
(1) *Stand Location*		
Philadelphia, Pa.	404	69.1
Pittsburgh, Pa.	327	63.0
Albany, N.Y.	304	57.9
Mercer, Pa.	279	77.8
Boston, Mass.	492	67.4
Detroit, Mich.	589	60.7
Newark, N.J.	432	59.1
Springfield, Mass.	288	65.0
Cleveland, Ohio	458	72.9
Bay City, Mich.	300	79.9
New York, N.Y.	425	47.8
La Porte, Ind.	299	73.3
Angola, Ind.	292	64.5
Cabarrus, N.C.	301	68.8
Los Angeles, Calif.	568	67.6
Savannah, Ga.	340	58.5
West Palm Beach, Fla.	501	72.6
Tucson, Ariz.	351	74.9
San Antonio, Tex.	603	75.9
Barbour, Ala.	596	72.5
Fresno, Calif.	340	82.4
Avayelles, La.	584	92.2
Columbia, S.C.	314	76.8
San Francisco, Calif.	570	79.7
Lamar, Miss.	432	78.4
New York, N.Y.	557	51.3
Callum, Wash.	306	83.9
St. Joseph, Mo.	302	75.3
Hartford, Conn.	311	79.7
Grant, Wash.	293	87.3
Chicago, Ill.	578	72.4

(cont.)

TABLE 20 (*cont.*)

Categories of demographic variables	*Number of sampled persons*	*Weighted response rate (%)*
Sussex, Del.	346	78.0
Boone, Iowa	323	89.2
Washington, D.C.	442	64.2
Milwaukee, Wis.	300	79.4
(2) *Age*		
1–5	1,860	80.4
6–19	2,939	80.0
20–44	4,587	68.6
45–64	2,121	66.1
65–74	2,640	62.1
(3) *Race*		
White	10,435	71.3
Black	3,562	77.8
Other	150	76.5
(4) *Sex*		
Male	5,936	72.5
Female	8,211	71.7
(5) *Income* ($)[a]		
<1,000	343	62.7
1,000–1,999	850	73.5
2,000–2,999	1,131	76.8
3,000–3,999	957	80.0
4,000–4,999	844	74.4
5,000–5,999	778	75.4
6,000–6,999	659	73.1
7,000–9,999	3,041	72.8
10,000–14,999	2,607	74.6
15,000–19,999	1,137	75.7
20,000–24,999	429	73.7
<25,000	399	63.1
Not obtained	972	48.0
(6) *Education*[a]		
None	199	80.5
1	67	59.0
2	116	88.8
3	265	80.1
4	292	70.2
5	301	79.2
6	498	79.3
7	481	69.9
8	1,519	72.2
9	837	70.8
10	1,035	71.4
11	845	71.3
12	4,085	73.8

TABLE 20 (*cont.*)

Categories of demographic variables	*Number of sampled persons*	*Weighted response rate (%)*
College		
1	468	72.0
2	736	70.5
3	255	78.2
4	774	73.5
5+	710	74.4
Not obtained	664	42.2
(7) *Urbanization*[a]		
Urbanized >3,000,000	3,306	61.5
Urbanized 1,000,000–2,999,999	2,349	72.9
Urbanized 250,000–999,999	1,050	67.3
Urbanized <250,000	1,709	74.1
Urban >25,000	283	78.1
Urban 10,000–24,999	526	81.0
Urban 2,000–9,999	743	74.1
Rural	4,181	75.9
(8) *SMSA*		
In SMSA, central city	5,846	68.8
Other SMSA	3,862	69.2
Not in SMSA	4,439	77.7
(9) *Region*		
Northeast	3,819	63.4
Midwest	3,272	74.5
South	3,441	71.6
West	3,615	79.1
(10) *Household size*[a]		
1	1,424	61.0
2	3,049	64.1
3	2,090	66.8
4	2,531	74.7
5	1,877	78.6
6	1,325	78.6
7	733	80.8
8	506	74.8
9	285	81.7
10	139	82.0
11	72	70.8
12	57	92.5
13	20	63.2
14	25	100.0
15	2	100.0
16	5	100.0
19	7	100.0

(*cont.*)

TABLE 20 (*cont.*)

Categories of demographic variables	*Number of sampled persons*	*Weighted response rate (%)*
(11) *Number sampled*		
1	6,183	69.7
2	5,462	73.4
3	2,002	75.2
4	419	73.4
5	74	73.6
7	7	100.0
(12) *Poverty index*[a]		
Less than 1.00	2,890	78.8
{ 1.00	39	60.0
{ Greater than 1.00	10,246	73.0
Not obtained	972	48.0

[a] Indicates a variable whose categories were collapsed for the variance and covariance calculations. Braces are drawn to indicate how the categories were combined.

APPENDIX B. AID TREE DIAGRAMS

This appendix gives the tree diagrams (Figs. 3–8) for six of the eleven AID analyses summarized in Table 2 of Section 2.2.2. The subgroup sizes are shown for each split, along with the name of the variable selected to make each split. The subgroups are numbered to identify the sequence of splitting. Furthermore, the number of final groups and the proportion of explained variation for all final groups combined are given in the legends.

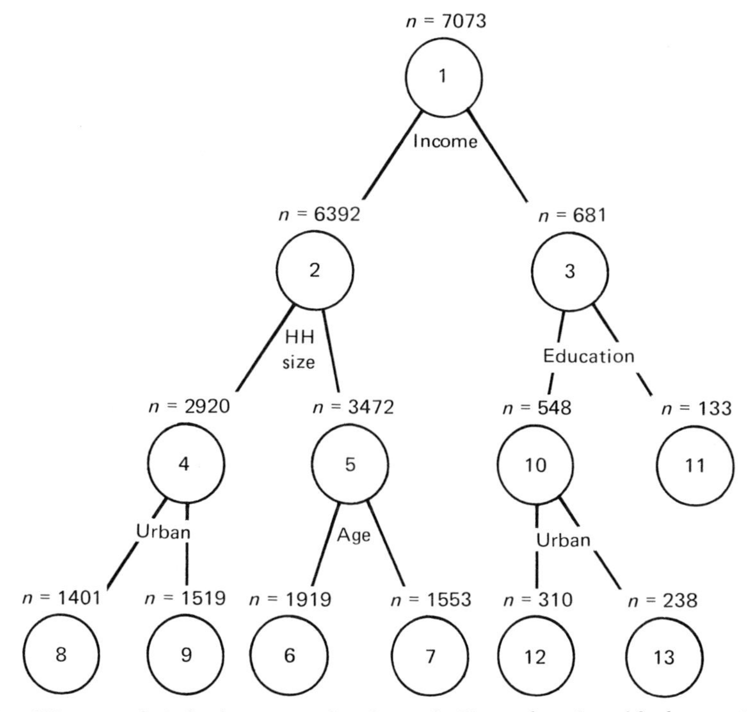

Fig. 3. *AID tree analysis for the even-numbered records. The total number of final groups is 11 and the percentage of total variation explained is 7.3.*

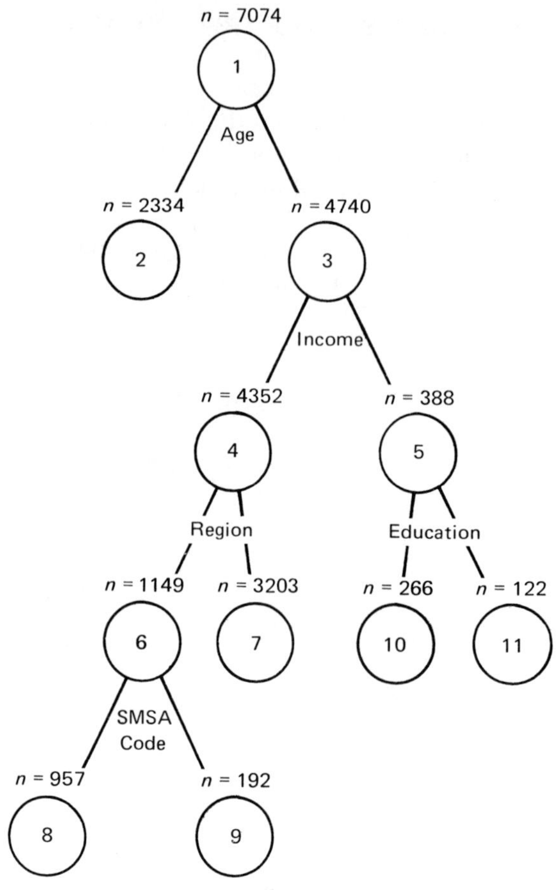

Fig. 4. *AID tree analysis for the odd-numbered records. The total number of final groups is 6 and the percentage of total variation explained is 6.0.*

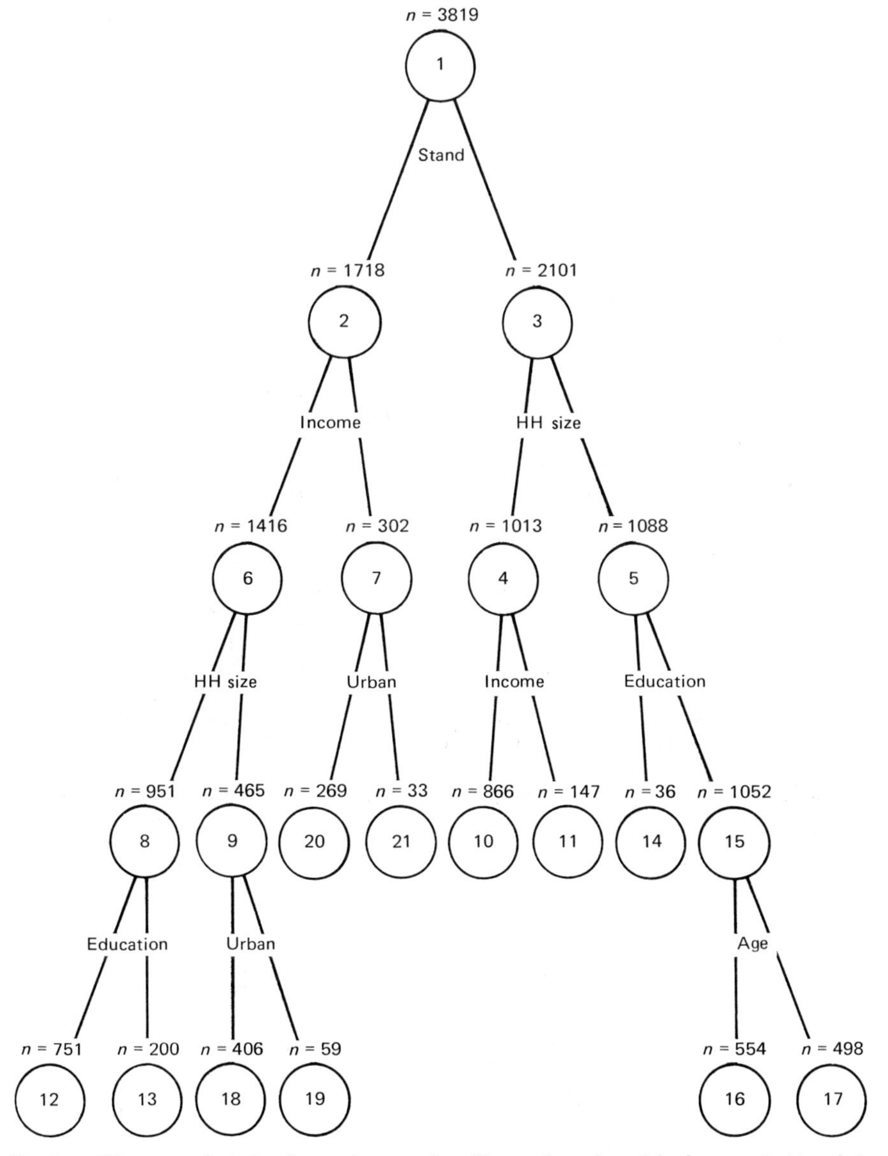

Fig. 5. *AID tree analysis for the northeast region. The total number of final groups is 11 and the percentage of total variation explained is 13.0.*

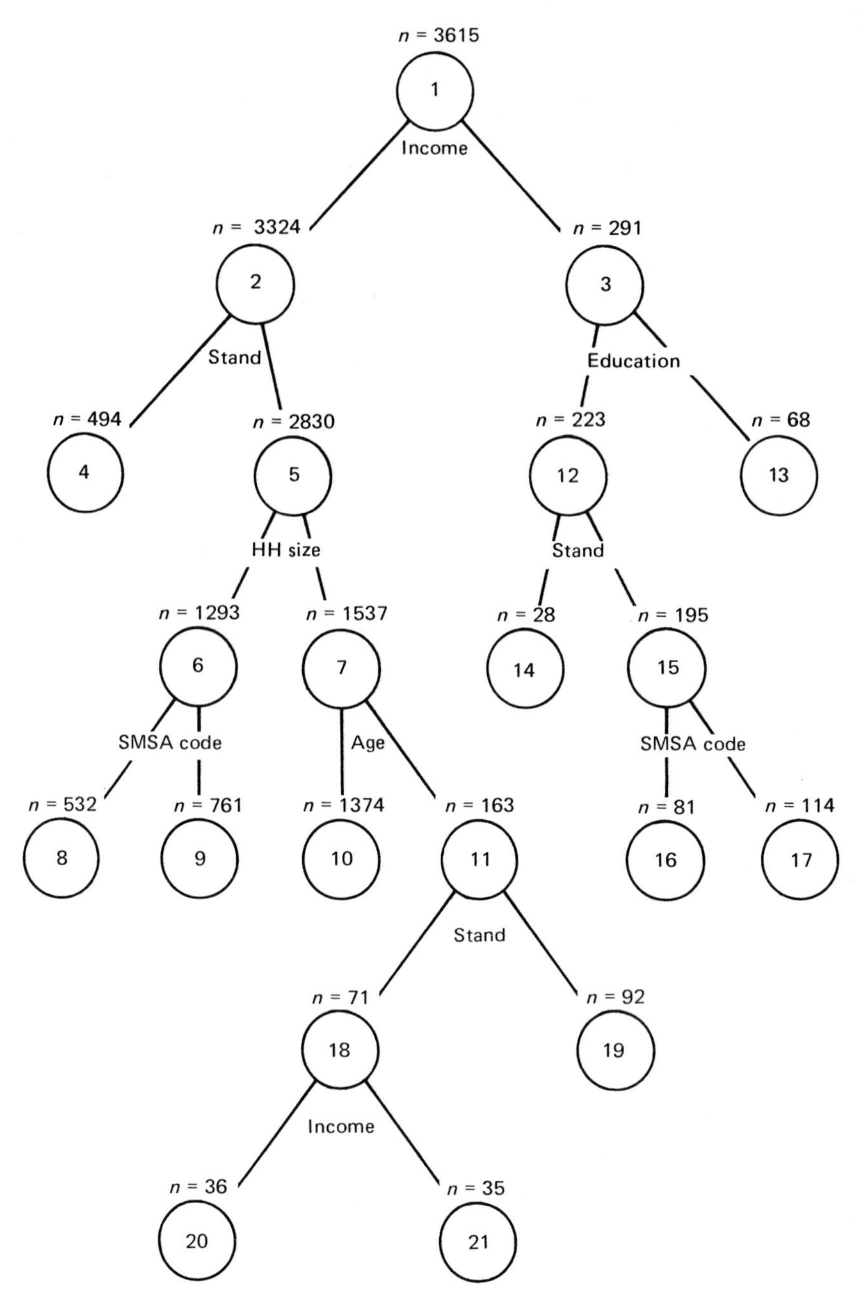

Fig. 6. *AID tree analysis for the west region. The total number of final groups is 11 and the percentage of total variation explained is 14.0.*

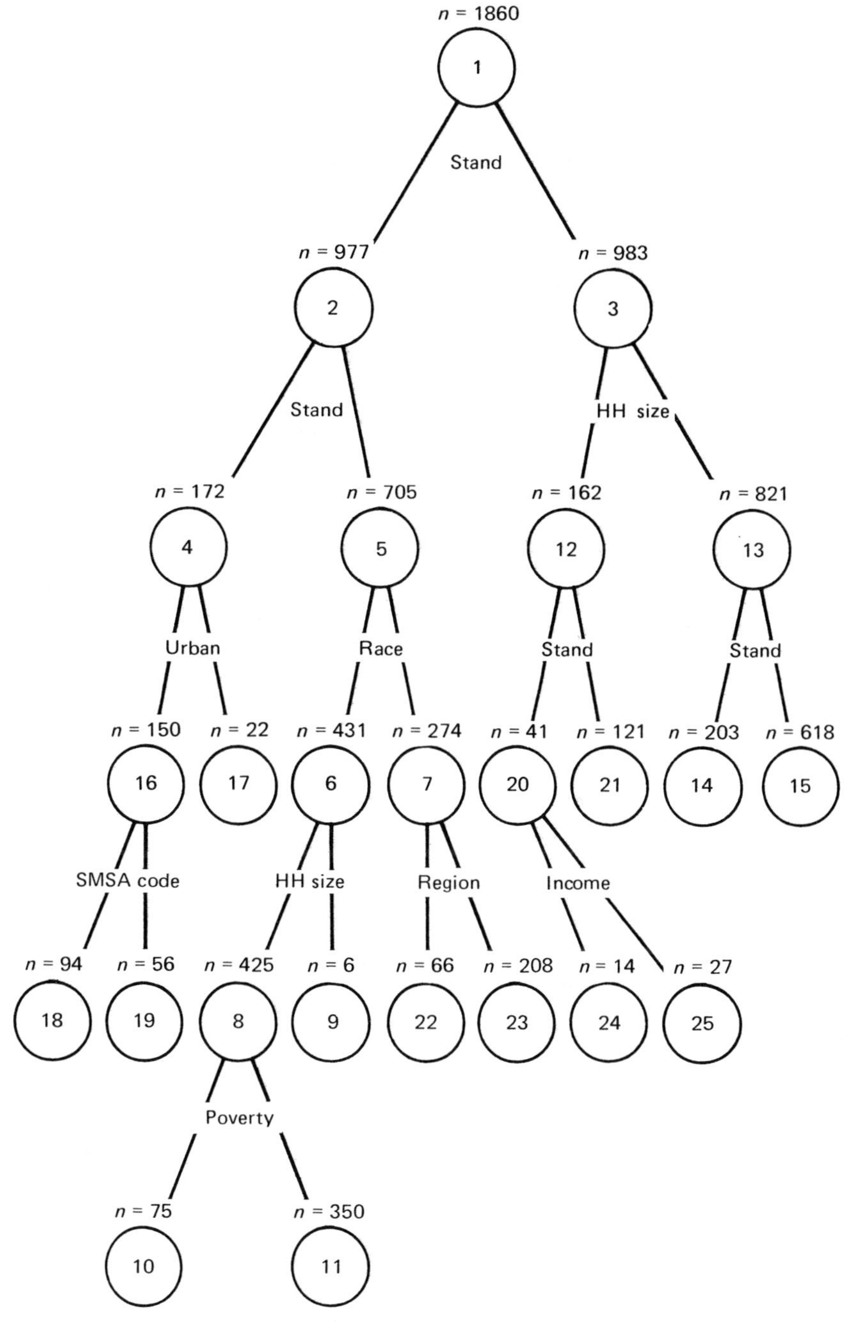

Fig. 7. *AID tree analysis for those sampled persons ages 1–5. The total number of final groups is 13 and the percentage of total variation explained is 15.4.*

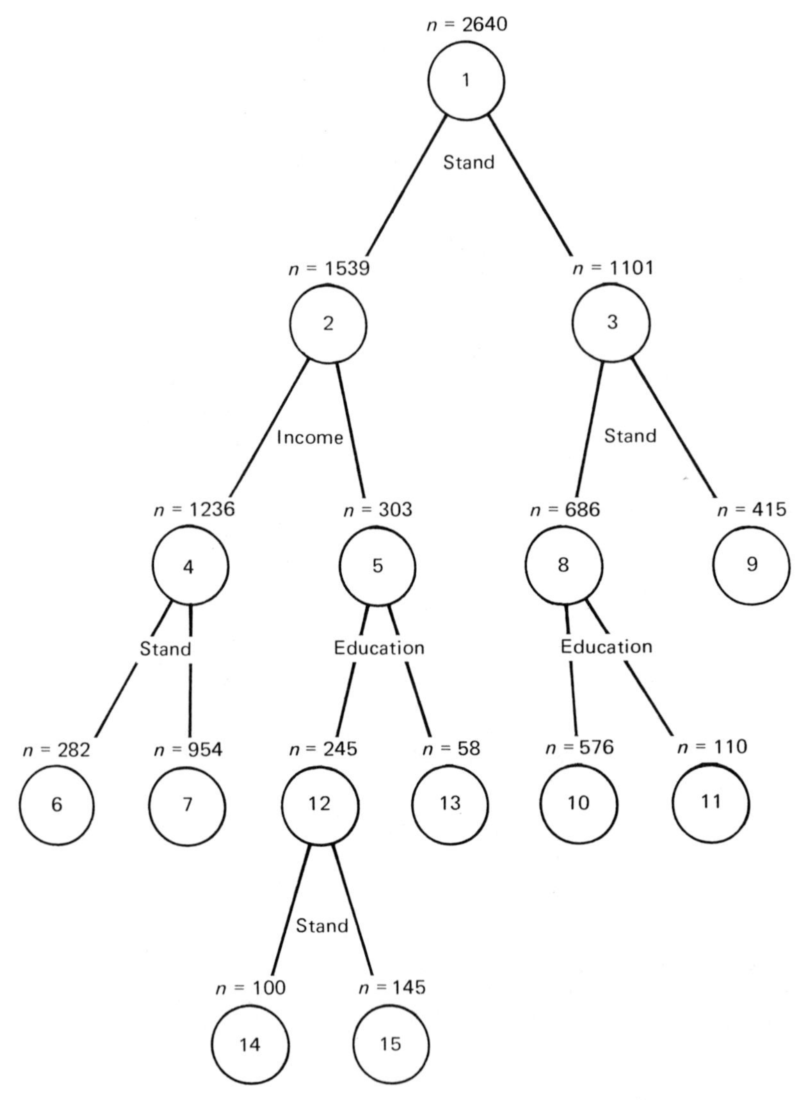

Fig. 8. *AID tree analysis for those sampled persons ages 65–74. The total number of final groups is 8 and the percentage of total variation explained is 11.5.*

APPENDIX C. IMPUTATION OF MISSING FAMILY INCOME AND EDUCATION OF HEAD OF HOUSEHOLD

TABLE 21

Imputed Family Incomes from Education of Household Head and Poverty Segment Designation

	Imputed level of family income	
Education reported	*Poverty segment*	*Nonpoverty segment*
Eighth grade or less	Less than $7,000	Less than $7,000
Some high school	Less than $7,000	$7,000–9,999
High school completed	$7,000–9,999	$10,000–14,999
Some college	$10,000–14,999	$15,000 or more
College graduate	$15,000 or more	$15,000 or more
Unknown	Less than $7,000	$7,000–9,999

TABLE 22

Imputed Education of Household Head from Family Income and Poverty Segment Designation

	Imputed education level of household head	
Income reported	*Poverty segment*	*Nonpoverty segment*
Less than $7,000	Eighth grade or less	Some high school
$7,000–9,999	Some high school	High school completed
$10,000–14,999	High school completed	High school completed
$15,000 or more	High school completed	Some college
Unknown	High school completed	High school completed

REFERENCES

Morgan, James N., and Sonquist, John A. (1963). *Journal of the American Statistical Association*, 58 (June): 514–535.

NCHS (1973). NCHS: Plan and Operation of the Health and Nutrition Survey, United States, 1971–1973. *Vital and Health Statistics*, Ser. 1, No. 10a, Washington, D.C.: U.S. Government Printing Office.

Schaible, Wesley (1973). Estimation weights for HANES, stands 01–35. Internal memorandum of DHES, March 27.

Author Index

Numbers in italic indicate the pages on which the complete reference is listed.

W

Subject Index

C

D

E

F

G

H